Metal Nanoparticles for Catalysis
Advances and Applications

RSC Catalysis Series

Series Editor:
Professor James J Spivey, *Louisiana State University, Baton Rouge, USA*

Advisory Board:
Krijn P de Jong, *University of Utrecht, The Netherlands*
James A Dumesic, *University of Wisconsin-Madison, USA*
Chris Hardacre, *Queen's University Belfast, Northern Ireland*
Enrique Iglesia, *University of California at Berkeley, USA*
Zinfer Ismagilov, *Boreskov Institute of Catalysis, Novosibirsk, Russia*
Johannes Lercher, *TU München, Germany*
Umit Ozkan, *Ohio State University, USA*
Chunshan Song, *Penn State University, USA*

Titles in the Series:
1: Carbons and Carbon Supported Catalysts in Hydroprocessing
2: Chiral Sulfur Ligands: Asymmetric Catalysis
3: Recent Developments in Asymmetric Organocatalysis
4: Catalysis in the Refining of Fischer–Tropsch Syncrude
5: Organocatalytic Enantioselective Conjugate Addition Reactions: A Powerful Tool for the Stereocontrolled Synthesis of Complex Molecules
6: N-Heterocyclic Carbenes: From Laboratory Curiosities to Efficient Synthetic Tools
7: P-Stereogenic Ligands in Enantioselective Catalysis
8: Chemistry of the Morita–Baylis–Hillman Reaction
9: Proton-Coupled Electron Transfer: A Carrefour of Chemical Reactivity Traditions
10: Asymmetric Domino Reactions
11: C-H and C-X Bond Functionalization: Transition Metal Mediation
12: Metal Organic Frameworks as Heterogeneous Catalysts
13: Environmental Catalysis Over Gold-Based Materials
14: Computational Catalysis
15: Catalysis in Ionic Liquids: From Catalyst Synthesis to Application
16: Economic Synthesis of Heterocycles: Zinc, Iron, Copper, Cobalt, Manganese and Nickel Catalysts
17: Metal Nanoparticles for Catalysis: Advances and Applications

How to obtain future titles on publication:
A standing order plan is available for this series. A standing order will bring delivery of each new volume immediately on publication.

For further information please contact:
Book Sales Department, Royal Society of Chemistry, Thomas Graham House, Science Park, Milton Road, Cambridge, CB4 0WF, UK
Telephone: +44 (0)1223 420066, Fax: +44 (0)1223 420247
Email: booksales@rsc.org
Visit our website at www.rsc.org/books

Metal Nanoparticles for Catalysis
Advances and Applications

Franklin Tao
University of Notre Dame, USA
Email: franklin.tao.2011@gmail.com

THE QUEEN'S AWARDS
FOR ENTERPRISE:
INTERNATIONAL TRADE
2013

RSC Catalysis Series No. 17

Print ISBN: 978-1-78262-033-4
PDF eISBN: 978-1-78262-103-4
ISSN: 1757-6725

A catalogue record for this book is available from the British Library

© The Royal Society of Chemistry 2014

All rights reserved

Apart from fair dealing for the purposes of research for non-commercial purposes or for private study, criticism or review, as permitted under the Copyright, Designs and Patents Act 1988 and the Copyright and Related Rights Regulations 2003, this publication may not be reproduced, stored or transmitted, in any form or by any means, without the prior permission in writing of The Royal Society of Chemistry or the copyright owner, or in the case of reproduction in accordance with the terms of licences issued by the Copyright Licensing Agency in the UK, or in accordance with the terms of the licences issued by the appropriate Reproduction Rights Organization outside the UK. Enquiries concerning reproduction outside the terms stated here should be sent to The Royal Society of Chemistry at the address printed on this page.

The RSC is not responsible for individual opinions expressed in this work.

Published by The Royal Society of Chemistry,
Thomas Graham House, Science Park, Milton Road,
Cambridge CB4 0WF, UK

Registered Charity Number 207890

For further information see our web site at www.rsc.org

Printed and bound in Great Britain by CPI Group (UK) Ltd, Croydon, CR0 4YY

Biography

Franklin (Feng) Tao, PhD, is a tenure-track assistant professor of chemistry. After receiving a PhD from Princeton University and being a postdoctoral fellow at UC-Berkeley and Lawrence Berkeley National Lab, he started his independent career in 2010. He was elected a Fellow of the Royal Society of Chemistry in 2013. Currently, he is leading a research group focusing on synthesis, evaluation of catalytic performance, and *in situ* and operando characterization of catalytic materials in catalytic reactions for chemical and energy transformations toward fundamental understanding of catalytic processes at the molecular level. He has published about 100 peer-reviewed publications and three books with Wiley and the Royal Society of Chemistry.

Acknowledgements

I would like to thank all the contributors who made the publication of this book possible, the Royal Society of Chemistry books team for their effort in publishing this book, and support from the Chemical Sciences, Geosciences and Biosciences Division, Office of Basic Energy Sciences, Office of Science, U. S. Department of Energy under the grant DE-FG02-12ER16353. An acknowledgement must also go to Dr Radha Narayanan for her time at the early stages of the development of this book.

Contents

Chapter 1	Introduction: Synthesis and Catalysis on Metal Nanoparticles *Franklin (Feng) Tao, Luan Nguyen and Shiran Zhang*		1
Chapter 2	Nanocatalysis: Definition and Case Studies *Choumini Balasanthiran and James D. Hoefelmeyer*		6
	2.1	Introduction	6
		2.1.1 Flash Synopsis of the History of Catalysis	7
		2.1.2 Reporting Turnover Frequency: the Common Denominator in Catalysis	8
	2.2	Factors Contributing to Structure Sensitivity in Catalysis	9
		2.2.1 Statistical Shape Analysis of Polyhedra Crystals and Relation to Catalysis	9
		2.2.2 Equilibrium Shapes of Nanocrystals	11
		2.2.3 Surface Restructuring	12
		2.2.4 Mobility of Surface Adsorbates	14
		2.2.5 Change in the Electronic Structure of Solids at the Nanometre Scale	14
		2.2.6 Example to Illustrate Anomalous TOF Behavior: Ethane Hydrogenolysis on Rhodium	15
	2.3	Synthesis and Properties of Well-defined Nanocrystals	16
	2.4	The Dawn of Nanocatalysis and Case Studies	18
		2.4.1 CO Oxidation on Au	18
		2.4.2 TiO_2 Nanocrystals with Reactive Facets	19
		2.4.3 Catalysis on Shape-controlled Pt Nanocrystals	21

		2.4.4	Advanced Templating Methods for Nanocrystal Synthesis	23
		2.4.5	Hybrid Nanocrystal Catalysis	23
	2.5	Conclusion		24
	References			25

Chapter 3 New Strategies to Fabricate Nanostructured Colloidal and Supported Metal Nanoparticles and their Efficient Catalytic Applications 30
Kohsuke Mori and Hiromi Yamashita

	3.1	Introduction		30
	3.2	New Route for the Preparation of Supported Metal Nanoparticle Catalysts		31
		3.2.1	A Photo-assisted Deposition Method Using a Single-site Photocatalyst	31
		3.2.2	A Microwave-assisted Deposition Method	34
		3.2.3	Deposition of Size-controlled Metal Nanoparticles as Colloidal Precursors	35
	3.3	Multifunctional Catalysts Based on Magnetic Nanoparticles		38
		3.3.1	Core–shell Magnetic FePt@Ti-containing Silica Spherical Nanocatalyst	38
		3.3.2	Water-soluble FePt Magnetic NPs Modified with Cyclodextrin	39
		3.3.3	FePd Magnetic NPs Modified with a Chiral BINAP Ligand	41
		3.3.4	Core–shell Nanostructured Catalyst for One-pot Reactions	42
	References			44

Chapter 4 Organometallic Approach for the Synthesis of Noble Metal Nanoparticles: Towards Application in Colloidal and Supported Nanocatalysis 47
Solen Kinayyigit and Karine Philippot

	4.1	Introduction			47
	4.2	Organometallic Synthesis of Noble Metal Nanoparticles			48
	4.3	Nanoparticles for Colloidal Catalysis			49
		4.3.1	Hydrogenation Reactions		50
			4.3.1.1	Ligand Stabilized Nanoparticles as Catalysts	50
			4.3.1.2	Water-soluble Nanoparticles as Catalysts	52

			4.3.1.3	Ionic Liquid Stabilized Nanoparticles as Catalysts	57
		4.3.2	Dehydrogenation Reactions of Amine-borane		60
		4.3.3	Carbon–carbon Coupling Reactions		63
			4.3.3.1	Pd Nanoparticles Stabilized by Chiral Diphosphite Ligands	63
			4.3.3.2	Pd Nanoparticles Stabilized by Pyrazole Ligands	65
		4.3.4	Hydroformylation Reactions		67
	4.4	Nanoparticles for Supported Catalysis			69
		4.4.1	Alumina as a Support for Hydrogenation and Oxidation Reactions		70
		4.4.2	Silica as a Support for Hydrogenation and Oxidation Reactions		71
			4.4.2.1	Hydrogenation of Olefins (Cyclohexene and Myrcene)	71
			4.4.2.2	Oxidation of Carbon Monoxide and Benzyl Alcohol	72
		4.4.3	Carbon Materials as Supports for Hydrogenation and Oxidation Reactions		75
			4.4.3.1	Hydrogenation of Cinnamaldehyde	75
			4.4.3.2	Oxidation of Benzyl Alcohol	76
			4.4.3.3	Versatile Dual Hydrogenation–oxidation Reactions	78
	4.5	Conclusion and Perspective			78
	References				79

Chapter 5 Nickel Nanoparticles in the Transfer Hydrogenation of Functional Groups — 83
Francisco Alonso

5.1	Introduction	83
5.2	Antecedents	84
5.3	Hydrogen-transfer Reduction of Alkenes	86
5.4	Hydrogen-transfer Reduction of Carbonyl Compounds	88
5.5	Hydrogen-transfer Reductive Amination of Aldehydes	94
5.6	Conclusions	96
References		96

Chapter 6 Ammonium Surfactant-capped Rh(0) Nanoparticles for Biphasic Hydrogenation — 99
Audrey Denicourt-Nowicki and Alain Roucoux

6.1	Introduction	99
6.2	Nanoparticles as Relevant Catalysts for Biphasic Hydrogenation	100

	6.3	Asymmetric Nanocatalysis: a Great Challenge	105
		6.3.1 Ethylpyruvate	106
		6.3.2 Prochiral Arenes	107
	6.4	Conclusions	108
	References	109

Chapter 7 Pd Nanoparticles in C–C Coupling Reactions   112
Dennis B. Pacardo and Marc R. Knecht

7.1	Introduction	112
7.2	Synthetic Scheme for the Fabrication of Pd Nanoparticles	115
7.3	C–C Coupling Reaction Mechanism for Pd Nanocatalysts	116
7.4	Pd Nanoparticles in the Stille Coupling Reaction	126
7.5	Pd Nanoparticles in the Suzuki Coupling Reaction	134
7.6	Pd Nanoparticles in the Heck Coupling Reaction	144
7.7	Summary and Conclusions	148
References	149

Chapter 8 Metal Salt-based Gold Nanocatalysts   157
Zhen Ma and Franklin (Feng) Tao

8.1	Introduction	157
8.2	Metal Salt-based Gold Nanocatalysts	158
	8.2.1 Metal Carbonate-based Gold Catalysts	158
	8.2.2 Metal Phosphate-based Gold Catalysts	160
	8.2.3 Hydroxyapatite-based Gold Catalysts	164
	8.2.4 Hydroxylated Fluoride-based Gold Catalysts	166
	8.2.5 Metal Sulfate-based Gold Catalysts	167
	8.2.6 Heteropolyacid Salt-based Gold Catalysts	168
8.3	Summary	169
Acknowledgments	169
References	169

Chapter 9 Catalysis with Colloidal Metallic Hollow Nanostructures: Cage Effect   172
Mahmoud A. Mahmoud

9.1	Introduction	172
9.2	Synthetic Approaches to Hollow Metallic Nanocatalysts	174
9.3	Assembling the Nanocatalysts on Substrates	176
9.4	Hollow Nanostructures are Different in Catalysis	179
	9.4.1 Hollow Nanostructures with a Catalytically Active Inner Surface and an Inactive Outer Surface	180

Contents xiii

		9.4.2	Comparing the Activity of Hollow and Solid Nanocatalysts of Similar Shapes	182
		9.4.3	Comparing the Activity of a Single Shell Hollow Nanocatalyst with a Double Shell Consisting of a Similar Inner Shell Metal	183
		9.4.4	Following the Optical Properties of Plasmonic Nanocatalysts During Catalysis	184
	9.5	Proposed Mechanism for Nanocatalysis Based on Spectroscopic Studies		185
	References			187

Chapter 10 Nanoreactor Catalysis   192
Kyu Bum Han, Curtis Takagi and Agnes Ostafin

	10.1	Introduction		192
	10.2	Steric and Structural Effects		193
		10.2.1	Dendrimers	193
		10.2.2	Microgels	194
		10.2.3	Polymer Core-shell Structures	195
		10.2.4	Hydrophobic–hydrophilic Structures: Micelle, Emulsion, and Liposome	195
	10.3	Absorbing Nanocatalyst Surface		198
		10.3.1	Micelle and Emulsion	198
		10.3.2	Carbon Nanotubes	198
	10.4	Conclusion		199
	References			199

Chapter 11 Nanoparticle Mediated Clock Reaction: a Redox Phenomenon   203
Tarasankar Pal and Chaiti Ray

	11.1	History		204
		11.1.1	Iodine Clock Reaction	204
		11.1.2	B–Z Reaction	205
		11.1.3	Bray-Liebhafsky Reaction	205
		11.1.4	Briggs–Rauscher Reaction	206
		11.1.5	The Blue Bottle Experiment	207
	11.2	Recent Work		207
		11.2.1	Clock Reaction of Methylene Blue	210
	11.3	Mechanistic Approach		215
		11.3.1	Eley–Rideal Mechanism	215
		11.3.2	Langmuir–Hinshelwood Mechanism	216
	11.4	Applications		216
		11.4.1	Water Purification	216
		11.4.2	Memory Facilitation by Methylene Blue and Brain Oxygen Consumption	217

	11.4.3	Novel UV-activated Colorimetric Oxygen Indicator	217
11.5	Conclusion		217
References			217

Chapter 12 Theoretical Insights into Metal Nanocatalysts — 219
Ping Liu

12.1	Introduction	219
12.2	Computational Method	220
12.3	Metal Nanocatalysts	220
	12.3.1 Copper Nanocatalysts for Water–Gas Shift Reactions: the Importance of Low-coordinated Sites	221
	12.3.2 Metal (Core)–Platinum Shell Nanocatalysts for Oxygen Reduction Reactions in Fuel Cells: the Essential Role of Surface Contraction	223
12.4	Supported Metal Nanocatalysts	227
12.5	Conclusions	230
Acknowledgements		231
References		231

Chapter 13 Porous Cryptomelane-type Manganese Oxide Octahedral Molecular Sieves (OMS-2); Synthesis, Characterization and Applications in Catalysis — 235
Saminda Dharmarathna and S. L. Suib

13.1	Introduction	235
13.2	Synthesis and Morphology Control	236
13.3	Catalysis	239
	13.3.1 Selective Oxidation and Fine Chemical Synthesis	239
	13.3.2 C–H Activation	241
	13.3.3 CO_2 Activation	243
	13.3.4 Environmental and Green Chemistry	244
13.4	Conclusion	246
Acknowledgments		247
References		247

Subject Index — 251

CHAPTER 1

Introduction: Synthesis and Catalysis on Metal Nanoparticles

FRANKLIN (FENG) TAO,* LUAN NGUYEN AND SHIRAN ZHANG

Department of Chemistry and Biochemistry, University of Notre Dame, Notre Dame, IN 46556, US
*Email: franklin.tao.2011@gmail.com

Heterogeneous catalysis is critical for chemical and energy transformations. It has played a cornerstone role in the chemical industry for more than one century. Many industrial catalysts were developed on the basis of trial-and-error. An industrial catalyst is typically a combination of a few or more components. From a materials science point of view, an industrial catalyst is very heterogeneous in terms of composition, structure, size, shape, and dispersion of catalyst particles on their support. In addition, a chemical reaction with a heterogeneous catalyst is performed on the surface of a catalyst particle at a high temperature while the catalyst particle is in a gaseous environment or in a liquid. An oxidizing and/or reducing reactant very likely restructures its surface and/or subsurface and/or bulk before a stable catalytic performance is obtained. In many cases, the chemistry and structure of a catalyst particle during a catalytic reaction could be different from those before catalysis. They could be very different from those of a catalyst under *ex situ* conditions after catalysis. Here, the *ex situ* condition is defined as a status at which the catalyst is at room temperature and all reactant gases are purged. Due to these potential differences between *in situ*

RSC Catalysis Series No. 17
Metal Nanoparticles for Catalysis: Advances and Applications
Edited by Franklin Tao
© The Royal Society of Chemistry 2014
Published by the Royal Society of Chemistry, www.rsc.org

and *ex situ* conditions and the heterogeneity of an industrial catalyst in chemistry and structure, understanding heterogeneous catalysis at a molecular level has been quite challenging. Despite the complications with catalyst materials and their reaction pathways, heterogeneous catalysts have played a cornerstone role in chemical and energy transformations, and heterogeneous catalysis has been one of the most important fields since the beginning of the last century.

The measured catalytic performance (activity, selectivity, and stability) is the outcome of many structural and chemical factors that interact with each other. However, a catalytic event is performed on a catalytic site which has a specific geometrical packing of catalyst atoms to give a suitable electronic structure for an appropriate molecular or dissociative adsorption with a subsequent coupling to form a product. Essentially, a catalytic event is determined by a molecular reaction with a catalytic site at a microscopic level. The correlation between the macroscopic catalytic performance and microcosmic "picture" of catalytic sites is lacking due to the heterogeneity of an industrial catalyst. Experimental and theoretical simulations of industrial catalysis with single crystal model catalysts have been an important approach to understanding how a specific site on the surface of a catalyst particle participates in a catalytic reaction at a solid (a catalyst)–gas or liquid (reactants) interface. This approach has provided a tremendous amount of information on catalytic reactions from a surface science point of view. It has been the cornerstone for understanding heterogeneous catalysis at a molecular level. However, due to the limited variability of structure, composition, and size, there exists a gap in materials between a well-defined single crystal model catalyst and an industrial catalyst with heterogeneous structure and chemistry.

In the last two decades, the significant advance in nanoscience and nanotechnology was partially driven by both the quantum effect of semiconductor nanoparticles with different sizes and plasmonic effect of some noble metal nanoparticles with different sizes. Accompanying this, spectacular achievements have been made in syntheses toward controlling the size, shape, or composition of nanomaterials. These achievements in the materials synthesis of nanomaterials, particularly the control of size, shape, structure, and composition of metal or bimetallic nanoparticles, have offered the possibility to bridge the gap between materials in the studies of heterogeneous catalysis. Colloidal chemistry allows tuning of the size, shape, structure, and composition readily. As metal atoms on different crystallographic faces of a metal nanoparticle pack differently, tuning the shape of metal nanoparticles to expose different crystallographic faces can offer different types of catalytic sites. For example, Pt atoms on a Pt nanocube pack into a (100) surface of a fcc lattice. An octahedral nanoparticle only exposes its (111) face. Both (100) and (111) faces can be found on a cubo-octahedral nanoparticle. A concave nanocube, in fact, offers a stepped surface with a high density of under-coordinated catalyst atoms. The capability of tuning size with synthesis provides a method to distinguish the sites of under-coordinated catalyst atoms at corners and edges from the site on the surface since there is a size-dependent density of these under-coordinated atoms at corners and/or edges.

Chapter 2 describes structural factors of catalyst nanoparticles, which could influence catalytic performances including catalytic activity and selectivity. In addition, it reviews correlations between these structural factors and catalytic performance, and discusses the potential restructuring of the surfaces of catalyst particles. A few examples are discussed from a structural point of view.

Chapter 3 reviews a few synthetic routes of metal nanoparticle catalysts. It presents three methods for the preparation of metal nanoparticles supported on a substrate including photo-assisted deposition methods using single-site photocatalysts, a microwave-assisted deposition method, and deposition of size-controlled metal nanoparticles as colloidal precursors. In addition, the syntheses of multi-functional catalysts are discussed.

In Chapter 4, syntheses of noble metal nanoparticles are reviewed from an organometallic point of view. It emphasizes the nature of the ligand stabilizing the nanocatalysts in solution and the role of the support for a supported catalyst. Syntheses of Pt, Rh, Ru, Ir and other metals are exemplified. Factors influencing the synthesis of metal nanoparticles and their performance in catalysis are discussed in detail.

After the review of syntheses in Chapters 2–4, Chapter 5 discusses the catalytic transfer hydrogenation of organic compounds. It focuses on the utilization of nanoparticle catalysts of the earth-abundant metal, Ni, a replacement for noble metal catalysts in hydrogen-transfer reductions of functional groups.

Chapter 6 reports the recent progress achieved in nanocatalysis, particularly use of quaternary ammonium salts as water-soluble capping agents of rhodium nanoparticles. Hydrogenation under biphasic liquid–liquid conditions and asymmetric catalysis including ethylpyruvate hydrogenation and prochiral arene hydrogenation are described.

As well as hydrogenations on Ni and Rh nanoparticles in Chapters 5 and 6, the catalysis of C–C coupling on Pd nanoparticles is reviewed in Chapter 7. The importance of Pd-catalyzed C–C coupling pioneered by Heck, Suzuki and Negishi was recognized by the 2010 Nobel prize in chemistry. Thanks to the advance in the synthesis of metallic Pd nanocatalysts with specific size and shape, efficient catalysis on Pd nanoparticles in contrast to traditional small molecules was revealed. Due to the limited space, only C–C coupling on selected examples are discussed in Chapter 7, though a large number of applications of Pd nanoparticles in C–C coupling have been reported in the literature.

Gold nanoparticles supported on oxide substrates exhibit exciting catalytic activities for many inorganic and organic reactions in contrast to the inert nature of macroscopic gold particles. Tremendous effort has been put into exploration of the origin of catalytic activity of Au nanoparticles following the discovery of activity for many reactions on them. Numerous papers on this topic have been published in the literature. In these reports, most of the Au catalyst particles were supported on an oxide. Other than metal oxides, salts have been used as supports for Au nanoparticles. Some salts are solid acids, which are used in acid catalysis. In fact, integration of the acidic support of a salt with gold nanoparticles forms a bi-functional catalyst that is highly active in organic reactions. Chapter 8 reviews the preparation and catalysis of salt-based Au catalysts. These salts in the Au/salt catalysts are carbonate, phosphate, hydroxyapatite, hydroxylated fluoride, metal sulfate, and heteropolyacid. A comparison between the roles of salts and oxides in gold catalysis is made.

All the metal nanoparticle catalysts discussed so far are metal nanoparticles with a solid core. From a structural point of view, a nanoparticle with a hollow or porous core could provide different catalytic activity. A porous metal nanoparticle consist of an external surface and an internal surface. Thus, the total surface area is significantly increased due to the existence of the internal surface in the hollow or porous core of the particle. More importantly, the density of catalyst atoms with a low coordination number in such a metal particle is in fact much higher than that of a metal nanoparticle with a solid core. In addition, its inner surface could be covered with much less capping molecules due to the limited space in a pore. Thus, the density of active sites could be significantly larger. In addition, the limited space at a scale of nanometres or a sub-nanometre, in fact increases the rate of collision of reactant molecules with catalyst atoms. These structural features could offer porous metal nanoparticles different catalytic activity and selectivity in contrast to a metal nanoparticle with a solid core. Chapter 9 presents the synthesis of porous metal nanocatalysts and discusses the catalysis of the nanoparticles.

As discussed in Chapter 9, the local structure of a catalyst nanoparticle is critical for its catalytic activity. Other than the localized structure of a metal nanoparticle, such as the surface of a catalyst particle with a solid core and the surface and inner surface of a porous nanoparticle, the external environment of a particle could influence catalytic shape to some extent. A metal particle together with its surrounding layers or shell can be considered as a single reactor at the nanoscale. The surroundings could be a scaffold of a dendrimer around the loaded metal nanoparticles, a cross-linked linear polymer of microgels, a polymer with a hydrophobic or hydrophilic substrate around a metal particle, or a shell of a metal particle. Chapter 10 describes several types of nanoreactors and discusses the effect of the surroundings on the catalysis of metal nanoparticles.

Other than these catalytic reactions, metal nanoparticles can also catalyze clock reactions. Chapter 11 reviews the clock reactions on metal and oxide particles, and discusses the mechanisms.

Computational chemistry is significant for a mechanistic understanding of chemical reactions. The application of a computational approach to catalysis studies is certainly the most successful synergy in chemical sciences. A tremendous effort has been made in this field in recent decades and spectacular achievements have been obtained. Theoretical studies are able to not only rationalize experimental findings and provide insights into reaction pathways and catalytic performance, they are also able to offer guidance for the design of new catalysts. Computation-aided screening of catalysts can largely narrow the coverage of potential composition and structure and accelerate the design and optimization of catalysts. Chapter 12 briefly introduces computational methods and presents examples to demonstrate how computational approaches were used to determine reaction pathways that could not be tracked with current experimental techniques. This chapter also reviews applications of the computational approach in studying catalytic reactions on metal nanoparticles supported on reducible oxides and demonstrates the essential role of the metal/oxide interface in promoting the catalytic performance of metal nanocatalysts.

Catalysis on manganese oxide octahedral molecular sieves exhibits high catalytic activity in many catalytic reactions for chemical and energy transformations. Chapter 13 reviews the synthesis of these molecular sieves and discusses catalytic reactions including selective oxidation and fine chemical synthesis, C–H activation and CO_2 activation, environmental remediation, and green chemistry.

CHAPTER 2

Nanocatalysis: Definition and Case Studies

CHOUMINI BALASANTHIRAN AND JAMES D. HOEFELMEYER*

Department of Chemistry, University of South Dakota, 414 E. Clark St., Vermillion, SD 57069, USA
*Email: jhoefelm@usd.edu

2.1 Introduction

Catalysis is the process of introducing a catalyst to a reaction to increase the rate of the reaction. The catalyst provides an alternate reaction pathway with a lower activation energy. It facilitates the reaction without being consumed and contributes to multiple turnovers of the catalytic reaction cycle at each catalytic site. Catalysis has enormous technological significance, being important in energy production, the chemical industry, and environmental technologies, and is a foundation of our modern way of life as well as of life itself.[1]

Early examples of catalysis in the chemical industry include the catalytic oxidation of sulfur dioxide to sulfur trioxide, which allowed the large-scale commercial production of sulfuric acid; ammonia synthesis from nitrogen and hydrogen (Haber process); and ammonia oxidation (Ostwald process).[2] Catalysis has been utilized extensively in the petroleum industry, for example in hydrodesulfurization and reforming. These uses arose from an increasing demand for high-octane gasoline. The long history of the growing demand for chemicals and fuel resulted in a parallel demand for better catalysts, and has triggered much research into understanding catalytic processes.

Most catalysis occurs between a substrate and a catalyst in which the catalyst hosts the substrate at a specific site. At this active site, the substrate undergoes a chemical transformation into a product. Finally, the product is released from the active site. There are some homogeneous reaction schemes that do not follow these guidelines, such as the catalytic decomposition of stratospheric ozone that occurs through free radical chain reactions.

In nature, highly pre-organized enzymes are optimized to catalyze specific biochemical reactions. Active sites in enzymes are highly specific in binding to substrates. The host–guest interactions are precise, and arise from the specific positioning of functional groups that participate in intermolecular interactions, such as hydrogen bonding or electrostatic interactions, with the substrate in order to achieve molecular recognition. Other amino acid residues present specific functional groups to activate the substrate and stabilize the transition state, which results in the minimization of the activation energy of the reaction. Finally, the product may have a low affinity for the active site and dissociate, or in some cases the product causes a change in the shape of the enzyme to assist in ejecting the product. Once the product is released, the enzyme, and its active site, is restored to its original state ready for another turnover of a catalytic cycle. The existence of such finely tuned structures for each of the intricately linked steps within the web of biochemical reactions is awe-inspiring.

2.1.1 Flash Synopsis of the History of Catalysis[3]

The utility of catalysts should be instantly recognizable, and should motivate interest in their development. Without any guiding principle to develop new catalysts for a reaction, the best approach may be an empirical one. Slowly, over time, some substances were observed to increase the rate of a chemical reaction. The catalyst may be homogeneous (dissolved catalyst) or heterogeneous (insoluble catalyst). It might follow that the next question would pertain to the relation between rate and mass of catalyst added, and it might quickly arise that different laboratories could report differing rates based on an identical mass of the same catalyst. The results could suggest that the rate is dependent on the mass of the catalyst as well as its morphology. For example, one gram of nickel shot *versus* one gram of freshly prepared, finely divided nickel precipitate may show remarkably different catalytic behavior in a heterogeneous system. Understanding this difference lies in understanding that the catalysis occurs on the surface of the nickel. This realization would certainly initiate a race to develop high surface area materials and the means to stabilize them. Thin films, porous materials, and finely divided materials are high surface area morphologies.

Without tools that enable detailed characterization of the catalyst morphology, the approach to develop new catalysts remains somewhat empirical. For instance, wetness impregnation of nickel onto a support followed by different calcination protocols could lead to catalysts with widely different activities. The use of electron microscopy and X-ray diffraction have proven enormously beneficial in providing answers, and have allowed further details about morphology to be elucidated. Key features include particle size, particle shape, and crystal phase. With this level of sophistication in the preparation and characterization of catalyst materials, it has become possible to realize the size- dependent behavior of matter manifest in catalysis.[4]

The focus of this chapter will be on catalysis on finely divided materials, and we have intentionally set aside the discussion of homogeneous catalysis with molecular entities, grafted catalysts, enzyme catalysis (nature's nanomachines), and soft materials that may spontaneously arrange to a shape that is an effective catalyst (artificial enzymes). Additionally, our emphasis is not on porous materials that include crystalline zeolites, metal–organic frameworks, ordered nanotube arrays, mesoporous materials, aerogels, and xerogels. While all of these materials have an important role in the emergent field of nanocatalysis,[5] we intend to emphasize the importance of surface science[6] and nanocrystal synthesis in the development of new heterogeneous catalyst systems.

2.1.2 Reporting Turnover Frequency: the Common Denominator in Catalysis

The activity of a catalyst should be defined in a way that allows comparison between measurements taken from various laboratories. The standard for comparing catalyst activity is to report the turnover frequency,[7] rate law,[8] conditions of the reaction, and active site of the catalyst. Reporting these data can require an enormous amount of effort; however, failure to define these characteristics does not allow a meaningful comparison of catalysis data. The correct reporting of catalytic data provides a link to understanding structure–activity relations over a wide variety of catalyst types including molecular, supported catalysts, and single crystals through calculating normalized turnover frequency according to the power rate law.[9] This important link has been vital in bridging the fields of homogeneous catalysis, heterogeneous catalysis, and surface science.

In heterogeneous catalysis, there are reactions that are structure-sensitive or insensitive. Structure-insensitive reactions exhibit nearly identical turnover frequencies independent of the structure of the catalyst. For example, ethylene hydrogenation catalyzed on platinum shows a turnover frequency of ~ 1–10 s^{-1} on Pt wire, Pt(111), or Pt supported on silica under normalized conditions of 10 torr C_2H_4, 100 torr H_2, 298 K.[10,11] Structure-insensitive reactions, therefore, can be useful probes for validating catalysis measurements. Due to the potential utility, it is imperative that catalytic data be reported with the utmost care in order to establish potential structure-insensitive reactions as probes. For example, one can measure catalyst particle size using electron microscopy, X-ray diffraction, chemisorption analysis, and turnover frequency of a structure-insensitive reaction. If electron microscopy data indicates small particles while chemisorption data and turnover frequency data of a structure-insensitive reaction indicate a larger particle size, then the disparate data may be diagnostic of a blockage of surface sites.

2.2 Factors Contributing to Structure Sensitivity in Catalysis

Here, the properties of matter responsible for structure sensitivity in catalysis will be outlined. In particular, it is instructive to describe the size- and shape-dependent changes in the properties of finely divided materials. The size range in which such changes are most dramatic is ~1–15 nm. Below this range, clusters tend to have molecular-like properties with discreet orbital energies; whereas above this size range, particles exhibit electronic properties similar to bulk materials. We note that control of the morphology of nanostructured materials above 15 nm (which is not the focus of this chapter) may have other implications, such as surface plasmon resonances, mass transport in porous structures, or directionalization of charge carriers or lasing in nanowires. It is also worth pointing out that the occurrence of facets in nanoparticles emerges at a size of ~4 nm; below which, shapes tend to be less well-defined due to a lack of extended facets. Finally, it is important to appreciate that the selection of a size–shape combination of a nanoparticle translates to a selection of an electronic structure–surface topology combination that has important implications in catalysis besides simply adjusting the surface area of the catalyst.

2.2.1 Statistical Shape Analysis of Polyhedra Crystals and Relation to Catalysis

It is rather intuitive to suppose that a nanocrystal with a cubic shape always has 8 corner atoms and that the ratio of those sites to the total number of atoms or total number of surface sites is dependent on the size of the cubic nanocrystal. The smallest cube would be composed of 8 atoms (2 × 2 faces), therefore the ratio of corner atoms to surface atoms is 1. The next size up, with 3 × 3 faces, is composed of 27 atoms, with 26 of them on the surface, and 8 being the corner atoms. Therefore the ratio of corner atoms to surface atoms is 8 : 26, or 0.31. The same ratio for a cube with 4 × 4 faces is 0.14, and so on. As the size increases, the corner and edge atoms become less significant as the surface atoms are increasingly made up of face sites. The surface topology of crystals >15 nm is dominated by atomically smooth crystal faces, which share chemical features with single crystal surfaces.

Statistical shape analysis to reveal the ratio of surface atom types for polyhedral nanocrystals was calculated by Van Hardeveld and Hartog.[12] For example, a cubo-octahedron is a common morphology for face-centered

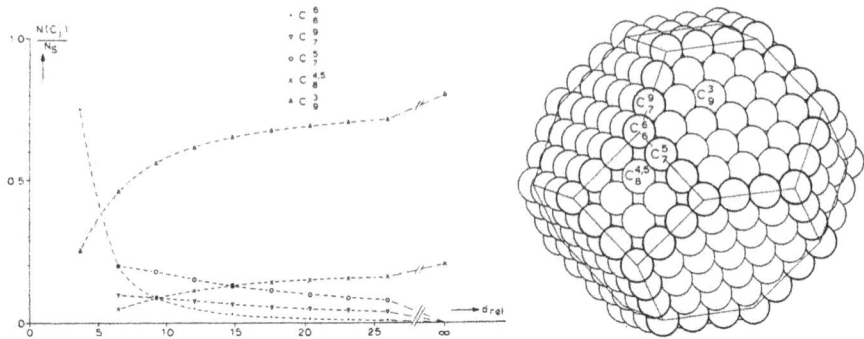

Figure 2.1 Statistical shape analysis of particles with cubo-octahedron shape. Reprinted from ref. 12 with permission from Elsevier.

cubic metals composed of five types of surface atoms: corners (C_6^6), atoms within {100} faces ($C_8^{4,5}$), atoms within {111} faces (C_9^3), edges between two {111} faces (C_7^9), and edges between {100} and {111} faces (C_7^5). The ratio of the surface atom types to total surface atoms calculated for crystals with n shells shows the rapidly diminishing importance of corner and edge atoms and dominance of facial atoms with increasing size (Figure 2.1).

This is an idealized model; whereas, nanoparticle samples have some distribution of size and shape that complicates the statistical shape analysis. In fact, the ability to precisely control the dimensions of nanocrystals in order to achieve minimum size and shape distribution can allow statistical shape analysis with smaller standard deviations. This seems to be an opportunity that defines the field of 'nanocatalysis'. Ideally, the nanocatalyst samples would be characterized and reported with quantitative details including statistical shape analysis. Perhaps this is an opportunity for mathematics to merge with chemistry. Some reports that include mechanistic details for the formation of special non-equilibrium nanocrystal shapes incorporate detailed qualitative shape analysis.[13] In special cases where homogeneous samples can be created, quantitative relations between surface topology and turnover frequency can be determined. This tends to work well for larger high-quality nanocrystals with small relative size distribution and uniform shape. For particles that have poorly defined shapes (size <4 nm), it becomes extremely challenging to develop quantitative relations, which interestingly is often the regime in which materials exhibit maximum turnover frequency!

Crystal domain size clearly contributes to structure sensitivity in catalysis.[14–21] As crystal domains become smaller, the ratio of corner and edge atoms to facial atoms increases, and the surface topology becomes increasingly disordered, or roughened. The open coordination spheres of corner and edge atoms should bind substrates actively. A reaction mechanism operating on corner and edge atoms is likely to be very different to one on an extended crystal face.

2.2.2 Equilibrium Shapes of Nanocrystals

It is already apparent from the discussion in section 2.1 that the shape of a nanocrystal in relation to its crystal structure gives rise to specific crystal facets. The example of a cubo-octahedron shape composed of atoms in a face-centered cubic arrangement gives rise to {100} and {111} low-index facets. In fact, the cubo-octahedron is one of many possible equilibrium shapes of nanocrystals. Each facet has a measurable surface tension or surface free energy such that the most stable facet has the lowest surface free energy. Low-index facets tend to have lower surface free energy and are often atomically smooth; whereas high-index facets are atomically rough and have high surface free energy. At high temperatures, the differences in the surface free energy of facets are lower than thermal energy, and the nanocrystal adopts the perfectly isotropic shape of a sphere. At lower temperatures, the differences are large enough to influence the equilibrium shape of the nanocrystal. The ratio of the surface free energy of two facets is proportional to the relative distance between the nanocrystal center and the face-centroids, which allows the theoretical construction of the equilibrium crystal shape—known as a Wulff construction.[22,23] If a nanocrystal shape deviates from the Wulff construct, then some other influence (for example, surface stabilizing agent or catalyst) leads to the non-equilibrium shape. This idea is fundamentally important in the design and synthesis of nanocrystals.

So, an origin of structure sensitivity is the arrangement of atoms on specific facets. For the face-centered cubic cubo-octahedron, the atoms on the {111} face are found within a hexagonal 2D array of atoms wherein each atom has six in-plane nearest neighbors and six three-fold hollow sites, which of course differs significantly from the atoms on the {100} face that are found within a square 2D array wherein each atom has eight in-plane nearest neighbors and four four-fold hollow sites. Geometric differences between surface facets can have enormous implications for their chemical properties. Examples include the extremely structure-sensitive nature of CO dissociation on rhodium,[24] or ammonia synthesis catalyzed on iron single crystal surfaces that differ in turnover frequency depending on the orientation of the surface plane.[25,26] Comparison of five iron single crystal surfaces showed the Fe(111) surface was ~500 times more active than the Fe(110) surface. The Fe(111) surface is rough compared with the Fe(110) surface; however, it also has a high ratio of C7 sites. The Fe(211) surface also shares these features and exhibits a high turnover frequency for ammonia synthesis.

One of the principal challenges defining nanocatalysis is the ability to engineer nanocrystals with uniform size and shape that possess characteristics optimized for specific catalytic reactions. Some recent advances in the literature include the synthesis of well-defined nanocrystals with high-index facets that have reactive surfaces.[27,28] High-index facets feature surface atoms at step-edges with lower coordination numbers that contribute to higher reactivity.

2.2.3 Surface Restructuring

The shapes of nanocrystals are not static. At high temperatures, a nanocrystal will morph into the thermodynamic equilibrium shape. The temperature at which a Wulff construct shaped nanocrystal becomes spherical is often stated as the nanocrystal melting point, and can be determined using transmission electron microscopy with high temperature stage capability. The melting point of nanomaterials generally decreases dramatically in comparison with bulk materials. This is because surface atoms, which make up a higher atom fraction of nanomaterials than bulk materials, are mobile and can desorb from the surface. At high temperatures, it is possible for interparticle atom migration to occur as thermal energy mobilizes atoms. On surfaces, this phenomenon leads to sintering of the nanoparticles characterized by an increase in average particle size. This effect reduces surface area, leads to loss of control in surface site types, and is detrimental to catalysis.

Surface restructuring and reshaping of nanocatalysts is a common phenomenon in heterogeneous catalysis to optimize bonding between adsorbed atoms and surface atoms. Different atomic termination and crystal planes revert to thermodynamically more stable structures at higher temperatures and pressures during catalysis. Surface restructuring has been studied extensively and can occur upon adsorption of molecules on the surface or formation of thin films. Surface science techniques including LEED, STM,[29] and XPS are powerful tools used to characterize such processes, and these studies have contributed enormously to the understanding of catalysis.[30–32]

Recent studies in the research of surface restructuring *via* oxide layer formation in bimetallic, metallic, and metal oxide nanocrystals have been described. *In situ* crystal plane controlled surface restructuring during CO oxidation on Cu_2O NPs has been investigated using XPS, TEM, HRTEM, SEM and H_2-TPR.[33] The *fcc* Cu_2O nanocrystals with octahedral or cubic shapes with exposed {111} and {100} facets, respectively, do not show any oxidation under ambient conditions. During CO oxidation, the Cu_2O octahedra and Cu_2O cubes show activation energies and conversion rates of 73.4 ± 2.6 kJ mol^{-1}, 91.5% and 110 ± 6.4 kJ mol^{-1}, 45.1%, respectively, under similar reaction conditions. During catalysis, XPS data show Cu(II) in both nanocrystals, indicating complete oxidation of the Cu(I) in the as-synthesized nanocrystals. Electron microscopy results show that the nanocrystals retain their original octahedral or cubic shape; however, with rounder corners and rougher edges on the surface. HRTEM shows that the inner region maintains the same lattice fringe structures as Cu_2O. Together, the data indicate an *in situ* formation of CuO thin films during catalytic oxidation of CO. Theoretical studies show that the faster rate of catalytic oxidation of Cu_2O octahedra over Cu_2O cubes is associated with the existence of coordinatively unsaturated Cu atoms in {111} planes that are more prone to surface oxidation.

High pressure STM experiments on Pt(110) single crystals show the flexible nature of the surface.[32] Exposure to 1.6 atm H_2 led to a 'missing row' reconstruction in which random rows of apex atoms were missing, resulting in slightly deeper valleys on the surface. After exposure of the surface to 1 atm O_2, the pits become deeper and result in (111) microfacets. Exposure to 1 atm CO led to regeneration of the (110) surface void of the (111) microfacets or the missing rows. In a separate investigation, Pt(110) single crystals were subjected to CO/O_2 atmospheres at high temperatures and studied within a combined UHV-high-pressure surface X-ray diffraction (SXRD) chamber.[34] Clean Pt(110) exhibited missing-row reconstruction after ion-bombardment and annealing cycles. Exposure to 100 mbar CO at 600 K led to regeneration of the bulk-terminated surface within minutes. The surface was brought to room temperature and exposed to 10 mbar CO and 500 mbar O_2. The sample was heated to 650 K, which led to the formation of a distorted PtO_2 layer. A PtO_2 layer forms over the missing row reconstructed surface, but with high disorder. The Pt(110) crystal with more ordered PtO_2 layer was exposed to 500 mbar O_2, then pulses of CO were introduced that instantly underwent oxidation to CO_2. Formation of CO_2 was simultaneous with the consumption of oxygen on the PtO_2 layer that caused increasing disorder. A sufficiently long pulse of CO led to complete consumption of the PtO_2 layer and regeneration of the bulk-terminated Pt(110) surface. When P_{CO}/P_{O2} was set to approximately 0.2, the catalyst showed its high activity again; however, the PtO_2 layer recovered after 20 minutes. It was found that the active surface in this case was Pt(110) with carbonate ions. Under O_2-rich conditions, CO oxidation was fast and PtO_2 or the Pt(110)–carbonate layer was found on the single crystal surface; whereas, under O_2-lean conditions, surface O was steadily depleted until the bulk-terminated Pt(110) surface was restored. This surface was 25 times less active for CO oxidation.

Structural and chemical changes on bimetallic nanoparticles in oxidizing and reducing environments at ambient pressure were demonstrated.[35] Ambient pressure X-ray photoelectron spectroscopy (APXPS) was used to analyze the *in situ* structure in the bimetallic system at different depths from the surface at gas pressures higher than usual limits. The reduction (CO, H_2), oxidation (NO, O_2) and catalytic reaction (NO, CO) environments led to surface restructuring on 15 ± 2 nm $Rh_{0.5}Pd_{0.5}$ and $Pt_{0.5}Pd_{0.5}$ bimetallic nanoparticles. As synthesized, the $Rh_{0.5}Pd_{0.5}$ nanoparticle surface is Rh-rich and the core is Pd-rich. During catalysis, $Rh_{0.5}Pd_{0.5}$ shows dramatic segregation and reversibility during catalytic reactions: (a) oxidation results in the formation of RhO_y on the surface due to its higher surface free energy compared to that of Pd; (b) reduction results in migration of Rh to the core and migration of Pd to the surface. On the other hand, $Pt_{0.5}Pd_{0.5}$ undergoes restructuring without substantial segregation and reversibility. As synthesized, the $Pt_{0.5}Pd_{0.5}$ nanoparticle surface is Pd-rich and the core is Pt-rich. During catalysis, $Pt_{0.5}Pd_{0.5}$ does not show segregation and reversibility during catalytic reactions. Pt has a higher surface energy and is less prone to oxidation. Thus, oxidation results in PdO_y on the surface while reduction leads to regeneration of Pd metal on the surface. The two contrasting behaviors are explained by the differences of surface energy in the metals and metal oxides.

Atomic restructuring was found to occur within a sample of 1.3 nm Rh particles exposed to hydrogen.[36] The particles were synthesized by reduction of [Rh(μ-Cl)(C$_2$H$_4$)$_2$]$_2$ with vanadocene in the presence of polyvinylpyrrolidinone. As synthesized, the Rh particles exhibit polytetrahedral atomic organization. After exposure to 6 bar H$_2$ for 1 day, the atoms restructure into single crystalline Rh nanoparticles (1.3 nm diameter) with the *fcc* structure.

2.2.4 Mobility of Surface Adsorbates

Platinum and rhodium catalyze ethylene hydrogenation with high TOF; however, introduction of CO poisons the catalyst with a large increase in activation energy. A high-pressure scanning tunneling microscopy experiment in which H$_2$, ethylene, and CO were co-adsorbed on Pt(111) or Rh(111) provided an important insight into the mechanism of CO poisoning.[37] Addition of ethylene and H$_2$ to the metal surfaces at room temperature is known to lead to the formation of strongly bound ethylidyne; however, it is impossible to resolve any surface structure with STM due to the high surface mobility. Upon introduction of CO, ordered structures form on the catalyst surface. The implication is that catalysis occurs when mobility of surface adsorbates is high. Immobile surface structures block surface sites, whether these are adsorbates on single crystals or nanoparticle catalysts (*i.e.* surfactants or polymers).

2.2.5 Change in the Electronic Structure of Solids at the Nanometre Scale

The interaction between light and matter (spectroscopy) is fundamentally important in the understanding of the electronic structure of materials. The theory of excitons was developed in the 1930s to explain the absorption of UV-visible light by solids.[38-42] In the 1950s, Dresselhaus described an effective mass approximation for excitons,[43] and theoretical efforts were attempted to reconcile Frenkel and Wannier type excitons.[44] In particular, the formulation of intermediate type excitons may have laid the foundation for later descriptions of confined excitons.[45] Experimental work with thin film semiconductor spectroscopy in the 1960s began to provide new data to test exciton theory. Detailed theoretical descriptions of confined excitons were put forth *ca.* 1980.[46-49] If the diameter of the semiconductor nanocrystal is less than the Bohr exciton radius, then there is quantum confinement in the system. Quantum confinement leads to a blue-shift of the absorption and emission spectra of a material, and is an example of how fundamental properties of nanomaterials differ from bulk materials.[50,51] In metal clusters, there is a transition from metallic to non-metallic behavior.[52-55]

Experimental verification of quantum confinement appeared when Ekimov observed a quantum size effect in the optical absorption spectra of CdS and CuCl nanocrystals.[56-58] The semiconductor crystallites were trapped within a glassy matrix. The optical spectra were shown to depend on crystallite size approaching the Bohr exciton radius of the semiconductor.

A major advance came with the development of synthesis methods to produce measurable quantities of uniform, processable nanocrystals.[59-61] The quantum dots were composed of CdSe and were synthesized *via* a nonhydrolytic molecular precursor route in which surfactants mediated nanocrystal growth and stabilized the nanocrystal surface. From this seminal discovery, in addition to the discovery of fullerenes and the ongoing study of colloids, was born the systematic investigation of uniform, high-quality nanocrystals from synthesis, characterization, and applications. These developments touched off a firestorm of research activity to develop easy, low-cost, robust methods for uniform, high-quality nanocrystals and other types of nanomaterials that have given rise to the fields of 'nanoscience' or 'nanotechnology'. While spherical quantum dots show size-dependent electronic structures, more recent studies have also shown shape-dependent electronic structures.[62]

2.2.6 Example to Illustrate Anomalous TOF Behavior: Ethane Hydrogenolysis on Rhodium

We end section 2.2 with an example from the catalysis literature that illustrates the size-dependent properties of matter. Yates and Sinfelt reported the results of ethane hydrogenolysis on Rh/SiO_2 catalysts.[63] The catalysts were synthesized by impregnation of $RhCl_3$ onto Cabosil, followed by H_2 reduction. Rhodium particle sizes of 11, 12, 20, 23, 41, and 127 Å were obtained. Hydrogen chemisorption was used to determine the Rh particle sizes. Unsupported Rh powder was found to have an average particle size of 2560 Å. The specific activity of catalytic ethane hydrogenolysis varied dramatically over this size range. Rh particle sizes of 11 and 12 Å exhibited a specific activity of 4 mmol C_2H_6 converted/hr/m^{2}Rh. The activity was about four-fold higher for Rh particle sizes of 20, 23, and 41 Å, and decreased 20 to 40-fold for large particle sizes. A peak in TOF was observed for Rh crystallite sizes close to 2 nm.

Schmidt and co-workers[64,65] prepared Rh/SiO_2 catalysts by wetness impregnation of $RhCl_3$ on Cabosil. The Rh loading was varied as 1, 5, and 15 wt% on SiO_2, and led to Rh particle sizes of 12, 32, and 44 nm, respectively. The pretreatment procedure had a significant influence on the catalyst structure and activity. Briefly, when Rh/SiO_2 samples were calcined in O_2, the particles were oxidized to Rh_2O_3. The samples were then reduced at different temperatures under hydrogen. Reduction at 300 °C in H_2 caused the oxide particles to fracture into smaller particles of 5–20 Å, separated by channels or cracks 5–10 Å wide. Increased reduction temperature caused the small particles to become annealed, and resulted in a reformation of the original Rh particle size at 600 °C. The particle structure was characterized with TEM. Intermediate states between fractured particles and fully annealed particles were observed, when reduction temperatures of 400 °C and 500 °C were used. Catalytic ethane hydrogenolysis on the Rh/SiO$_2$ materials was studied. The rate was recorded for the annealed particles, oxidized particles, fractured particles, and partially annealed particles. Interestingly, the rate was highest for the samples annealed at 300 °C, and decreased with increasing annealing temperature. It is also worth noting that the fully annealed particles (600 °C) with sizes of 12, 32, and 44 nm exhibited nearly identical TOFs for ethane hydrogenolysis. The low activity observed for large Rh particle sizes is consistent with the results of Yates and Sinfelt, discussed earlier.

Datye and co-workers examined alkane hydrogenolysis over supported Rh particles and Rh single crystals.[66,67] The Rh(111) and Rh(100) surfaces were selected, and annealed under hydrogen at 773 K. A study of *n*-pentane hydrogenolysis showed that these surfaces exhibited a large difference in the TOFs and activation energies. Supported Rh particles on silica were annealed at identical conditions, and exhibited a TOF that was similar to the Rh(111) crystal. Oxidative pretreatment (O$_2$, 773 K) causes the crystal, and particle, surfaces to roughen through the formation of polycrystalline Rh$_2$O$_3$. Mild reduction (H$_2$, <573 K) resulted in the conversion of the Rh$_2$O$_3$ to Rh, and a highly disordered, non-equilibrium structure was retained. The roughened surfaces were much more active than highly ordered surfaces, which were annealed under H$_2$ at 773 K.

2.3 Synthesis and Properties of Well-defined Nanocrystals

Many industrial catalysts are composed of high surface area porous oxide supports with highly dispersed (or finely divided) metals deposited on the surface. The dispersed metal can be formed on addition of an aqueous metal ion to the support followed by high temperature reduction. The metal particles formed from 'wetness impregnation' typically have a broad size or shape distribution that makes it difficult to assign quantitative relations between properties of matter and catalytic activity. In nanocatalysis, there is an added level of rational design with careful control of uniform, high-quality nanocrystals that allows detailed structure–activity relationships to be documented. The synthesis of nanoparticles can be achieved using a variety of methods, which have been organized within several important review articles.[68–70] The careful investigation of nanocrystal synthesis has yielded numerous important findings.

Nanoparticles can be amorphous, polycrystalline, (multiply)twinned crystals, or single crystals with various degrees of quality in purity and defects such as stacking faults or point defects. The theory of nanocrystal synthesis shows that there is a nucleation event along the reaction coordinate followed by growth.[71,72] The nucleation event should be confined to a single, narrow time domain followed by controlled growth to form monodisperse nanocrystals. During the growth phase, the precursor concentration should be kept low, but non-zero, to achieve size-focusing; otherwise the sample tends to drift to a broader size distribution.[73] At high temperature and without preferential stabilization of facets, isotropic monodisperse spherical nanocrystals form. Remarkable control has been demonstrated such that spherical metal nanocrystals with size distributions within ±5% can be routinely synthesized. Such samples self-organize into colloidal crystal thin films or 3D colloidal crystals.[74]

Studying the kinetics and mechanism of nanocrystal nucleation, growth, and aggregation/agglomeration is exceptionally challenging due to the difficulty of monitoring the reaction *in situ* in a well-defined system. To address this, a platform for investigation of nucleation and growth of metal nanoparticles from molecular precursors was devised that includes a 'reporter reaction' that provides real-time data. The work stemmed from the important observation that late transition metal molecular homogeneous catalysts (polyoxoanion-supported Rh(I) or Ir(I); $(Bu_4N)_5Na_3[(1,5\text{-}COD)MP_2W_{15}Nb_3O_{62})]$ under reducing conditions (H_2) gave rise to metal nanoparticles.[75] Hydrogenation of cyclohexene was utilized as a convenient *in situ* probe for the kinetics of metal cluster formation.[76] The reaction was characterized by a significant delay in the onset of activity that gives rise to a sigmoidal kinetics curve. The product mixture after hydrogenation contained monodisperse metal nanocrystals with a magic number of metal atoms. The formation of the metal particles and the sigmoidal kinetics curve could be modeled precisely with a 2-step nucleation (A → B) and autocatalytic growth mechanism (A + B → 2B).[77] The 2-step mechanism was demonstrated for Rh(I),[78] Ir(I),[77] and likely operates initially for reduction of Pd(II), Pt(IV), Ru(III), Rh(III), Ag(I), Au(III), Cu(II), and Ir(III) in the presence of a base.[79] After further investigation, a 4-step mechanism became apparent that included agglomeration (B + B → C) and autocatalytic agglomeration (B + C → 1.5 C).[80,81] The 4-step mechanism was demonstrated for Pt(II),[80,81] Ir(I),[82] and Au(III).[83] In the case of Ir, the size of cluster B ($Ir_{\sim 900}$) and C ($\geq Ir_{\sim 2000}$) could be estimated.[82] Most recently, the 2-step mechanism was shown to operate for the reduction of supported metal complexes of Ir or Pt on Al_2O_3.[84–85] The kinetics studies provide unprecedented insight into the formation of nanocrystals and supported nanocrystals, and could potentially lead to rational synthesis of nanocatalysts by design.

A high degree of shape control of nanocrystals is possible.[86,87] Surfactants, or other agents such as metal ions, can be chosen that preferentially stabilize a facet of the nanocrystal. In this case, an isotropic shape can form that deviates from the equilibrium spherical shape. Many examples exist of cubes, tetrahedral, octahedral, cubo-octahedra, icosohedra, tetrahexahedra, and other shapes. Another strategy to achieve isotropic non-equilibrium shapes is to passivate a facet upon adsorption of a metal ion. At lower temperatures, anisotropic growth of the nanocrystal can occur due to the difference in activation energy of growth in various crystallographic directions within the nanocrystal, which can lead to nanorods or multipods. It has been shown that twin planes in nanocrystal seeds can be used to control the shape of nanocrystals. Pt tripod nanocrystals form from seeds with a twin plane and stable {111} facets. Overgrowth occurs at the reactive corners in the <2 − 1 − 1> directions.[88]

In some cases, the presence of surfactant on the nanocrystal can induce disruptions of the crystalline order. High quality Ru nanocrystals with weakly bonding acetate stabilizing ligands were synthesized.[89] After exchange with an alkylthiol ligand, the nanoparticles were highly amorphous. Ru nanocrystals stabilized with alkylamine ligands that bind more strongly than acetate and less strongly than alkylthiol were prepared.[90] The nanocrystals show much weaker ligand-induced effects, with only a few stacking faults. Alkylamine stabilized Ru hourglass nanocrystals were found to form through a highly disordered intermediate that crystallizes with time.[91]

2.4 The Dawn of Nanocatalysis and Case Studies

Nanocatalysis involves an intentional imbuement or perturbation of catalytic properties (*via* selection of an electronic structure–surface topology combination) upon synthesis of specific well-defined nanocrystals. Based on the well-developed synthetic methods for nanocrystals, a wide range of tunable morphological properties can be achieved and set as independent variables in carefully designed experiments, such as a catalytic reaction. Therefore, scientists can establish the dependent variables (catalytic activity or selectivity) as a function of morphology in three-dimensional high surface area model catalyst systems. This is the birth of nanocatalysis, which has a foundation built from the convergence of many fundamental principles encapsulated within a vast pyramid of investigation and offers the potential to design optimized catalysts.[92–94] We now highlight some important work in the area of nanocatalysis.[95]

2.4.1 CO Oxidation on Au

Gold clusters can be grown on the surface of rutile TiO_2 (110) single crystal surfaces under ultra-high vacuum conditions. A tungsten filament wrapped with high purity Au wire provides Au atoms with well-calibrated flux. The Au atoms are deposited on the TiO_2 surface at 300 K followed by annealing at 850 K for 2 minutes. With this method, Au clusters (1–6 nm diameter) with relatively narrow size distributions can be formed in which size depends on the initial Au coverage.[96]

The clusters were characterized using scanning tunneling microscopy and scanning tunneling spectroscopy to establish their size and electronic structure. A quantum effect was clearly observed for the set of Au particles. Particles with diameter x height greater than 3.5 nm×1.0 nm showed metallic properties. Below this size, the Au clusters exhibit an increasing band-gap. Clusters with band-gaps in the 0.2–0.6 eV range were found in the 2.5–3.0 nm range.

The Au clusters supported on TiO$_2$ (110) were exposed to CO : O$_2$ mixtures and resulted in CO oxidation at 350 K. The reaction was highly structure-sensitive, and the turnover frequency of the reaction peaked for clusters with diameter ~3.0 nm. The results correlate with data obtained for supported Au-TiO$_2$ catalysts prepared by wetness impregnation of TiO$_2$ with HAuCl$_4$.[97]

2.4.2 TiO$_2$ Nanocrystals with Reactive Facets

Most anatase TiO$_2$ crystals are dominated by the thermodynamically stable {101} facets.[98] Several studies have focused on the preparation of anatase TiO$_2$ with reactive facets, such as {100} and {001}. This has been achieved significantly by the selection of capping agents, solvents and other experimental conditions. Computational investigation found that fluoride-termination led to a reversal in the stability such that the {001} facet was more stable than the {101} facet. The theoretical prediction was followed by experimental synthesis of anatase TiO$_2$ in the presence of fluoride that showed a high percentage of {001} facets.[99] Liu et al developed a modified hydrothermal synthesis of anatase TiO$_2$ nanocrystals with exposed {001} and {110} facets.[100] The system shows enhanced photocatalytic activity for degradation of methylene blue under UV irradiation. High thermal stability allowed the removal of fluorine from the surfaces without altering morphology and the fluorine free samples are used to estimate catalytic activity.

Shape and size controlled uniform anatase TiO$_2$ nanocuboids enclosed by active {100} and {001} facets were synthesized using the ionic liquid 1-butyl-3-methylimidazolium tetrafluoroborate ([bmim][BF$_4$]) as the capping agent.[101] The fluoride ion preferentially stabilized the {001} facet while [bmim]$^+$ stabilized the {100} facet. The obtained nanocrystals exhibit high thermal stability and enhanced photocatalytic activity. The nanocrystals were sufficiently stable to allow removal of the surface fluoride, which was critical to achieve the high photoactivity that exceeded the performance of Degussa P25.

A fluorine-free strategy was developed in which TiO$_2$ nanoparticles with a tetragonal bipyramidal shape enclosed by {101} facets were transformed to truncated tetragonal bipyramids enclosed by {001} and {101} facets, to sixteen-faceted polyhedral particles with exposed {103} and {101} facets, and finally to low aspect ratio tetragonal bipyramidal particles enclosed exclusively by {102} facets.[102] The activity via degradation of methylene blue was compared and the order of reactivity was {001}>{102}≈{103}>{101}. The {102} and {103} facets are similar in surface structure.

The photocatalytic activity of anatase nanocrystals with varying ratios of {101}, {001}, and {010} facets to generate •OH and H$_2$ under UV irradiation was reported.[103] Similar rates of evolution of •OH radicals or H$_2$ were observed in all three samples with fluorine-terminated surfaces. Removal of surface fluorine revealed photoactivity (in both reactions) to be in the order of {010}>{101}>{001}. Both {010} and {001} facets possess 100% Ti$_{5c}$ surface sites that were considered important to photocatalytic activity. Analysis of the electronic band structure led the authors to conclude that the {010}

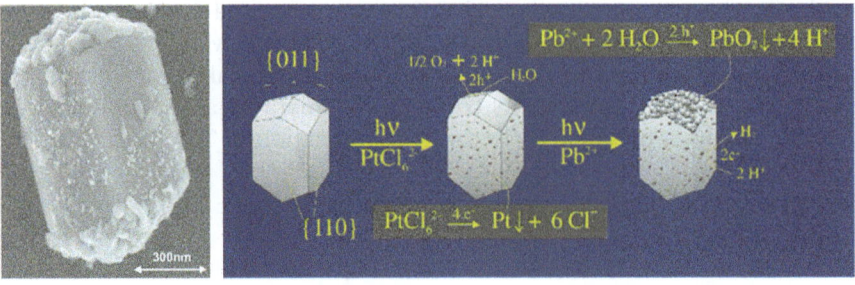

Figure 2.2 Selective growth of PbO$_2$ on the {011} facets of rutile modified with platinum on the {110} facet.
Reprinted from ref. 94 with permission from the Royal Society of Chemistry.

terminated crystal has higher activity due to the slightly larger band-gap; though, the differences are only on the order of a few tens of millivolts.

Exposed facets of nanocrystals influence the trapping and interfacial transfer of photogenerated carriers. UV irradiation of rutile nanocrystals with {011} and {110} facets in aqueous hexachloroplatinate led to deposition of platinum on the {110} facets while water oxidation occurred on the {011} facets. UV irradiation of the Pt-TiO$_2$ powder in aqueous Pb^{2+} led to the growth of PbO$_2$ on the {011} facets (Figure 2.2).[104] Similar behavior was observed with anatase TiO$_2$; however, platinum deposition required addition of isopropanol as a sacrificial oxidant. The platinum was preferentially deposited on the {101} facet of anatase, and PbO$_2$ growth occurred on the {001} facet. The results indicate separation of electrons and holes on different facets of the nanocrystal during irradiation. The effect was stronger for rutile particles than for the anatase particles.

A similar result was found when anatase TiO$_2$ nanocrystals with {101} and {001} facets were evaluated for mineralization of aqueous phenol in the presence of O$_2$. Charge trapping centers on the catalyst surface (Ti^{3+}, O$^-$, and O$_2^-$) that formed upon UV irradiation were quantitatively evaluated using EPR spectroscopy.[105] During exposure to vacuum conditions, the trapped holes (O$^-$ centers) increased with higher proportions of {001} facets; while the Ti^{3+} sites were associated with the {101} surfaces. Such a mechanism can allow for separation of the self-trapped polarons, which could provide a big advantage in photocatalysis. Otherwise, their proximity on the same facet could make recombination more competitive. The photooxidation of phenol was found to increase with nanocrystals that had higher proportions of {001} facets. Because of the correlation between trapped holes and increased activity with increasing ratio of the {001} facet, the mechanism for photooxidation appeared to be driven by trapped holes. The Ti^{3+} sites act as reduction centers that form superoxide in the presence of O$_2$ and contribute to the mineralization of phenol. It remains to be elucidated whether the improved photooxidation of nanocrystals with {100} facets was due to separation of charge-trapped centers that allow simultaneous oxidation *via* O⁻ traps and O_2^- (from Ti^{3+} traps) or if the improvement is only dependent on the concentration of O⁻ traps. These investigations have important implications for future work in nanophotocatalysis.

2.4.3 Catalysis on Shape-controlled Pt Nanocrystals

Electron transfer between hexacyanoferrate(III) and thiosulfate on Pt nanocrystals with tetrahedral, cubic, or spherical shape was studied.[106] The tetrahedral particles are bound by {111} facets whilst cubic particles are bound by {100} facets. The 'near spherical' particles are likely a statistical distribution of truncated octahedra that feature both facets. The reaction was conducted near room temperature and, consequently, there was little change in the shape of the nanocrystals during catalysis. The activation energy of catalytic electron transfer was 14.0 kJ mol⁻¹, 22.6 kJ mol⁻¹, and 26.4 kJ mol⁻¹ for tetrahedral, spherical, and cubic particles, respectively. The result shows that electron transfer was facet- (and therefore shape-) dependent, which is consistent with established structural effects in electrocatalysis on Pt single crystal electrodes.[107]

Monodisperse PVP-stabilized platinum nanocrystals with three shapes—cubes, cubo-octahedra, and octahedra—were synthesized.[108] Shape control was achieved with careful addition of an Ag⁺ ion that enhances the crystal growth rate along [100]. The PVP stabilizer, however, is difficult to remove from the Pt surface and blocks adsorbates thereby lowering catalytic activity. Cetyltrimethylammonium bromide (CTAB) stabilized Pt nanocrystals were synthesized with cubic, cubo-octahedral, and dendritic shapes.[109] CTAB–Pt particles were an order of magnitude more active than PVP–Pt particles for ethylene hydrogenation. Benzene hydrogenation was found to be shape-dependent on the CTAB–Pt nanocrystals.[110] Cyclohexane and cyclohexene were formed on cubo-octahedral nanoparticles; whereas, only cyclohexane was produced on cubic nanoparticles. The results are consistent with data obtained from single crystal studies.[111] These findings implicate nanocrystal morphology as a way to control selectivity in catalysis.

Turnover frequency of ethane hydrogenolysis on platinum supported on mesoporous silica was studied. Monodisperse PVP-stabilized platinum nanoparticles were synthesized with sizes 1.7 – 7.1 nm and added to mesoporous silica supports.[112,113] This model catalyst system differs from a conventional catalyst in its synthesis approach. The Pt nanoparticles were synthesized first using precise chemical methods that yielded monodisperse particle samples. A conventional approach, such as wetness impregnation, can yield much larger particle size distributions. This is a distinction between nanocatalysis and a more conventional catalysis study. Two routes were used for the addition of Pt particles to the support. In the capillary inclusion method, the PVP–Pt particles were combined with SBA-15 and sonicated at room temperature for 3 hours; whereas, in the nanoparticle encapsulation method, the PVP–Pt particles were added to the triblock copolymer surfactant mixture in which mesoporous silica condenses.

The latter method was devised to yield a catalyst with more even distribution of particles throughout the support. The catalysts were calcined at 623–723 K to remove the PVP stabilizer. Chemisorption analysis with H_2, O_2, CO, and H_2–O_2 titration were performed to measure dispersion of the catalyst. The activity for ethylene hydrogenation was measured as a means to calculate dispersion due to its structure-insensitive nature. The dispersions measured from chemisorption were compared with data from TEM and powder XRD measurements, which establish the number of surface sites. The correct tabulation of surface sites is required for the calculation of turnover frequency that enables a true comparison of the catalysts. Ethane hydrogenolysis, known to be a structure-sensitive reaction, was studied on the Pt/SBA-15 catalysts. The turnover frequency increased with decreasing particle size. This was attributed to the higher surface roughness in the smaller particles; however, an electronic structure effect cannot be ruled out.

The catalytic hydrogenation of pyrrole was studied on several Pt catalysts.[114] A series of uniform Pt nanocrystals with a cubic shape with sizes of 5, 6, 7, and 9 nm were prepared, as well as rounded polyhedra with sizes of 3.5 and 5 nm. The nanocrystals were supported on MCF-17 large-pore mesoporous silica. The activity of the catalysts for ethylene hydrogenation exhibited a turnover frequency of $\sim 6 \text{ s}^{-1}$ (10 Torr C_2H_4, 100 Torr H_2, and 298 K) that is similar to Pt(111), and demonstrated active catalyst materials and the structure-insensitive nature of the reaction. Pyrrole hydrogenation can proceed step-wise to form pyrrolidine, *n*-butylamine, and butane/ammonia wherein the latter products arise from C–N bond scission. The reaction was studied in the temperature range 385–415 K. Over this temperature range, cubic Pt nanocrystals showed high selectivity for *n*-butylamine. Interestingly, at lower temperatures, the rounded polyhedra Pt catalyzed formation of *n*-butylamine/pyrrolidine in approximately 2 : 1 ratio that became increasingly selective to *n*-butylamine at higher temperatures. SFG investigation of pyrrole hydrogenation on Pt(111) and Pt(100) suggested a preference for *n*-butylamine on the Pt(100) surface. In comparison, hydrogenation of pyrrole on ~ 1 nm Pt particles over the same temperature range exhibited very different selectivity, and pyrrolidine was the major product.[115] The ~ 1 nm particles were formed from a dendrimer template that confines the size of the metal domain within a very narrow size range. The small particles do not feature any long range surface facets, so the preferential stabilization of *n*-butylamine on Pt(100) was not possible.

Particle shape was found to be important in controlling the *cis–trans* isomerism of olefins.[116] The interactions of *cis*- and *trans*-2-butene were compared for Pt(111), Pt(557), Pt(100), and Pt(110)-(2 × 1) surfaces. Except in the case of Pt(110), *cis*-2-butene desorbs from Pt surfaces at a higher temperature than the *trans* isomer, which indicates a higher stability of the *cis* isomer on the surface. DFT calculations show deformation of the Pt(111) surface on adsorption of 2-butene. The deformation involves movement of a Pt atom out of plane toward the olefin, and is more severe (and energetically unfavorable) for the *trans* isomer. The difference in surface deformation is likely the reason for the slightly higher stability of the *cis* isomer. The data suggest that olefin isomerization on Pt should lead to conversion to the *cis* isomer, as long as the catalyst surfaces exhibit the preferential stabilization of the *cis* isomer. To this end, tetrahedral Pt nanocrystals were prepared and deposited on silica xerogel. The Pt–xerogel SiO_2 catalysts were calcined at 475 K, 525 K, and 575 K. At 475 K, the tetrahedral shapes were retained; however, at 575 K, the particles had become spherical in shape. Isomerization of 2-butene at 375 K was studied over the three catalysts with tetrahedral, intermediate, and spherical shapes. The catalyst with tetrahedral Pt particles exhibited selectivity for the *cis* isomer that greatly diminished with increasing population of spherical particles.

2.4.4 Advanced Templating Methods for Nanocrystal Synthesis

The use of dendrimers has emerged as a tool for the highly precise preparation of nanoparticles in the 1–2 nm range.[117] Dendrimers are somewhat like highly pre-organized micelles. They are polymers that grow radially from a starting molecular block and repeating units that branch. Due to the geometric expansion of the dendrimer with successive generations, their size tends to be limited yet uniform. Functional groups, typically amines, on the repeating units can coordinate metal ions. In this way, the core of a dendrimer can be filled with metal ions, and subsequent reduction leads to formation of a nanoparticle within the dendrimer. Due to the high uniformity of the dendrimer, the nanoparticles are monodisperse. More importantly, it is straightforward to synthesize well-defined bimetallic nanoparticles simply by adjusting the metal ion ratio in the dendrimer.[118] Such control is typically not possible in untemplated nanoparticle synthesis. For example, random alloys of Pt/Cu, Pd/Cu, Pd/Au, Pt/Au, and Au/Cu DENs were synthesized.[119] These materials were used to catalyze reduction of *p*-nitrophenol to *p*-aminophenol by $NaBH_4$. Theoretically, combinations of metals should exhibit unique binding energy to *p*-nitrophenol that governs the reaction rate. Experiments showed this to be true for metal combinations with matching lattice constants; however, metals with lattice mismatch showed bimodal distribution of binding energies. This is an excellent example of synergism between theory and experiment in the design and testing of model nanocatalysts. Whereas dendrimers are limited in size, star-like block copolymers offer the advantages of micellular pre-organization over a larger size range.[120] Both of these phenomena are made possible through the use of uniform, high-quality polymers, and there is an interesting parallel between monodisperse polymers and the ability to organize matter at the nanoscale.

2.4.5 Hybrid Nanocrystal Catalysis

Hybrid nanocrystals are nanoparticles that feature multiple crystal phases, and are synthesized by a seeded growth mechanism wherein the seed and growing nanocrystal are of different compositions and crystal phases. Reactive facets or surface defects on the seed crystal serve as nucleation sites to form the hybrid nanocrystal. Given the seed is a well-defined nanocrystal, it is theoretically possible to construct the hybrid nanocrystal through a rational approach upon selective nucleation of the new phase at specific positions on the seed. Hybrid nanocrystals that consisted of CdSe nanorods with Au tips were synthesized according to this principle.[121] Other examples have appeared in the literature, and excellent reviews can be found in the literature.[122] An attractive feature of hybrid nanocrystals is the ability to join completely different materials at small length scales. For instance, one may combine a semiconductor particle with a magnetic particle to build a multifunctional nanocrystal. From a catalysis perspective, there is an enormous opportunity to capitalize on studying the effect of material combinations in catalytic reactions.

Metal–semiconductor hybrid nanocrystals have attracted attention as photocatalyst materials.[123] Excitation of the semiconductor component results in an exciton that can become separated as the electron becomes trapped on the metal domain. Careful selection of the metal and semiconductor domain in which they catalyze reduction and oxidation half-reactions, respectively, should give rise to an enhanced photocatalytic effect. This idea was the impetus for the study of modified titania colloids that could achieve water splitting.[124] New metal–semiconductor hybrid nanocrystals based on semiconductor nanorods can offer the specific advantages of high longitudinal mobility of charge carriers and well-defined sites for selective nucleation of catalyst particles. Ultrafast charge separation in matchstick type Au–CdS hybrid nanocrystals was studied using an optical time-resolved pump probe technique and probing the light-induced change of their optical response.[125] Electron–hole pairs photoexcited in the semiconductor part of the nanohybrids are shown to undergo rapid charge separation with the electron transferred to the metal part on a sub-20 fs time scale.

2.5 Conclusion

Catalysis began with a tradition of empirical work that evolved into a rational design of materials to attain improved performance. The development of model systems has been critical to the understanding of catalytic processes. The study of chemistry on surfaces has had profound implications, such as identification of surface species, the role of crystal facets in adsorption energy or reaction rates, restructuring processes, and surface mobility. Model three-dimensional systems were lacking until it became possible to synthesize monodisperse nanocrystals. Advances in nanocrystal synthesis methods extend the limits of model catalyst systems, and will surely open the door to new performance possibilities in catalysis. The ability to precisely control matter at the nanoscale, therefore its electronic structure and surface features, is at the heart of nanocatalysis.

While numerous opportunities for study exist to evaluate the catalytic performance of existing series of nanocrystals with tunable morphological properties, there is a parallel opportunity to learn how to arrange the nanocrystal components in space and develop multifunctional materials. The focus of this chapter was to illustrate size- and shape-effects in nanocatalysis; however, studies in this field should not be limited to this idea.

References

1. I. Chorkendorff and J. W. Niemantsverdriet, *Concepts of Modern Catalysis*, Wiley-VCH, Weinheim, Germany, 1st edn, 2003.
2. J. H. Sinfelt, *Surf. Sci.*, 2002, **500**, 923–946.
3. More detailed historic accounts can be found in:(a) A. J. B. Robertson, *Platinum Met. Rev.*, 2002, **19**, 64–69; (b) J. K. Smith, *History of Catalysis*, in *Encyclopedia of Catalysis*, John Wiley & Sons, Inc., 2002; (c) J. N. Armor, *Catal. Today*, 2011, **163**, 3–9.
4. M. Che and C. O. Bennett, *Adv. Catal.*, 1989, **36**, 55–172.
5. D. R. Rolison, *Science*, 2003, **299**, 1698–1701.
6. P. L. J. Gunter, J. W. Niemantsverdriet, F. H. Ribiero and G. A. Somorjai, *Catal. Rev.: Sci. Eng.*, 1997, **39**, 77–168.
7. M. Boudart, *Chem. Rev.*, 1995, **95**, 661–666.
8. It is important to point out that reaction order can be temperature dependent!.
9. T. Kwan, *J. Phys. Chem.*, 1956, **60**, 1033–1037.
10. H. Song, R. M. Rioux, J. D. Hoefelmeyer, R. Komor, K. Niesz, M. Grass, P. Yang and G. A. Somorjai, *J. Am. Chem. Soc.*, 2006, **128**, 3027–3037.
11. R. M. Rioux, H. Song, J. D. Hoefelmeyer, P. Yang and G. A. Somorjai, *J. Phys. Chem. B*, 2005, **109**, 2192–2202.
12. R. Van Hardeveld and F. Hartog, *Surf. Sci.*, 1969, **15**, 189–230.
13. S. Maksimuk, X. Teng and H. Yang, *J. Phys. Chem. C*, 2007, **111**, 14312–14319.
14. M. J. Yacamán, S. Fuentes and J. M. Dominguez, *Surf. Sci.*, 1981, **106**, 472–477.
15. M. Gillet and A. Renou, *Surf. Sci.*, 1979, **90**, 91–101.
16. B. Coq and F. Figueras, *Coord. Chem. Rev.*, 1998, **178**, 1753–1783.
17. A. S. McLeod and L. F. Gladden, *J. Catal.*, 1998, **173**, 43–52.
18. L. F. Gracia and E. E. Wolf, *Chem. Eng. J.*, 2001, **82**, 291–301.
19. S. Ladas, *Surf. Sci.*, 1986, **175**, L681–L686.
20. R. B. Greegor and F. W. Lytle, *J. Catal.*, 1980, **63**, 476–486.
21. J. J. Burton, *Catal. Rev.*, 1974, **9**, 209–222.
22. G. Wulff, *Z. Kristallogr.*, 1901, **34**, 449–530.
23. (a) L. D. Marks, *Surf. Sci.*, 1985, **150**, 358–366; (b) L. D. Marks, *Rep. Prog. Phys.*, 1994, **57**, 603–649.
24. M. Mavrikakis, M. Baumer, H. J. Freund and J. K. Norskov, *Catal. Lett.*, 2002, **81**, 153–156.

25. N. D. Spencer, R. C. Schoonmaker and G. A. Somorjai, *J. Catal.*, 1982, **74**, 129–135.
26. D. R. Strongin, J. Carrazza, S. R. Bare and G. A. Somorjai, *J. Catal.*, 1987, **103**, 213–215.
27. Z. Quan, Y. Wang and J. Fang, *Acc. Chem. Res.*, 2013, **46**, 191–202.
28. L. Zhang, W. Niu and G. Xu, *Nano Today*, 2012, 7, 586–605.
29. F. Besenbacher, *Rep. Prog. Phys.*, 1996, **59**, 1737–1802.
30. C. R. Henry, *Surf. Sci. Rep.*, 1998, **31**, 235–325.
31. F. Gao and D. W. Goodman, *Ann. Rev. Phys. Chem.*, 2012, **63**, 265–286.
32. G. A. Somorjai, *Appl. Surf. Sci.*, 1997, **121**, 1–19.
33. H. Bao, W. Zhang, Q. Hua, Z. Jiang, J. Yang and W. Huang, *Angew. Chem., Int. Ed.*, 2011, **50**, 12294–12298.
34. M. D. Ackermann, T. M. Pedersen, B. L. M. Hendriksen, O. Robach, S. C. Bobaru, I. Popa, C. Quiros, H. Kim, B. Hammer, S. Ferrer and J. W. M. Frenken, *Phys. Rev. Lett.*, 2005, **95**, 255505(1–4).
35. F. Tao, M. E. Grass, Y. W. Zhang, D. R. Butcher, J. R. Renzas, Z. Liu, J. Y. Chung, B. S. Mun, M. Salmeron and G. A. Somorjai, *Science*, 2008, **322**, 932–934.
36. R. Choukroun, D. de Caro, B. Chaudret, P. Lecante and E. Snoeck, *New J. Chem.*, 2001, **25**, 525–527.
37. D. C. Tang, K. S. Hwang, M. Salmeron and G. A. Somorjai, *J. Phys. Chem. B*, 2004, **108**, 13300–13306.
38. J. Frenkel, *Phys. Rev.*, 1931, **37**, 17–44.
39. J. Frenkel, *Phys. Rev.*, 1931, **37**, 1276–1294.
40. R. E. Peierls, *Ann. Physik.*, 1932, **13**, 905–952.
41. J. Frenkel, *Physik. Z. Sowjetunion*, 1936, **9**, 158–186.
42. G. H. Wannier, *Phys. Rev.*, 1937, **52**, 191–197.
43. G. Dresselhaus, *J. Phys. Chem. Solids*, 1956, **1**, 14–22.
44. R. S. Knox, *Theory of Excitons*, Academic Press, New York, US, 1963.
45. (a) H. Haken and W. Schottky, *Z. Physik. Chem.*, 1958, **16**, 218–244; (b) F. Englert, *Phys. Chem. Solids*, 1959, **11**, 78–91.
46. D. C. Mattis and G. Beni, *Phys. Rev. B*, 1978, **18**, 3816–3819.
47. Al. L. Efros and A. L. Efros, *Fiz. Tekh. Poluprovodn.*, 1982, **16**, 1209–1214.
48. L. E. Brus, *J. Chem. Phys.*, 1984, **80**, 4403–4409.
49. Y. Kayanuma, *Phys. Rev. B*, 1988, **38**, 9797–9805.
50. H. Weller, *Angew. Chem., Int. Ed.*, 1993, **32**, 41–53.
51. A. P. Alivisatos, *J. Phys. Chem.*, 1996, **100**, 13226–13239.
52. C. P. Vinod, G. U. Kulkarni and C. N. R. Rao, *Chem. Phys. Lett.*, 1998, **289**, 329–333.
53. Y. Jinlong, F. Toigo and W. Kelin, *Phys. Rev. B*, 1994, **50**, 7915–7924.
54. C. H. Chien, E. Blaisten-Barojas and M. R. Pederson, *J. Chem. Phys.*, 2000, **112**, 2301–2307.
55. R. L. Johnston, *Philos. Trans. R. Soc., A*, 1998, **356**, 211–230.
56. A. I. Ekimov and A. A. Onushchenko, *Sov. Phys. Semicond.*, 1982, **16**, 775–778.
57. A I. Ekimov and A. A. Onushchenko, *JETP Lett.*, 1984, **40**, 1136–1139.

58. A. I. Ekimov and A. A. Onushchenko, *ZhETF Pis ma Redaktsiiu*, 1981, **34**, 363–366.
59. A. P. Alivisatos, A. L. Harris, N. J. Levinos, M. L. Steigerwald and L. E. Brus, *J. Chem. Phys.*, 1988, **89**, 4001–4011.
60. M. L. Steigerwald, A. P. Alivisatos, J. M. Gibson, T. D. Harris, R. Kortan, A. J. Muller, A. M. Thayer, T. M. Duncan, D. C. Douglass and L. E. Brus, *J. Am. Chem. Soc.*, 1988, **110**, 3046–3050.
61. C. Murray, D. J. Norris and M. G. Bawendi, *J. Am. Chem. Soc.*, 1993, **115**, 8706–8715.
62. J. Li and L. W. Wang, *Nano Lett.*, 2003, **3**, 1357–1363.
63. D. J. C. Yates and J. H. Sinfelt, *J. Catal.*, 1967, **8**, 348–358.
64. C. Lee and L. D. Schmidt, *J. Catal.*, 1986, **101**, 123–131.
65. S. Gao and L. D. Schmidt, *J. Catal.*, 1988, **111**, 210–219.
66. A. D. Logan, K. Sharoudi and A. K. Datye, *J. Phys. Chem.*, 1991, **95**, 5568–5574.
67. D. Kalakkad, S. L. Anderson, A. D. Logan, J. Peña, E. J. Braunschweig, C. H. F. Peden and A. K. Datye, *J. Phys. Chem.*, 1993, **97**, 1437–1444.
68. A. Roucoux, J. Schulz and H. Patin, *Chem. Rev.*, 2002, **102**, 3757–3778.
69. A. R. Tao, S. Habas and P. D. Yang, *Small*, 2008, **4**, 310–325.
70. Y. Xia, Y. Xiong, B. Lim and S. E. Skrabalak, *Angew. Chem., Int. Ed.*, 2009, **48**, 60–103.
71. C. B. Murray, C. R. Kagan and M. G. Bawendi, *Annu. Rev. Mater. Sci.*, 2000, **30**, 545–610.
72. C. D. Dushkin, S. Saita, K. Yoshie and Y. Yamaguchi, *Adv. Colloid Interface Sci.*, 2000, **88**, 37–78.
73. J. Park, J. Joo, S. G. Kwon, Y. Jang and T. Hyeon, *Angew. Chem., Int. Ed.*, 2007, **46**, 4630–4660.
74. C. B. Murray, C. R. Kagan and M. G. Bawendi, *Science*, 1995, **270**, 1335–1338.
75. J. A. Widegren and R. G. Finke, *J. Mol. Catal. A*, 2003, **198**(1–2), 317–341.
76. Y. Lin and R. G. Finke, *Inorg. Chem.*, 1994, **33**, 4891–4910.
77. M. A. Watzky and R. G. Finke, *J. Am. Chem. Soc.*, 1997, **119**(43), 10382–10400.
78. J. D. Aiken and R. G. Finke, *Chem. Mater.*, 1999, **11**(4), 1035–1047.
79. J. A. Widegren, J. D. Aiken, S. Ozkar and R. G. Finke, *Chem. Mater.*, 2001, **13**(2), 312–324.
80. C. Besson, E. E. Finney and R. G. Finke, *J. Am. Chem. Soc.*, 2005, **127**(22), 8179–8184.
81. C. Besson, E. E. Finney and R. G. Finke, *Chem. Mater.*, 2005, **17**(20), 4925–4938.
82. L. S. Ott and R. G. Finke, *Chem. Mater.*, 2008, **20**(7), 2592–2601.
83. E. E. Finney, S. P. Shields, W. E. Buhro and R. G. Finke, *Chem. Mater.*, 2012, **24**(10), 1718–1725.
84. J. E. Mondloch, X. H. Yan and R. G. Finke, *J. Am. Chem. Soc.*, 2009, **131**(18), 6389–6396.
85. J. E. Mondloch and R. G. Finke, *ACS Catal.*, 2012, **2**(2), 298–305.

86. Y. W. Jun, J. H. Lee, J. S. Choi and J. Cheon, *J. Phys. Chem. B*, 2005, **109**, 14795–14806.
87. Y. W. Jun, J. W. Seo, S. J. Oh and J. Cheon, *Coord. Chem. Rev.*, 2005, **249**, 1766–1775.
88. S. Maksimuk, X. Teng and H. Yang, *J. Phys. Chem. C*, 2007, **111**, 14312–14319.
89. G. Viau, R. Brayner, L. Poul, N. Chakroune, E. Lacaze, F. Fievet-Vincent and F. Fievet, *Chem. Mater.*, 2003, **15**, 486–494.
90. M. W. Brink, M. A. Peck, K. L. More and J. D. Hoefelmeyer, *J. Phys. Chem. C*, 2008, **112**, 12122–12126.
91. J. Watt, C. Yu, S. L. Y. Chang, S. Cheong and R. D. Tilley, *J. Am. Chem. Soc.*, 2013, **135**, 606–609.
92. G. A. Somorjai, F. Tao and J. Y. Park, *Top. Catal.*, 2008, **47**, 1–14.
93. R. Schlogl and S. B. Abd Hamid, *Angew. Chem., Int. Ed.*, 2004, **43**, 1628–1637.
94. G. A. Somorjai and J. Y. Park, *Top. Catal.*, 2008, **49**, 126–135.
95. This is not a comprehensive account. We note an excellent special issue 'Nanoparticles for Catalysis' in *Acc. Chem. Res.*, 2013, **46**(8), 1671 that includes recent major developments.
96. M. Valden, X. Lai and D. W. Goodman, *Science*, 1998, **281**, 1647–1650.
97. G. R. Bamwenda, S. Tsubota, T. Nakamura and M. Haruta, *Catal. Lett.*, 1997, **44**, 83–87.
98. U. Diebold, *Surf. Sci. Rep.*, 2003, **48**, 53–229.
99. H. G. Yang, C. H. Sun, S. Z. Qiao, J. Zou, G. Liu, S. C. Smith, H. M. Cheng and G. Q. Lu, *Nature*, 2008, **453**, 638–641.
100. M. Liu, L. Piao, L. Zhao, S. Ju, Z. Yan, T. He, T. He, C. Zhou and W. Wang, *Chem. Commun.*, 2010, **46**, 1664–1666.
101. X. Zhao, W. Jin, J. Cai, J. Ye, Z. Li, Y. Ma, J. Xie and L. Qi, *Adv. Funct. Mater.*, 2011, **21**, 3554–3563.
102. X. Han, B. Zheng, J. Ouyang, X. Wang, Q. Kuang, Y. Jiang, Z. Xie and L. Zheng, *Chem.–Asian J.*, 2012, 7, 2538–2542.
103. J. Pan, G. Liu, G. Q. M. Lu and H. M. Cheng, *Angew. Chem., Int. Ed.*, 2011, **50**, 2133–2137.
104. T. Ohno, K. Sarukawa and M. Matsumura, *New J. Chem.*, 2002, **26**, 1167–1170.
105. M. D'Arienzo, J. Carbajo, A. Bahamonde, M. Crippa, S. Polizzi, R. Scotti, L. Wahba and F. Morazzoni, *J. Am. Chem. Soc.*, 2011, **133**, 17652–17661.
106. R. Narayanan and M. A. El-Sayed, *Nano Lett.*, 2004, **4**, 1343–1348.
107. For example:(a) J. Clavilier, R. Parsons, R. Durand, C. Lamy and J. M. Leger, *J. Electroanal. Chem.*, 1981, **124**, 321–326; (b) C. Lamy, J. M. Leger, J. Clavilier and R. Parsons, *J. Electroanal. Chem.*, 1983, **150**, 71–77.
108. H. Song, F. Kim, S. Connor, G. A. Somorjai and P. D. Yang, *J. Phys. Chem. B*, 2005, **109**, 188–193.
109. H. Lee, S. E. Habas, S. Kweskin, D. Butcher, G. A. Somorjai and P. D. Yang, *Angew. Chem., Int. Ed.*, 2006, **45**, 7824–7828.

110. K. M. Bratlie, H. Lee, K. Komvopoulos, P. D. Yang and G. A. Somorjai, *Nano Lett.*, 2007, **7**, 3097–3101.
111. (a) K. M. Bratlie, L. D. Flores and G. A. Somorjai, *J. Phys. Chem. B*, 2006, **110**, 10051–10057; (b) K. M. Bratlie, C. J. Kliewer and G. A. Somorjai, *J. Phys. Chem. B*, 2006, **110**, 17925–17930.
112. R. M. Rioux, H. Song, J. D. Hoefelmeyer, P. D. Yang and G. A. Somorjai, *J. Phys. Chem. B*, 2005, **109**, 2192–2202.
113. H. Song, R. M. Rioux, J. D. Hoefelmeyer, R. Komor, K. Niesz, M. Grass, P. D. Yang and G. A. Somorjai, *J. Am. Chem. Soc.*, 2006, **128**, 3027–3037.
114. C. K. Tsung, J. N. Kuhn, W. Huang, C. Aliaga, L. I. Hung, G. A. Somorjai and P. Yang, *J. Am. Chem. Soc.*, 2009, **131**, 5816–5822.
115. W. Huang, J. N. Kuhn, C. K. Tsung, Y. Zhang, S. E. Habas, P. Yang and G. A. Somorjai, *Nano Lett.*, 2008, **8**, 2027–2034.
116. I. Lee, F. Delbecq, R. Morales, M. A. Albiter and F. Zaera, *Nat. Mater.*, 2009, **8**, 132–138.
117. R. M. Crooks, M. Zhao, L. Sun, V. Chechik and L. K. Yeung, *Acc. Chem. Res.*, 2001, **34**, 181–190.
118. X. H. Peng, Q. M. Pan and G. L. Rempel, *Chem. Soc. Rev.*, 2008, **37**, 1619–1628.
119. Z. D. Pozun, S. E. Rodenbusch, E. Keller, K. Tran, W. Tang, K. J. Stevenson and G. Henkelman, *J. Phys. Chem. C*, 2013, **117**, 7598–7604.
120. (a) X. Pang, L. Zhao, M. Akinc, J. K. Kim and Z. Lin, *Macromolecules.*, 2011, **44**, 3746–3752; (b) X. C. Pang, L. Zhao, W. Han, X. K. Xin and Z. Q. Lin, *Nat. Nanotechnol.*, 2013, **8**, 426–431.
121. T. Mokari, E. Rothenberg, I. Popov, R. Costi and U. Banin, *Science*, 2004, **304**, 1787–1790.
122. (a) R. Costi, A. E. Saunders and U. Banin, *Angew. Chem., Int. Ed.*, 2010, **49**, 4878–4897; (b) P. D. Cozzoli, T. Pellegrino and L. Manna, *Chem. Soc. Rev.*, 2006, **35**, 1195–1208; (c) C. D. Donega, *Chem. Soc. Rev.*, 2011, **40**, 1512–1546; (d) L. Carbone and P. D. Cozzoli, *Nano Today*, 2010, **5**, 449–493.
123. S. Rawalekar and T. Mokari, *Adv. Energy Mater.*, 2013, **3**, 12–27.
124. E. Borgarello, J. Kiwi, E. Pelizzetti, M. Visca and M. Gratzel, *Nature*, 1981, **289**, 158–160.
125. D. Mongin, E. Shaviv, P. Maioli, A. Crut, U. Banin, N. Del Fatti and F. Vallee, *ACS Nano*, 2012, **6**, 7034–7043.

CHAPTER 3

New Strategies to Fabricate Nanostructured Colloidal and Supported Metal Nanoparticles and their Efficient Catalytic Applications

KOHSUKE MORI[a,b] AND HIROMI YAMASHITA*[a,b]

[a] Division of Materials and Manufacturing Science, Graduate School of Engineering, Osaka University, 2-1 Yamada-oka, Suita, Osaka 565-0871, Japan; [b] ESICB, Kyoto University, Katsura, Kyoto 615-8520, Japan
*Email: yamashita@mat.eng.osaka-u.ac.jp

3.1 Introduction

Among the various catalyst materials, metal NPs (1 to 100 nm) in particular are gaining increasing attention because of their existence on the borderline molecular states with discrete quantum energy levels.[1–3] Thus, NP-based catalysts are considered to bridge the gap between mononuclear metal complexes and heterogeneous bulk catalysts; this system is often referred to as a "quasihomogeneous system".[4] Their large surface area-to-volume ratio and the high concentration of low coordination sites and surface vacancies allow effective utilization of expensive catalyst metals.

The general routes to NP synthesis are based on the chemical reduction of metal precursors with the appropriate reducing reagents in the presence of stabilizing organic ligands. Control of the primary structures, such as size, composition and morphology, is widely utilized to tune their catalytic activities and selectivities. Protective ligands can also impart their selective solubility, resistance to aggregation and derivatization with functional groups. Promising synthetic methods providing suitable NPs responsible for target catalytic reactions are extremely important. The NP-based catalysts are key components of catalytic activities, but they often encounter deactivation caused by sintering and/or particle agglomeration during the catalytic reactions. Moreover, their applications in liquid suspensions are limited due to the difficulties with product separation and catalyst recycling. In order to overcome these drawbacks, catalytic NPs have generally been immobilized on solid supports (*e.g.*, carbon, metal oxides and zeolites) or stabilized by capping with large polymers.[5,6]

Supported NP catalysts are widely recognized as an important class of industrial catalysts that are closely related with versatile key technologies in the petrochemical industries, in the conversion of automobile exhausts, in chemical sensors, and in the manufacturing of fine and specialty chemicals. There are a number of criteria for practical supported metal catalysts including specific catalyst–support interaction, resistance to agglomerization, site isolation, good accessibility of substrate molecules, mechanical robustness and low synthetic cost, *etc*. The most convenient methods for attaining the immobilization of metal NPs on high surface area support materials are the incipient wetness technique and the ion-exchange method, in which the support material is impregnated with metal precursors in the solution phase, followed by activation under a reducing atmosphere.[7,8] Owing to its inherent simplicity, this is successful for the large scale production of catalysts, but often results in non-uniform distribution of the active phase on the support. A novel and radically different approach for the generation of supported metal catalysts is desirable in order to develop efficient and practical catalysts.

In this chapter, we highlight our recent advancement in the design and architecture of the NP-based catalysts and their efficient application. Owing to the dramatic growth of this research field, we sincerely hope that this review can offer a helpful overview of this fascinating area to the readers.

3.2 New Route for the Preparation of Supported Metal Nanoparticle Catalysts

3.2.1 A Photo-assisted Deposition Method Using a Single-site Photocatalyst

Single-site photocatalysts, exemplified by Ti-containing zeolites and Ti-containing mesoporous silicas, have been shown to exhibit exceptional photocatalytic activities compared to bulk TiO_2.[9–11] These differences can be explained by considering their photo-excitation mechanism. In the case of

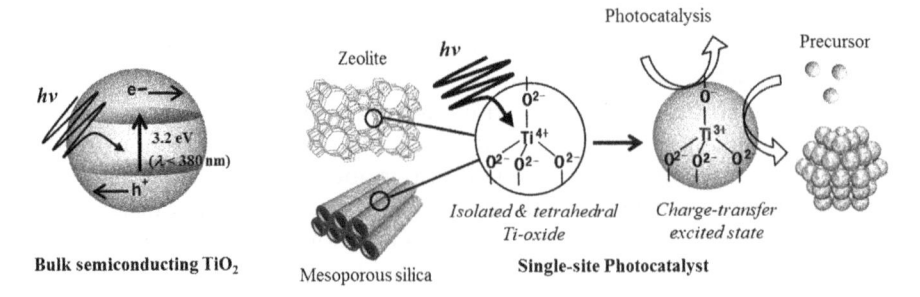

Figure 3.1 (A) Excitation by UV light and charge separation in the semiconducting bulk TiO$_2$. (B) Formation of charge-transfer excited state in the single-site photocatalyst under UV light irradiation and its application for the synthesis of nano-sized metal particles.

bulk TiO$_2$, electrons and holes in the conduction and valence bands formed *via* photoexcitation contribute to the reduction and oxidation reaction, respectively (Figure 3.1A).[12] On the other hand, in the case of single-site photocatalysts, excitation by light at around 230–270 nm brings about an electron transfer from the oxygen to Ti^{4+} ions, resulting in the formation of pairs of trapped hole centers (O$^-$) and electron centers (Ti^{3+}).[13,14] These charge transfer excited states, *i.e.* the electron hole pair state which localize quite near to each other compared with the electrons and holes generated in TiO$_2$, play a crucial role in photocatalytic reactions (Figure 3.1B). It can be expected that the metal precursor species can be easily deposited on the excited state of a single-site photocatalyst to form well-controlled metal NPs from the mixture of single-site photocatalysts in the aqueous solution of metal precursors. Recently, we have successfully utilized single-site photocatalysts as a platform for the synthesis of nano-size metal catalysts by the advanced photo-assisted deposition (PAD) method.[15–20]

Under UV-light irradiation of a slurry of Ti-containing zeolites (TS-1, Pd/Ti = 0.6) in an aqueous PdCl$_2$ solution, the Pd metal can be successfully deposited on the TS-1. No Pd deposition was observed both on the silicalite zeolite without Ti-oxide under UV-light irradiation and the TS-1 zeolite without UV-light irradiation. These results suggest that the Pd precursor underwent anchoring on the single framework Ti^{4+} center of TS-1 under UV-light irradiation. The subsequent calcination at 723 K for 5 h and reduction with H$_2$ at 473 K for 1 h generates nano-sized Pd metal particles (PAD-Pd/TS-1).

The TEM images showed that the PAD-Pd/TS-1 exhibits nano-sized Pd metals with a mean diameter of *ca.* 2.1 nm and has a narrow size distribution, while the imp-Pd/TS-1, which was prepared by a conventional impregnation method, had Pd metal particles with various sizes within the range of 2–15 nm. These results clearly suggest that the size of the Pd metal particles depends on the preparation method, and that the smaller Pd metal particles are formed on the photo-deposited catalyst rather than on the impregnated catalyst.

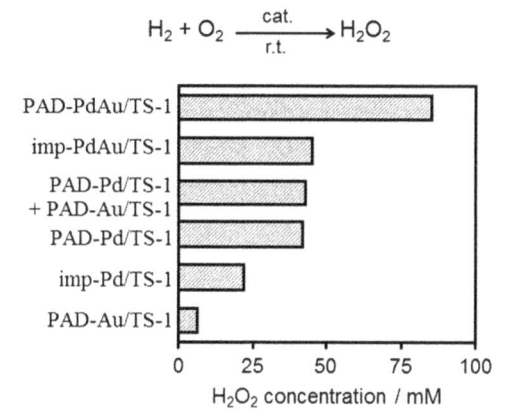

New Strategies to Fabricate Nanostructured Colloidal and Supported Metal NPs 33

$$H_2 + O_2 \xrightarrow[\text{r.t.}]{\text{cat.}} H_2O_2$$

Figure 3.2 The effect of catalysts on concentration of H_2O_2. Reaction conditions: catalyst (0.1 g), 0.01 M HCl (50 mL), 293 K, 6 h, H_2 and O_2 (80 ml min^{-1}, $H_2 : O_2 = 1 : 1$).

The nano-sized Pd catalysts can be applied to catalytic reactions with high efficiency. With the flow of H_2 and O_2 into an aqueous slurry of Pd/TS-1 under atmospheric pressure at room temperature, H_2O_2 can be produced. The PAD-Pd/TS-1 can exhibit higher activity than the imp-Pd/TS-1 (Figure 3.2). The high dispersion of Pd metal particles in the PAD-Pd/TS-1 is preferable for the formation of H_2O_2.

The PAD method also provides PdAu bimetallic nanoparticles from an aqueous solution of a mixture of PdCl$_2$ and HAuCl$_4$.[21] The TEM image of PAD-PdAu/TS-1 exhibited larger particles that ranged in size from 12–15 nm with a composition of 80 mol% Au – 20 mol% Pd. Pd K-edge and Au L$_{III}$-edge XAFS results support the existence of Au-rich bimetallic particles by the present PAD method. For the structural model of the PdAu nanoparticles, it is reasonable to suggest that most of the Au atoms are preferentially located in the core region, while the Pd atoms are preferentially located in the shell region, since the reduction potentials of Au ions are more positive than Pd ions ($E^0(Pd^{2+}/Pd^0) = +0.63$ V, $E^0(Au^{3+}/Au^0) = +1.0$ V vs. NHE).[22]

As shown in Figure 3.2, PAD-PdAu/TS-1 exhibited the highest H_2O_2 formation rate, which was significantly better than that of imp-PdAu/TS-1 and PAD-Pd/TS-1. The reaction rate of pure PAD-Au/TS-1 was extremely low under identical reaction conditions. The enhancement effect was almost negligible when the physical mixture of pure PAD-Pd/TS-1 and PAD-Au/TS-1 was employed. These results indicate that a synergic effect exists between Au and Pd atoms for the TS-1-supported catalysts prepared by photo-irradiation. The main origin of the enhanced catalytic activity can be explained by the higher selective formation of H_2O_2 due to the electronic promotion of Pd by the addition of Au.

The PAD method also enables the preparation of nano-sized Pd-Ni bimetal particles with a narrow size distribution on Ti-containing mesoporous silica

(Ti-HMS).[23] The PAD-PdNi/Ti-HMS catalyst exhibited specifically higher hydrogenation activity than imp-PdNi/Ti-HMS prepared by conventional methods. In particular, it was proven that the PAD-PdNi/Ti-HMS with Ni/Pd = 1.5 mol% exhibited a strong synergistic effect. These studies verified a promising approach for the preparation of supported metal catalysts, and we expect that this synthetic strategy will be extendable in the preparation of other bimetallic nanoparticles as unique catalysts and/or photocatalysts.

3.2.2 A Microwave-assisted Deposition Method

Microwave dielectric heating has been attracting a great deal of attention as a new and promising method in the organic and inorganic synthesis fields. It is widely accepted that microwave irradiation enables rapid, uniform, and energy efficient heating, and it is also applicable to the simple preparation of metallic nanostructured materials.[24,25] As a supported system, carbon-supported and polymer stabilized colloids have been synthesized[26]. Although much studies dealing with the preparation of metal colloids and clusters are reported, the microwave-assisted synthesis of metal particles supported on silica based porous materials is unexplored so far.

We have found that Ti-containing mesoporous silica (Ti-HMS) also acts as a promising platform to deposit uniform and small metal particles under microwave irradiation (500 W, 2450 ± 30 MHz).[27] The precursor of Pt metal can be successfully deposited on the Ti-HMS under microwave irradiation of a slurry of Ti-HMS in an aqueous H_2PtCl_6 solution. On the other hand, the Pt metal precursor could not be deposited on the HMS support without Ti-oxide moieties under identical conditions. Without microwave irradiation conditions, the deposition hardly occurred on the Ti-HMS support. These results suggested that the existence of the tetrahedrally coordinated Ti-oxide moieties and microwave irradiation were indispensable for attaining the deposition of Pt precursors. Characterization by CO adsorption, XAFS, and TEM analysis revealed that the size of metal particles depends on the preparation methods and that smaller metal particles were formed on the microwave-assisted metal catalysts compared to the conventionally prepared impregnated catalysts. These nano-sized metal catalysts are useful as efficient catalysts for various reactions such as the hydrogenation of nitrobenzene and oxidation of CO.

The Mw heating method was also found to be applicable to semiconducting anatase-type TiO_2 to exploit the highly efficient photocatalyst for the formation of hydrogen (H_2) and nitrogen (N_2) gases from aqueous ammonia (NH_3) as harmful nitrogen-containing chemical waste.[28] The effect of the Pt deposition method on the photocatalytic activity was examined using 0.3 wt% Pt-TiO_2 prepared by various methods (microwave: Mw, PAD, equilibrium adsorption: EA, imp). As shown in Figure 3.3, PAD-, EA- and imp-Pt-TiO_2 also exhibited stable H_2 and N_2 formation activities and stoichiometries as those of Mw-Pt-TiO_2, but the H_2 and N_2 formation activities decreased in

$$2NH_3 \xrightarrow[h\nu]{Pt/TiO_2} 3H_2 + N_2$$

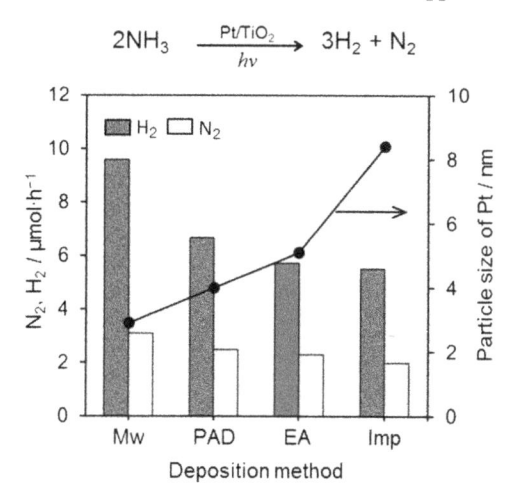

Figure 3.3 Effect of preparation method of 0.3 wt% Pt-TiO$_2$ and Pt metal average particle size on the photocatalytic decomposition of aqueous NH$_3$ under UV-light irradiation.

the order of Mw > PAD > imp ≈ EAD. It should be noted that the order of gas formation activity was in close agreement with the order of decreasing Pt nanoparticle size calculated using pulsed CO adsorption measurements. The small and highly dispersed Pt nanoparticles loaded on TiO$_2$ exhibited a high surface area and low photonic shielding effect by the Pt particles in the photocatalytic reaction, resulting in a more efficient reaction.

Moreover, in the presence of surface-modifying ligands possessing different carbon chain lengths such as sodium laurate (C, 12) and sodium stearate (C, 18), size-controlled Cu nanoparticles with average particle sizes of 4.5, 6.3, and 7.3 nm can be prepared on ZnO/SBA-15 mesoporous silica by microwave-assisted alcohol reduction[29] and the efficient hydrogenation reaction of p-nitrophenol to p-aminophenol was achieved by Cu/ZnO/SBA-15 with Cu nanoparticles. It should be noted that the order of hydrogenation activity was in close agreement with the order of decreasing Cu nanoparticle size.

3.2.3 Deposition of Size-controlled Metal Nanoparticles as Colloidal Precursors

A rational synthetic method of supported metal catalysts has also been realized by employing Ag NPs with a mean diameter of ca. 10 nm stabilized by 3-mercaptopropionic acid (3-MPA) as a colloidal precursor solution.[30] The synthetic strategy is illustrated in Figure 3.4. In solution at higher pH values, the proton of the carboxyl group at the opposite side to the thiol group was electrolytically dissociated to form a carboxylate anion (−COO$^-$) and negative electric potential was formed on the surface of the Ag NPs. Such Ag NPs

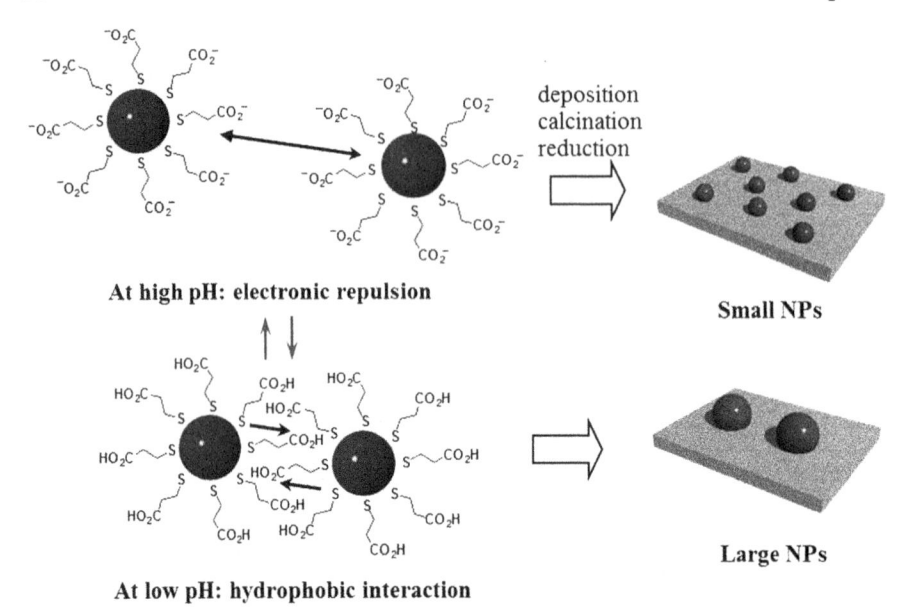

Figure 3.4 Schematics of the assembly-dispersion mechanism of the colloidal Ag NPs with 3-MPA at different pH regions and their deposition onto an Al$_2$O$_3$ support.

can demonstrate a high dispersion state without aggregation because of the electrostatic repulsion between Ag NPs. At a lower pH, the Ag NP surface becomes electrostatically neutral to lose the driving force of dispersion and the hydrophobic interactions between Ag NPs predominate, because the proton of the weak acid carboxyl group of 3-MPA cannot be ionized. It can be envisioned that such pH-triggered assembly-dispersion properties of the Ag NPs provide a possibility to control the particle size of the supported metal catalysts.

As expected, several colloidal Ag NP aqueous solutions, whose pH was adjusted with HCl (pH = 2.9, 3.2, 3.4 and 3.7), were smoothly deposited on an Al$_2$O$_3$ surface by electrostatic attraction while keeping their dispersion state. After deposition of Ag NPs, the samples were calcined at 823 K to remove the surface-attached 3-MPA and further treated with H$_2$ at 473 K. UV-vis diffuse reflectance spectra showed the characteristic plasmon absorption peak of Ag NPs at around 450 nm. TEM images revealed that the Ag/Al$_2$O$_3$(3.7) showed smaller Ag NPs, where the average diameter was determined to be 12.1 nm. The average diameter gradually increased with decreasing the pH value in the preparation sequence; the average diameters at pH = 3.4, 3.2 and 2.9 were estimated to be 15.5, 19.2 and 26.3 nm, respectively. HR-TEM images did not show significant grain boundary, revealing that the Ag NPs deposited on Al$_2$O$_3$ were not the assemblages of the original small NPs but the single crystal.

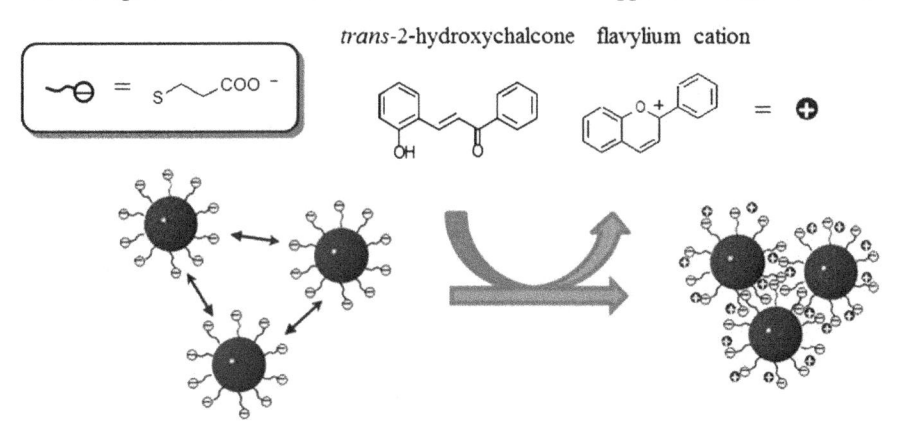

Figure 3.5 Schematics of the assembly-dispersion mechanism of the colloidal Ag NPs with a flavylium cation generated under light irradiation.

The reaction rate increased with increasing the pH value of the prepared colloidal solution, and the Ag/Al$_2$O$_3$ (3.7) with the smallest Ag NPs exhibited higher activity in the reduction of 4-nitrophenol in the presence of NaBH$_4$, because the higher dispersion of Ag metal NPs is preferable for the catalytic reactions.

We also exploited the size controlled deposition of Ag nanoparticles on Al$_2$O$_3$ with the assistance of a photo-induced chromic reaction.[31] 2-Hydroxychalcon derivatives are known to be one of the organic photochromic compounds. By light irradiation to *trans*-2-hydroxychalcon derivatives under slightly acidic conditions, flavylium cations are photochemically produced *via cis*-2-hydroxychalcon and flav-3-en-2-ol, as shown in Figure 3.5.[32] In the presence of *trans*-2-hydroxychalcone, the photo-irradiation gradually neutralized the negative electric charge of 3-MPA-stabilized Ag NPs owing to the formation of flavylium cations, which induced the assembly of Ag NPs. We demonstrated that such photo-induced assembly-dispersion control of Ag NPs enables a size selective deposition on an Al$_2$O$_3$ support by varying the photo-irradiation time while keeping their inherent aggregated form. After calcination and reduction processes, the assemblage of Ag NPs transforms into monocrystalline Ag particles with different diameters. From the TEM image, the average diameter gradually increased with increasing the irradiation time in the preparation sequence; the average diameters after 0, 2, 4, and 8 hours of irradiation times were estimated to be 10.2, 12.3, 15.9, and 20.5 nm, respectively. It can be concluded that these preparation methods based on the precise control of the assembly-dispersion state of the colloidal NPs enable a strong protocol to create supported metal catalysts with different diameters. The present strategy provides great flexibility in the selection of the kind of metals as well as primary NP sizes. Introduction of these features into the catalyst design enables the achievement of desired supported metal catalysts for the target catalytic reactions.

3.3 Multifunctional Catalysts Based on Magnetic Nanoparticles

3.3.1 Core–shell Magnetic FePt@Ti-containing Silica Spherical Nanocatalyst

It is widely accepted that the activity of a solid catalyst suspended in a liquid phase can benefit greatly from the use of smaller catalyst particles (<1 μm) to avoid mass-transfer limitations. However, the difficulties in recovering small particles from the reaction mixture severely circumvent their industrial applications. Although NP-based catalysts possessing extremely large surface areas are also key components of catalytic activity, the separation step becomes a more troublesome issue as the size of the particles decreases to the nanometre scale. In order to overcome these drawbacks, the separation of suspended magnetic catalyst bodies from the liquid system using an external magnetic field is a promising strategy.[33–36] We have developed core–shell type Ti-containing silica (Ti-SiO$_2$) encapsulating superparamagnetic fcc FePt NPs (FePt@Ti-SiO$_2$). The inner magnetic FePt cores are shielded from the external environment by an impermeable coating, which prevents sintering and agglomeration of metal NPs under catalytic conditions. The catalytically active Ti-oxide moieties are located on the external surface of the coating and are considered to be highly dispersed at the atomic level.

The following synthetic procedure for the FePt@Ti-SiO$_2$ was used (Figure 3.6). Uniformly distributed FePt NPs were synthesized by decomposition of Fe(CO)$_5$ and reduction of Pt(acac)$_2$ in the presence of oleic acid and oleylamine. Next, (3-mercaptopropyl)-triethoxysilane (3-MPTS) was added to the hexane solution of as-synthesized FePt NPs. Ligand exchange with 3-MPTS made the FePt NPs dispersible in the alcoholic solution. Finally, precursors of Ti-SiO$_2$ (TEOS and TPOT) were added to the FePt NPs dispersed in ethanol and stirred at room temperature, affording FePt@Ti-SiO$_2$.

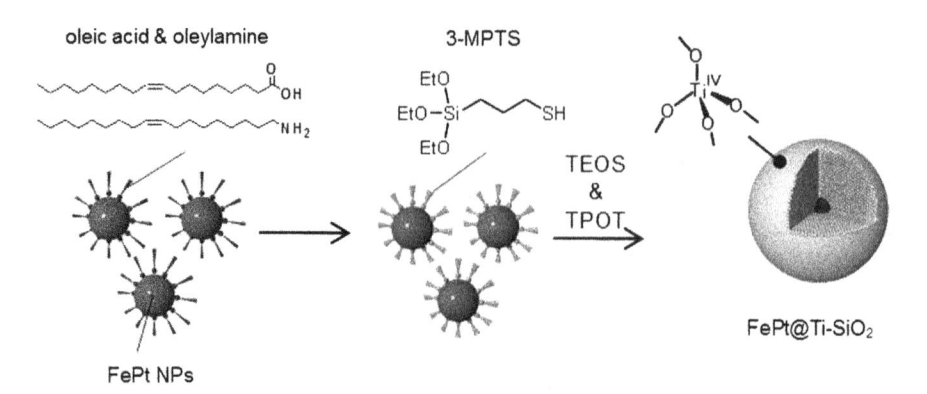

Figure 3.6 Schematics of the procedure for the synthesis of FePt@Ti-SiO$_2$.

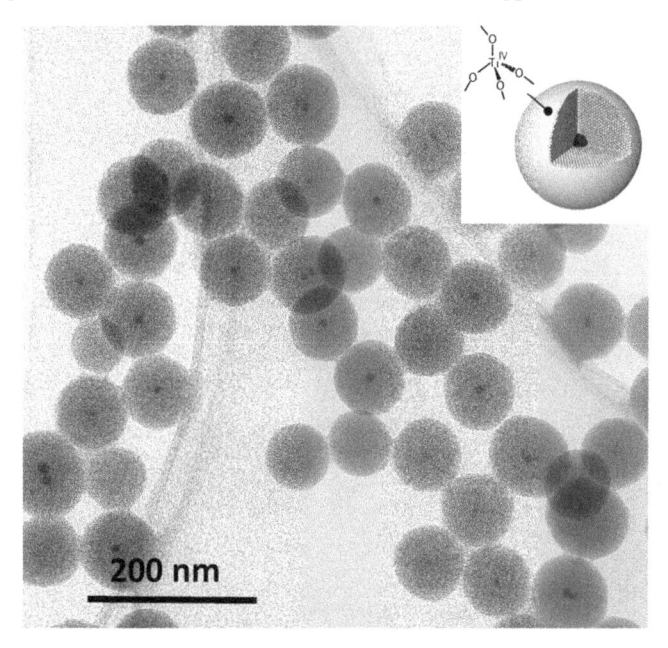

New Strategies to Fabricate Nanostructured Colloidal and Supported Metal NPs 39

Figure 3.7 TEM image of FePt@Ti-SiO$_2$.

TEM images show that the spherical particles have a 45–50 nm thick Ti-SiO$_2$ shell (Figure 3.7). Each Ti-SiO$_2$ sphere encapsulates one FePt NPs, although a small fraction of the particles possesses either two or zero FePt NPs. The average diameter of FePt NPs was determined to be *ca.* 6 nm. Ti K-edge XAFS spectra also confirm the formation of isolated and tetrahedrally-coordinated Ti-oxide moieties within the silica shell framework.

The catalytic ability of the FePt@Ti-SiO$_2$ was investigated in the epoxidation of cyclooctene using 30% H$_2$O$_2$ as a simple test reaction. The results of this reaction indicated that the corresponding cyclooctene oxide was obtained in >99% selectivity and the turnover number (TON) based on Ti approached 48 after 24 h. A more important advantage of the FePt@Ti-SiO$_2$ is the facile recovery from the reaction mixture and the high reusability. Upon completion of the oxidation reaction, the magnetic properties of the FePt@Ti-SiO$_2$ can afford a straightforward means to isolate the catalyst from the colloidal solution. By applying an external permanent magnet, the catalyst was attracted. The recovered catalyst could then be recycled in the epoxidation of cyclooctene at least three times while maintaining identical inherent activity to the initial run.

3.3.2 Water-soluble FePt Magnetic NPs Modified with Cyclodextrin

The synthesis of cyclodextrin (CD)-stabilized metal NPs represents research on the functionalization of NPs. CDs are a group of cyclic oligosaccharides

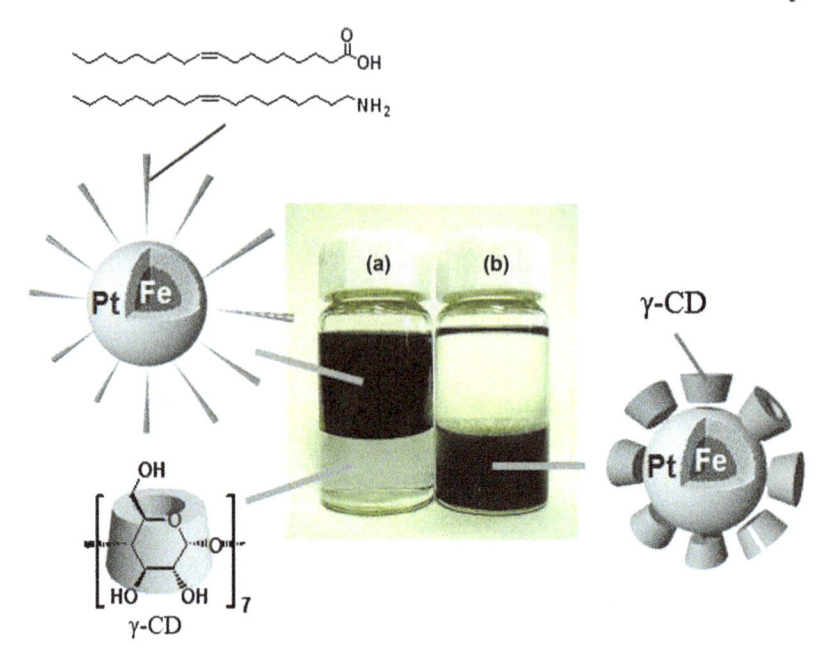

Figure 3.8 Photograph of a two-phase mixture of FePt NPs stabilized by (a) oleic acid/oleylamine and (b) γ-CD. The top layer is hexane and the bottom layer is aqueous solution.

composed of six (α-CD), seven (β-CD), or eight (γ-CD) α(1,4)-linked glucopyranose units. The cavity exhibits a hydrophobic character, which allows CDs to form inclusion complexes with hydrophobic molecules that fit into the cavity.[37] We present a CD-stabilized FePt magnetic NP exhibiting catalytic activity under aqueous conditions,[38] in which the composition of FePt NPs was controlled to form Fe$_{core}$Pt$_{shell}$ by the first decomposition of Fe(CO)$_5$ to form the Fe core, followed by the successive reduction of Pt(acac)$_2$ in the presence of oleic acid and oleylamine. Elemental analysis confirmed that the average composition of the NPs was Fe$_{48}$Pt$_{52}$.

As shown in Figure 3.8, the as-synthesized FePt NPs can be dispersed in nonpolar hexane solvent due to the presence of oleic acid and oleylamine. A mixture of a hexane suspension of as-synthesized FePt NPs and an equal volume of γ-CD aqueous solution was stirred vigorously at room temperature. After stirring for 24 h, the top hexane layer becomes colorless, confirming the existence of water-compatible γ-CD as host molecules (Figure 3.8.). It can be said that the surface properties of the NPs have been modified through the formation of an inclusion complex between surface-bound organic molecules and γ-CD. The surface coverage of γ-CD on the FePt NPs was roughly estimated to be *ca.* 60 % by CHN elemental analysis.

The aqueous reaction of allyl alcohol using Fe$_{core}$Pt$_{shell}$ NPs capped with γ-CD proceeded smoothly to give the corresponding *n*-propanol. It should be noted that the reaction using the Fe$_{shell}$Pt$_{core}$ NPs as well as Fe NPs stabilized by γ-CD hardly occurred because Fe atoms do not accelerate the aforementioned hydrogenation reaction, confirming that our FePt NPs consists of an Fe-rich core and a Pt-rich shell. More interestingly, the reaction rate performed with FePt NPs capped with γ-CD in water was three times higher than hydrophobic FePt NPs stabilized by oleic acid/oleylamine in THF. Similar results were observed in other organic solvents, such as toluene and hexane. We suppose that the host–guest complexation of organic substrate (allyl alcohol) with γ-CD on FePt NPs is the major driving force for the enhancement of the catalytic efficiency under aqueous conditions. Such a driving force is lost in the case of the hydrophobic FePt NPs stabilized by oleic acid/oleylamine in an organic solvent as the reaction medium, resulting in a low catalytic activity.

3.3.3 FePd Magnetic NPs Modified with a Chiral BINAP Ligand

With a similar strategy for catalyst design, the FePd NP with an Fe-rich core and a Pd-rich shell have been modified with chiral 2,2′-bis(diphenylphosphino)-1,1′-binaphthyl (BINAP) through the simple ligand exchange procedure (Figure 3.9).[39] In the FT-IR spectrum, the peaks due to the methylene ν(C–H) stretch vibration of oleic acid and/or oleylamine disappeared, accompanied with the appearance of the aromatic ring ν(C–H) stretch vibration of BINAP ligand at around 2800–3000 cm^{-1}. The circular dichroism (CD) spectra of (S)-BINAP-modified NPs in CHCl$_3$ showed a negative cotton effect, while (R)-BINAP-modified NPs showed a positive cotton effect. For the structural model of the FePd NPs, the XAFS results suggested that most of the Pd atoms were preferentially located in the shell region, while the Fe atoms were preferentially located in the core region.

It was found that the FePd NPs modified with (S)-BINAP served as a chiral nanocatalyst in the Suzuki–Miyaura coupling reaction of 1-bromo-2-methylnaphthylene (**1**) and naphthylboronic acid (**2**) afford the corresponding (S)-binaphthalene (**3**) in >99% yield with a moderate enantioselectivity of 48% (Figure 3.9). Since Fe atoms do not accelerate these coupling reactions, our FePd NPs are thought to consist of an Fe-rich core and a Pd-rich shell. In the case of (R)-BINAP, (R)-binaphthalene was obtained with the same level of chemical yield and enantioselectivity, revealing that the absolute configuration of the coupling products is determined by the employed ligands. Periodic monitoring of the coupling reaction showed no induction period or degradation: the enantioselectivity appeared to be independent of the conversion level and remained unchanged during the reaction, suggesting that racemization of the product did not take place.

Figure 3.9 Ligand exchange procedure of FePd NP from oleic acid/oleylamine to chiral BINAP and asymmetric Suzuki–Miyaura coupling reaction. The red blue core is Fe rich and the yellow shell is Pt rich.

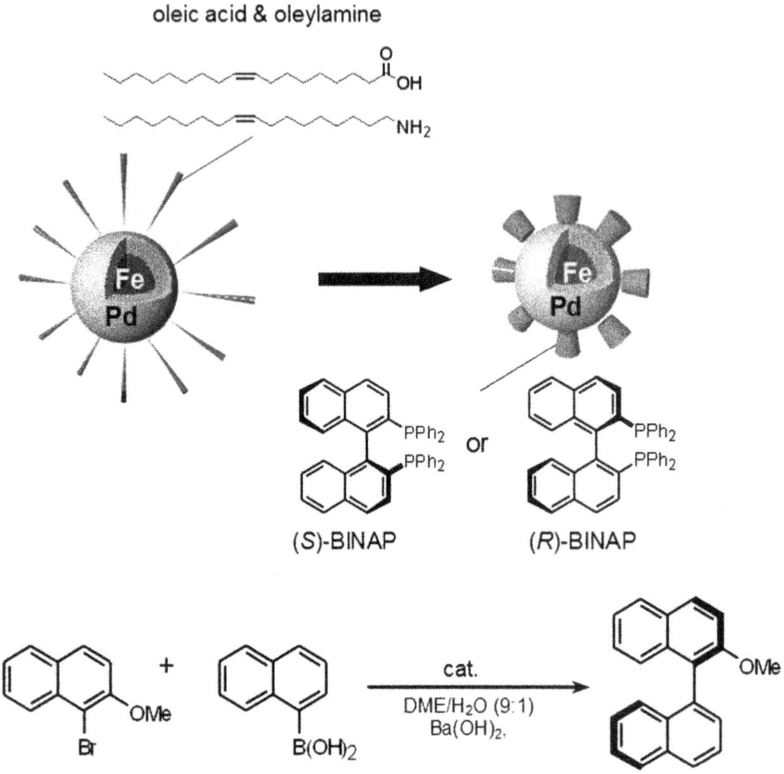

	Conv. (%)	Ee. (%)
Pd/Fe_2O_3-(S)-BINAP	>99	48
Pd/Fe_2O_3-(S)-BINAP	>99	46

Figure 3.9 Ligand exchange procedure of FePd NP from oleic acid/oleylamine to chiral BINAP and asymmetric Suzuki–Miyaura coupling reaction. The dark grey core is Fe rich and the light grey shell is Pt rich.

3.3.4 Core–shell Nanostructured Catalyst for One-pot Reactions

As mentioned earlier, the direct synthesis of H_2O_2 from H_2 and O_2 over supported Pd-based catalysts has attracted significant attention. The key problem is stabilizing the resulting H_2O_2, because H_2O_2 simultaneously undergoes hydrogenation and/or decomposition to water by the same catalysts employed for its formation. This undesirable feature causes a slow H_2O_2 formation rate and prevents the accumulation of high H_2O_2 concentration. In order to overcome this problem, the utilization of *in situ* generated H_2O_2 from H_2 and O_2 to subsequently oxidize organic reactants in the same reaction vessel (one-pot oxidation reaction) could be an alternative strategy. It is possible to use the unstable H_2O_2 immediately without isolation/purification steps, which would contribute to saving energy and time as well as avoiding the risk of transportation of the concentrated H_2O_2.

We predict that the key to constructing a one-pot oxidation system is the efficient dispersion of H_2O_2 generated on Pd NP sites to the neighboring catalytically active oxidation sites before any undesirable conversion to water can occur. Therefore, the primary requirement is the precise architecture of the two types of active sites. Considering this, we present a new type of core–shell structured catalyst to enable the one-pot oxidation, in which a uniform SiO_2 core supporting Pd NPs was covered with a Ti-containing mesoporous silica shell (Pd/SiO$_2$@TiMSS).[40] Although the inner Pd NPs are located at the boundary of the core–shell structure, reactants can penetrate to the Pd NPs through the mesoporous shell. The Ti-oxide moieties located within the mesoporous region are catalytically active toward oxidation with H_2O_2 and are considered to be highly dispersed at the atomic level.

To elucidate the effect of the location of the Pd NPs on the enhancement of one-pot oxidation using *in situ* generated H_2O_2, two reference samples, SiO$_2$@Pd(R)/TiMSS and SiO$_2$@Pd(S)/TiMSS, were synthesized. In the case of the SiO$_2$@Pd(R)/TiMSS, where (R) indicates random, the Pd NPs are randomly distributed within the mesoporous structure. In the case of SiO$_2$@Pd(S)/TiMSS, where (S) indicates the surface, most of the Pd NPs are deposited on the outer surface of the mesopores. The HR-TEM images of all samples show that the spherical NPs consist of nonporous SiO_2 cores, Ti-containing mesoporous shells with channels oriented perpendicular to the core surface, and Pd NPs. The thickness of the shell was determined to be approximately 30 nm. The Pd NPs were apparently present at the intended sites; the boundary between the SiO_2 core and mesoporous silica shell, randomly distributed within the mesopores, and the surface of mesoporous shell in Pd/SiO$_2$@TiMSS, SiO$_2$@Pd(R)/TiMSS, and SiO$_2$@Pd(S)/TiMSS, respectively. The average size of the Pd NPs is 3.5 nm and no significant differences in the Pd NP size are observed in the three samples.

It should be emphasized that, as expected, the relative position between the H_2O_2 generation sites and the oxidation sites has a significant effect on the catalytic activity in the one-pot oxidation of methyl phenyl sulfide into methyl phenyl sulfoxide. Pd/SiO$_2$@TiMSS is the most effective catalyst among those investigated; the reaction rate for Pd/SiO$_2$@TiMSS is almost three times higher than those for SiO$_2$@Pd(S)/TiMSS and SiO$_2$@Pd(R)/TiMSS (Figure 3.10). The high catalytic activity of Pd/SiO$_2$@TiMSS can be ascribed to the efficient utilization of H_2O_2 for the subsequent sulfide oxidation, where *in situ* generated H_2O_2 within the mesoporous channels can promptly react at the neighboring Ti sites before H_2O_2 can be dispersed into the solvent. On the contrary, H_2O_2 is mainly produced at the outer surface of SiO$_2$@Pd(S)/TiMSS and it is difficult for the generated H_2O_2 to enter the mesoporous channels to interact with the Ti species; therefore most of the H_2O_2 is decomposed by the Pd catalyst used in its formation. In the case of SiO$_2$@Pd(R)/TiMSS, the random distribution of the Pd NPs within the mesoporous channels may inhibit the diffusion of reactants to both types of

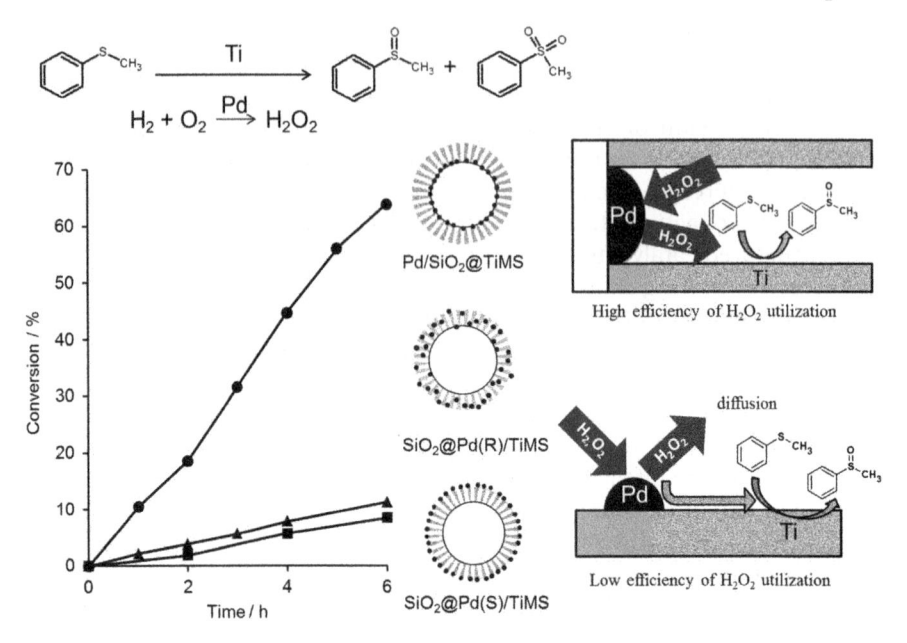

Figure 3.10 Kinetics of methyl phenyl sulfide oxidation using *in situ* generated H$_2$O$_2$.

active site. With further optimization of the pore diameter and thickness of the Ti-MS shell, enhancement of the catalytic activity can be attained by a factor of 6.[41] Therefore, the core–shell catalyst is considered to be a promising structure for one-pot reactions.

References

1. J. P. Wilcoxon and B. L. Abrams, *Chem. Soc. Rev.*, 2006, **35**, 1162–1194.
2. J. D. Mackenzie and E. P. Bescher, *Acc. Chem. Res.*, 2007, **40**, 810–818.
3. S. G. Kwon and T. Hyeon, *Acc. Chem. Res.*, 2008, **41**, 1696–1709.
4. D. Astruc, F. Lu and J. R. Aranzaes, *Angew. Chem., Int. Ed.* 2005, **44**, 7852–7872.
5. J. Sun and X. Bao, *Chem.–Eur. J.*, 2008, **14**, 7478–7488.
6. M. Zahmakran and S. Ozkar, *Nanoscale*, 2011, **3**, 3462–3481.
7. K. Hayek, R. Kramer and Z. Paál, *Appl. Catal. A*, 1997, **162**, 1–15.
8. C. -m. Yang, P. -h. Liu, Y. -f. Ho, C. -y. Chiu and K. -j. Chao, *Chem. Mater.*, 2002, **15**, 275–280.
9. C. T. Kresge, M. E. Leonowicz, W. J. Roth, J. C. Vartuli and J. S. Beck, *Nature*, 1992, **359**, 710–712.
10. Y. Tao, H. Kanoh, L. Abrams and K. Kaneko, *Chem. Rev.*, 2006, **106**, 896–910.
11. W. Fan, R.-G. Duan, T. Yokoi, P. Wu, Y. Kubota and T. Tatsumi, *J. Am. Chem. Soc.*, 2008, **130**, 10150–10164.

12. A. Fujishima and K. Honda, *Nature*, 1972, **238**, 37–38.
13. H. Yamashita, K. Mori, S. Shironita and Y. Horiuchi, *Catal. Surv. Asia*, 2008, **12**, 88–100.
14. H. Yamashita and K. Mori, *Chem. Lett.*, 2007, **36**, 348–353.
15. H. Yamashita, Y. Miura, K. Mori, T. Ohmichi, M. Sakata and H. Mori, *Catal. Lett.*, 2007, **114**, 75–78.
16. S. Shironita, K. Mori, T. Shimizu, T. Ohmichi, N. Mimura and H. Yamashita, *Appl. Surf. Sci.*, 2008, **254**, 7604–7607.
17. K. Mori, T. Araki, S. Shironita, J. Sonoda and H. Yamashita, *Catal. Lett.*, 2009, **131**, 337–343.
18. K. Mori, T. Araki, T. Takasaki, S. Shironita and H. Yamashita, *Photochem. Photobiol. Sci.*, 2009, **8**, 652–656.
19. S. Shironita, K. Mori, T. Ohmichi, E. Taguchi, H. Mori and H. Yamashita, *J. Nanosci. Nanotech.*, 2009, **9**, 557–561.
20. S. Shironita, M. Goto, T. Kamegawa, K. Mori and H. Yamashita, *Catal. Today*, 2010, **153**, 189–192.
21. K. Mori, Y. Miura, S. Shironita and H. Yamashita, *Langmuir*, 2009, **25**, 11180–11187.
22. M.-L. Wu, D.-H. Chen and T.-C. Huang, *Langmuir*, 2001, **17**, 3877–3883.
23. K. Fuku, T. Sakano, T. Kamegawa, K. Mori and H. Yamashita, *J. Mater. Chem.*, 2012, **22**, 16243–16247.
24. M. Tsuji, M. Hashimoto, Y. Nishizawa, M. Kubokawa and T. Tsuji, *Chem.-Eur. J.*, 2005, **11**, 440–452.
25. J.-F. Zhu and Y.-J. Zhu, *J. Phys. Chem. B*, 2006, **110**, 8593–8597.
26. M. Cano, A. Benito, W. K. Maser and E. P. Urriolabeitia, *Carbon*, 2011, **49**, 652–658.
27. S. Shironita, T. Takasaki, T. Kamegawa, K Mori and H. Yamashita, *Catal. Lett.*, 2009, **129**, 404–407.
28. K. Fuku, T. Kamegawa, K. Mori and H. Yamashita, *Chem.-Asian J*, 2012, **7**, 1366–1371.
29. K. Fuku, S. Takakura, T. Kamegawa, K. Mori and H. Yamashita, *Chem. Lett.*, 2012, **41**, 614–616.
30. K. Mori, A. Kumami, M. Tomonari and H. Yamashita, *J. Phys. Chem. C*, 2009, **113**, 16850–16854.
31. K. Mori, A. Kumami and H. Yamashita, *Phys. Chem. Chem. Phys.*, 2011, **13**, 15821–15824.
32. F. Pina, M. J. Melo, M. Maestri, R. Ballardini and V. Balzani, *J. Am. Chem. Soc.*, 1997, **119**, 5556–5561.
33. K. V. S. Ranganath and F. Glorius, *Catal. Sci. Tech.*, 2011, **1**, 13–22.
34. K. Mori, Y. Kondo, S. Morimoto and H. Yamashita, *Chem. Lett.*, 2007, **36**, 1068–1069.
35. K. Mori, S. Kanai, T. Hara, T. Mizugaki, K. Ebitani, K. Jitsukawa and K. Kaneda, *Chem. Mater.*, 2007, **19**, 1249–1256.
36. K. Mori, Y. Kondo, S. Morimoto and H. Yamashita, *J. Phys. Chem. C*, 2007, **112**, 397–404.

37. J. Liu, W. Ong, E. Román, M. J. Lynn and A. E. Kaifer, *Langmuir*, 2000, **16**, 3000–3002.
38. K. Mori, N. Yoshioka, Y. Kondo, T. Takeuchi and H. Yamashita, *Green Chem.*, 2009, **11**, 1337–1342.
39. K. Mori, Y. Kondo and H. Yamashita, *Phys. Chem. Chem. Phys.*, 2009, **11**, 8949–8954.
40. S. Okada, K. Mori, T. Kamegawa, M. Che and H. Yamashita, *Chem.-Eur. J.*, 2011, **17**, 9047–9051.
41. S. Okada, S. Ikurumi, T. Kamegawa, K. Mori and H. Yamashita, *J. Phys. Chem. C*, 2012, **116**, 14360–14367.

CHAPTER 4

Organometallic Approach for the Synthesis of Noble Metal Nanoparticles: Towards Application in Colloidal and Supported Nanocatalysis

SOLEN KINAYYIGIT[a,b] AND KARINE PHILIPPOT*[a,b]

[a] Laboratoire de Chimie de Coordination, CNRS, LCC, BP 44099, 205 Route de Narbonne, F-31077 Toulouse cedex 4, France; [b] Université de Toulouse, UPS, INPT, LCC, F-31077 Toulouse, France
*Email: karine.philippot@lcc-toulouse.fr

4.1 Introduction

Catalysis is an important area of chemistry in which metal nanoparticles (MNPs) are considered as promising substitutes of catalysts; both for homogeneous (molecular complexes) and heterogeneous (bulk metals on supports) ones.[1–4] MNPs with their small size are highly interesting systems due to their high proportion of surface atoms that offer numerous active sites and to their unique electronic properties at the frontier between the molecular and metallic states.[5] Furthermore, the surface chemistry of MNPs can be tuned by addition of a stabilizer or by using a support. Polymers, dendrimers, ionic liquids, surfactants, ligands *etc.* used as capping agents can transfer their own properties such as solubility or chirality whereas widely used supports such as ceria, titania, alumina, silica, or carbon materials can give a synergy to the nanoparticle systems to orientate a catalytic reaction.[6,7] There are some examples in literature that focus on both aspects by combining the ligands with supports. With intensive studies over the past two decades, nanocatalysis has emerged as a modern domain of catalysis at the borderline between homogeneous and heterogeneous ones. However, from an industrial point of view regarding sustainable and economical concerns, the use of MNPs as catalysts is still hampered by two important drawbacks: the stability of the particles and the recovery/recyclability of the catalyst.[8,9] To overcome these problems, several approaches have been developed, such as the design of appropriate stabilizing agents or the immobilization of MNPs in a magnetic solid support, which allows an easy separation of the nanocatalyst from the reaction mixture. Regardless of the method followed for the preparation, promising alternatives to increase stability and recycling of the nanocatalysts have been reported.

4.2 Organometallic Synthesis of Noble Metal Nanoparticles

Nowadays, two important concepts are being considered in nanocatalysis in order to develop efficient catalytic systems, namely the bottom-up approach for the synthesis of nanoparticles (NPs) well-controlled in size/shape and the molecular approach to obtain more selective nanocatalysts. The bottom-up strategy allows the building of metallic NPs from the monometallic species.[10,11] Intensive research is currently devoted to this approach with the belief that carefully designed NPs with well-defined size, shape, composition and surface chemistry should be able to display the benefits of both homogeneous and heterogeneous catalysts in terms of high efficiency and selectivity.[12] The introduction of ligands as NP stabilizers is of special interest as it focuses on the precise molecular definition of the catalytic surface. This strategy potentially allows optimization of the parameters that govern efficiency in catalysis, including enantioselectivity and solubility in organic or aqueous media.[13,14] Since ligands are expected to modulate both the electronic and the steric environment at the surface of the particles, it is of key importance to analyze the influence of ligands on the stabilization of NPs and on their surface properties in order to develop more active and selective nanocatalysts. Therefore, significant efforts are presently directed towards the preparation of metallic NPs stabilized by ligands of various nature to study their influence on catalytic performances.

In this context of modern nanocatalysis, we are using the tools of molecular chemistry in order to develop an organometallic approach for the synthesis of MNPs. Our methodology that was first introduced by Chaudret in the 90s is based on the decomposition of metalloorganic complexes as the metal source. The main advantage of these precursors is that their decomposition can be easily achieved under the mild conditions of solution

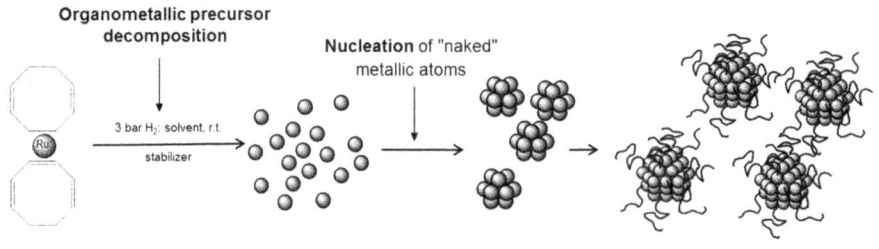

Organometallic Approach for the Synthesis of Noble Metal Nanoparticles 49

Figure 4.1 Schematic representation of the organometallic approach for the synthesis of MNPs by hydrogen decomposition of a metalloorganic complex.

chemistry. The decomposition step is performed through reduction or ligand displacement from the metal coordination sphere, most often in tetrahydrofuran (THF) solution and in the presence of a stabilizer (see Figure 4.1).[15–17] This allows the control of particle size, shape and surface state in order to reach a monodisperse assembly of NPs with the desired properties. This method can also be applied to bimetallic systems.[18,19] The choice of the stabilizing agent is fundamental as it will govern the growth, the stability and the catalytic performance of the MNPs.[20] Various functional organic molecules such as polymers,[21,22] alcohols,[23,24] ionic liquids[25–27] and various ligands[28–30] can be used for this purpose. Immobilization of NPs using supports like alumina, silica[31] or carbon materials is also possible and they allow the easy orientation of the catalytic reaction and facilitate the recovery of the nanocatalysts. During these efforts, simple spectroscopic methods derived from molecular chemistry, such as infrared (IR) and nuclear magnetic resonance (NMR), both in solution and in the solid state, aid in precisely characterizing the metallic surface.[32–37]

In the following parts of this chapter, we will present an overview of the recent developments in the organometallic synthesis of noble metal nanoparticles for application as catalysts in solution or supported conditions. This review is not an exhaustive report but aims to show the progress obtained recently. The results are reported taking into account the targeted catalytic reaction, with the emphasis on either the nature of the stabilizing ligand (ionic liquids, organic solvent- or water-soluble ligands) for nanocatalysis in solution or of the support for supported catalysis.

4.3 Nanoparticles for Colloidal Catalysis

For the past few years, our research has widely focused on the preparation of well-defined metallic NPs stabilized by ligands of different nature.[20] Our aim was to better understand the influence of the ligand on the size and the stability of the NPs as well as on their surface chemistry, and consequently how they affect the catalytic performance of NPs, depending on their interaction with the metallic surface. Another interest for us in using ligands was their potential to orientate the course of a catalytic reaction by their selectivity, a well-known phenomenon in homogeneous catalysis. Numerous efforts have been devoted to the use of simple ligands but also to more sophisticated ones as these parameters are major for the development of nanocatalysts with improved catalytic performances in comparison to known homogeneous and heterogeneous catalysts.

Studies in the improvement of the NP synthesis mainly involved ligands containing nitrogen (aminoalcohols, oxazolines), phosphorus (phosphines, diphosphites) or carbon (carbenes) as coordinating atoms to the metal surface. Ionic liquids were also proven to be very efficient stabilizers to obtain size-controlled MNPs, and the coaddition of ligands could offer new perspectives in this way. Some of our NP systems have been applied in catalysis, mainly in hydrogenation reactions, but also in the dehydrogenation of amine-boranes, carbon–carbon coupling and hydroformylation reactions.

4.3.1 Hydrogenation Reactions

A high number of papers describe the preparation of various MNPs in suspension for their application as catalysts in the hydrogenation of substrates mainly of olefin and arene derivatives in addition to the carbonyl substrates such as ketones.[38] The asymmetric hydrogenation of prochiral substances is also a domain explored with nanocatalysts.

4.3.1.1 Ligand Stabilized Nanoparticles as Catalysts

4.3.1.1.1 Aminoalcohol- and Oxazoline-stabilized Ru Nanoparticles. In a collaboration with Gomez *et al.*, we used chiral aminoalcohol- and oxazoline-stabilized ruthenium NPs as catalysts in the hydrogenation of unsaturated prochiral substrates such as acetophenone, *ortho*- and *para*-anisoles and dimethyl itaconate.[39] The nanocatalysts were prepared by decomposition of the [Ru(COD)(COT)] precursor under mild conditions (3 bar H_2; r. t.) and in the presence of optically pure ligands, namely β-aminoalcohols, amino(oxazolines), hydroxy(oxazolines) and bis(oxazolines) to give stable RuNPs with a mean size between 1.6 and 2.5 nm. Their catalytic behaviour was examined in comparison with that of Ru molecular complexes prepared *in situ* with the same ligands (see Figure 4.2). For the hydrogen transfer of acetophenone, only the Ru/amino(oxazoline) colloidal system gave asymmetric induction with a modest value of *ca.* 10% ee. In contrast to the molecular system, no enantioselectivity was achieved in the hydrogenation of dimethyl itaconate with the same nanocatalyst despite observing a 100% conversion. Finally, when the same catalytic system was applied in the reduction of *ortho*- and *para*-anisoles (methanol; 50 °C; 40 bar H_2; 6 h), cyclohexanes were detected as the main products (*ca.* 57% for *ortho*-anisole and 87% for *para*-anisole) while the corresponding molecular

Figure 4.2 aminooxazoline-stabilized RuNPs and their catalytic behavior in hydrogenation of *ortho*- and *para*-anisoles.[39]

catalyst was found to be inactive. In both cases, the *trans*-isomer was favoured, with a *trans–cis* ratio up to 19. The very low asymmetric induction encountered with this colloidal catalyst suggested a fluxional behaviour of the ligands at the surface of RuNPs, as previously described for Ru/amine colloids.[22]

4.3.1.1.2 Diphosphite-stabilized Ru, Rh and Ir Nanoparticles. In a collaboration with the groups of Castillón, Claver and Roucoux, 1,3-diphosphites derived from carbohydrates were used as stabilizers for the preparation of Ru, Rh and Ir NPs for their investigation in asymmetric hydrogenation reactions.[40,41] These ligands allowed the formation of very small Ru, Rh and Ir NPs in mild conditions (THF; 3 or 6 bar H_2; r.t.) from [Ru(COD)(COT)], [Rh(η^3-C_3H_5)$_3$] and [Ir(COD)Cl]$_2$ complexes, respectively. Structural modifications in the diphosphite backbone were found to influence the NP size, dispersion and catalytic activity. In the hydrogenation of *ortho*- and *meta*-methylanisoles, the RhNPs were shown to display the highest activity while the IrNPs presented the lowest one. In contrary with the results previously described for the aminooxazoline-stabilized RuNPs, the hydrogenation of *o*-anisole gave total selectivity for the *cis*-product in all cases. However, the ee of the product was always less than 6%. A maximum of 81% *cis*-selectivity was obtained for the hydrogenation of *m*-anisole, but with no enantioselectivity.

4.3.1.1.3 Carbene-stabilized Ru Nanoparticles. In a collaborative work with van Leeuwen *et al.*, RuNPs stabilized by *N*-heterocyclic carbenes (NHC), namely *N*,*N*-di(*tert*-butyl)imidazole-2-ylidene (I^{t}Bu) and 1,3-bis(2,6-diisopropylphenyl)imidazole-2-ylidene (IPr),[42] were used as catalysts in the hydrogenation of several substrates under various reaction conditions (solvent, substrate concentration, substrate–metal ratio, temperature). The Ru/NHC NPs appeared as active catalysts in the hydrogenation of aromatics and showed an interesting ligand effect; Ru/IPr NPs were generally

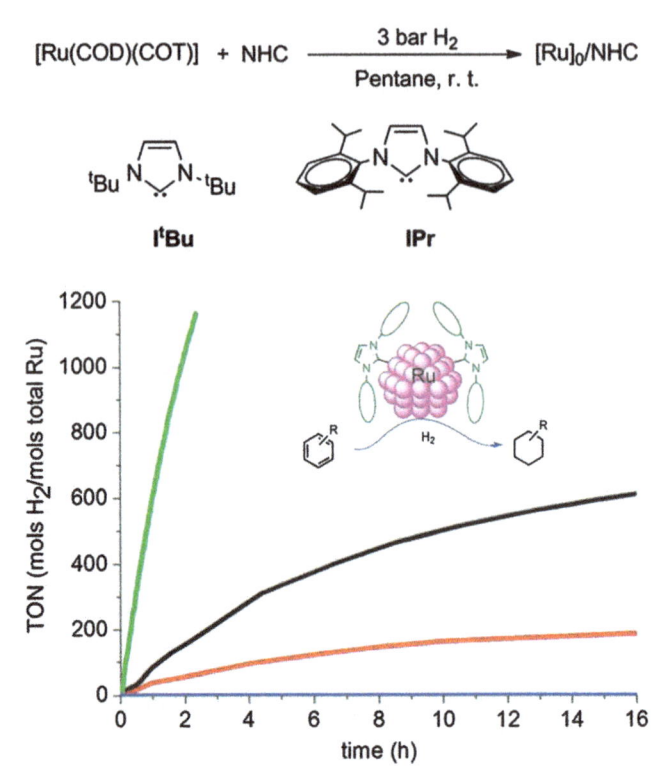

Figure 4.3 Synthesis of Ru/NHC NPs and their catalytic performance in hydrogenation of *ortho*-methylanisole (0.5 M) in different solvents (Ru/IPr0.2 (0.3% Ru); 40 bar H$_2$; 298 K : pentane (black), THF (red), methanol (blue), no-solvent (green, 0.1% Ru).[43]

more active than Ru/I^{t}Bu NPs.[43] The influence of the reaction parameters (changes in solvent, temperature, *etc.*) was also studied (see Figure 4.3).

4.3.1.2 Water-soluble Nanoparticles as Catalysts

Due to concerns about sustainability issues, we have been interested in the synthesis of water-soluble MNPs for their use as catalysts in water or under biphasic liquid–liquid conditions. For this purpose, we took inspiration from organometallic catalytic systems in water and considered common ligands employed to stabilize complexes such as 1,3,5-triaza-7-phosphaa-damantane (PTA) and sulfonated diphosphines.

4.3.1.2.1 PTA-stabilized Pt and Ru Nanoparticles. The synthesis of PTA-stabilized Ru and PtNPs was carried out by decomposition of [Ru(COD)(COT)] and [Pt(CH$_3$)$_2$(COD)] precursors in the presence of 0.8 equiv. of PTA (P(H$_2$) = 3 bar; THF; 70 °C) (see Figure 4.4).[44] The resulting NPs displayed a spherical shape and low size dispersity. They were purified by washing

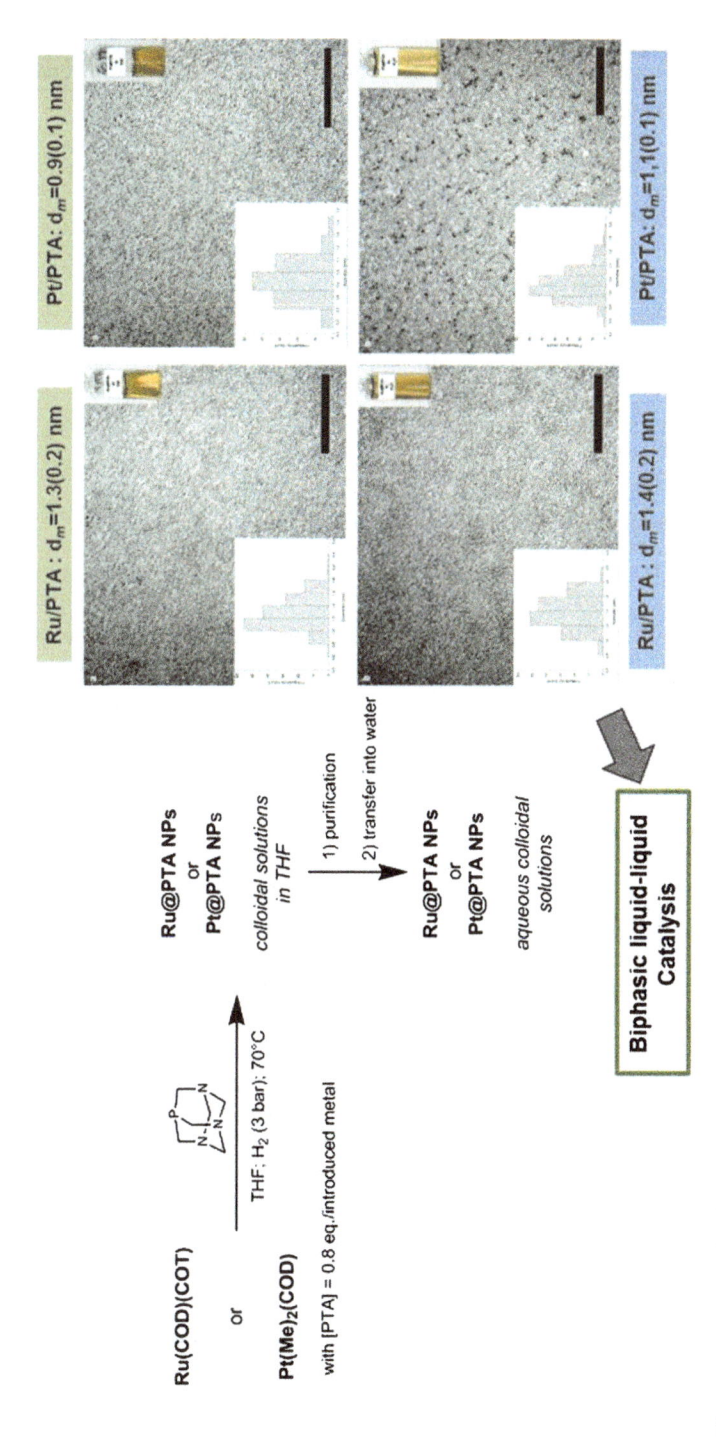

Organometallic Approach for the Synthesis of Noble Metal Nanoparticles 53

Figure 4.4 Synthesis of PTA-stabilized NPs and TEM images in THF and in water for Ru NPs (left) and Pt NPs (right).[44]

Table 4.1 Hydrogenation of olefins and aromatic derivatives with aqueous Ru/PTA and Pt/PTA colloidal solutions.[a]

Catalyst	Substrate	P H$_2$ (bar)	Time (h)	Conversion (%)[f]
Pt/PTA[b]	Octene	1	2	100
Pt/PTA[b]	Octene	1	2	68[g]
Pt/PTA[c]	Toluene	10	2	8
Pt/PTA[c]	Toluene	10	16	100
Pt/PTA[c]	m-Methylanisole	10	2	15
Pt/PTA[c]	m-Methylanisole	10	16	100
Ru/PTA[d]	Dodecene	1	5	100[h]
Ru/PTA[d]	Octene	10	1	100
Ru/PTA[e]	Toluene	10	16	100
Ru/PTA[e]	m-Methylanisole	10	16	60

[a]Reaction conditions: [substrate]-[metal] = 100, $T = 20°C$ stirred at 1500 mn^{-1}.
[b]Pt (0.0119 mmoL); substrate (1.19 mmoL).
[c]Pt (0.0122 mmoL); substrate (1.22 mmoL).
[d]Ru (0.0159 mmoL); substrate (1.59 mmoL).
[e]Reaction conditions: Ru (0.020 mmoL); substrate (2 mmoL).
[f]Substrate conversion determined by gas chromatography analysis.
[g]Recycling of the aqueous suspension of entry 1.
[h]Determined by ^{1}H and ^{13}C NMR.

with pentane and filtration, and were further dissolved in water without any change in dispersion and in their mean diameter, with a value of ~1.4 nm and ~1.1 nm for the RuNPs and the PtNPs, respectively. Aqueous suspensions of these NPs were stable for weeks when kept under an argon atmosphere. ^{1}H, ^{13}C, ^{31}P solution and solid-state NMR studies showed the strong coordination of PTA at the surface of the particles by the phosphorous atom.

Biphasic liquid–liquid hydrogenation was investigated with model olefins and aromatic substrates using aqueous colloidal solutions of Ru/PTA and Pt/PTA NPs as catalysts, with Roucoux et al. (see Table 4.1).[45] Octene and dodecene were totally converted into the corresponding alkanes (r.t.; 1 bar H$_2$), with moderate activities. An increase in the hydrogen pressure (P(H$_2$) = 10 bar) was not detrimental for the colloidal suspension stability. Complete hydrogenation of toluene into cyclohexane was observed overnight with Ru/PTA NPs whereas 60% of m-methoxymethylcyclohexane was formed from methoxymethylanisole. In comparison, the reduction with Pt/PTA NPs was achieved after 16 h even if very low conversions were obtained after 2 h with 8% and 15% cyclohexyl derivatives, respectively. In summary, these PTA-stabilized Ru and Pt NPs were active in the hydrogenation of olefinic and aromatic compounds under mild conditions despite the change of environment that they underwent after their dissolution into water.

4.3.1.2.2 Sulfonated Diphosphine-stabilized Ru Nanoparticles. By applying the same procedure as previously described with the PTA ligand (see Figure 4.5), we used sulfonated diphosphines to stabilize RuNPs.[46] In this collaborative work with Roucoux et al. and Monflier et al., the RuNPs

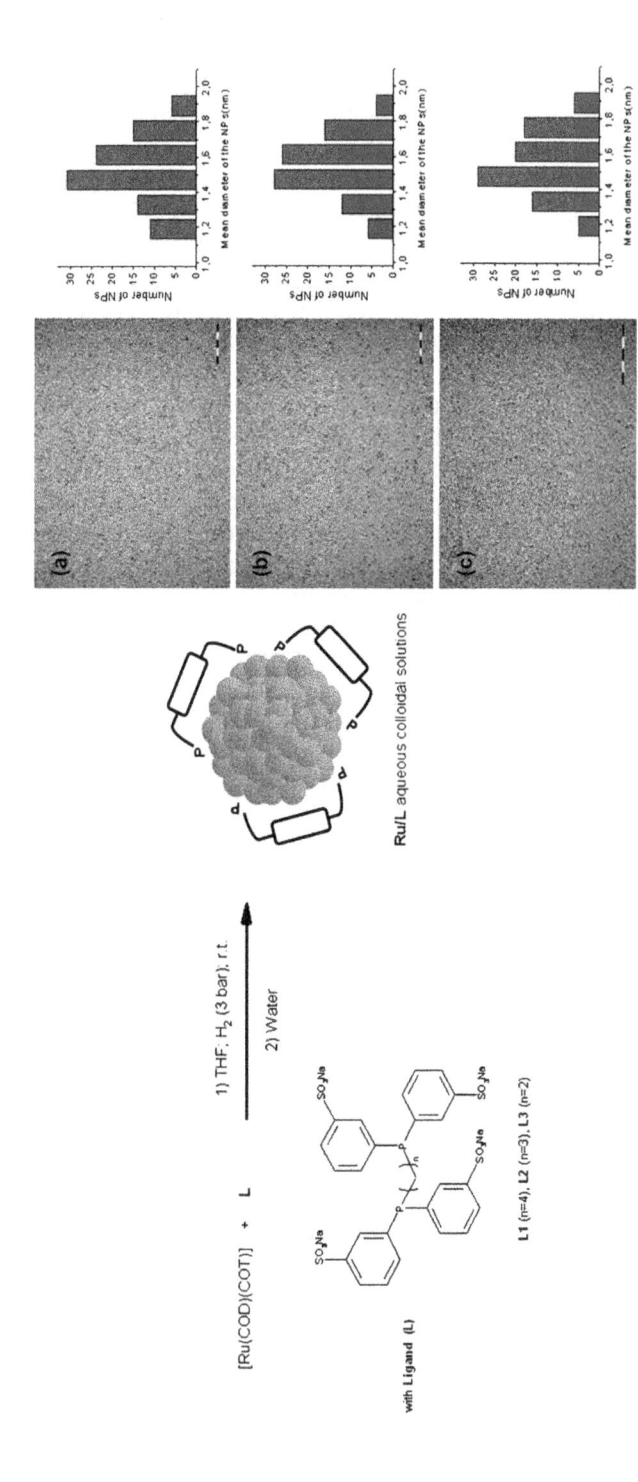

Organometallic Approach for the Synthesis of Noble Metal Nanoparticles   55

Figure 4.5 Synthesis and TEM images of sulfonated diphosphine-stabilized RuNPs in water with (a) dppb-TS, (b) dppp-TS, and (c) dppe-TS with [L]-[Ru] = 0.1.[46]

were prepared from [Ru(COD)(COT)] and the diphosphines in THF (3 bar H_2; r.t.). Different diphosphines, (1,4-bis[(di-m-sulfonatophenyl)phosphine]-butane = dppb-TS, 1,4-bis[(di-m-sulfonatophenyl)phosphine]propane = dppp-TS, 1,4-bis[(di-m-sulfonatophenyl)phosphine]ethane = dppe-TS), and ligand–Ru ratios were employed in order to analyze the effect of the backbone and the diphosphine concentration on the stability and the size of the NPs and thus on their catalytic properties. Depending on the ligand amount, well-crystallized RuNPs in a mean size range of 1.2–1.5 nm were formed. The coordination of the sulfonated diphosphines at the surface of the RuNPs allowed their further dispersion in water giving rise to very homogeneous and stable aqueous colloidal solutions (up to several months) without any change in their mean sizes.

The catalytic behaviour of these aqueous colloidal solutions showed promising results in terms of reactivity when tested for the hydrogenation of unsaturated substrates (tetradecene, styrene and acetophenone) in biphasic liquid–liquid conditions. Interestingly, minor structural differences in the diphosphine ligands, such as the alkyl chain length, influenced the catalytic activity in styrene hydrogenation significantly, in addition to the positive effect of an increase in temperature (from 20 °C to 50 °C) or pressure (from 1 to 10 bar H_2) (see Table 4.2). As the [ligand]–[Ru] ratio increased, conversion and selectivity (expressed as ethylbenzene (EB)–ethylcyclohexane (EC) ratio)

Table 4.2 Hydrogenation of styrene Ru/sulfonated diphosphine aqueous colloidal solutions.[a]

			Product selectivity (%)		
Nanocatalysts	T (°C)	Time (h)	ST	EB	EC
Ru/dppb-TS	20	1	25	75	0
	20	20	0	45	55
	20	40	0	3	97
Ru/dppp-TS	20	1	59	40	1
	20	20	0	47	53
	20	40	0	2	98
Ru/dppe-TS	20	1	75	24	1
	20	20	0	41	59
	20	40	0	1	99
Ru/dppb-TS	50	1	10	90	0
	50	2	0	82	18
	50	3	0	0	100

[a]Reaction conditions: [ligand]–[Ru] = 0.1; ruthenium (3.9 × 10^{-5} mol), styrene (3.9 × 10^{-3} mol), 1 bar H_2, water (10 mL).

also increased at short reaction times. The best results were obtained with the dppb-TS ligand with the longest alkyl chain, giving rise to 75% EB after 1 h compared to 40% EB and 20% EB with dppp-TS and dppe-TS, respectively. As all the RuNPs display similar mean sizes, the observed differences in their catalytic properties were correlated to the difference in the flexibility of the alkyl chain. Due to having the highest number of carbon atoms, the dppb-TS ligand had the highest flexibity and therefore favoured a better diffusion of the substrate towards the metal surface. Although the mean size (1.25 nm) of the RuNPs did not alter with the increase of [dppe-TS]–[Ru] from 0.2 to 0.5, the variation in selectivity was explained by a limited access of the aromatic substrate to the NP surface due to increased steric hindrance when more ligands were coordinated. A molar ratio of 0.1 appeared to be a good compromise between the stabilization and the catalytic activity of the NPs. Finally, as also observed for Ru/PTA NP systems, preliminary results of recycling were encouraging for the recovery of these water-soluble Ru/sulfonated diphosphine nanocatalysts.

4.3.1.3 Ionic Liquid Stabilized Nanoparticles as Catalysts

Providing more environmentally-friendly conditions than traditional solvents, the use of ionic liquids (ILs), acting as both the solvent and the stabilizer, is emerging as an alternative for the preparation of nanocatalysts.[9,47,48] Both experiments and simulations reported in literature have evidenced the segregation phenomenon that exists between polar and non-polar domains in imidazolium-based ILs.[49,50] In collaboration with Santini *et al.*, we investigated the stabilization of RuNPs in the presence of imidazolium-derived ionic liquids.[25–27,51–53] The synthesis of RuNPs was first performed in 1-butyl-3-methylimidazoliumbis(trifluoromethanesulfonyl)imide ([RMIm][NTf$_2$]; R=C$_4$H$_9$).[25] The nanoparticles were prepared by decomposition of [Ru(COD)(COT)] under 4 bar of H$_2$ without adding any further stabilizer, but varying the reaction temperature (0 °C and 25 °C) and the stirring rate, to study their influence on the NPs formed. At 25 °C under stirring, the RuNPs were homogeneously dispersed with a mean size of ∼2.4 nm. At 0 °C, the RuNPs had a smaller size of ∼0.9 nm but in agglomerations of big clusters of 2–3 nm. When no stirring was applied at 0 °C, the NPs displayed a slightly larger mean size, ∼1.1 nm, but without any agglomeration. These results showed the influence of the temperature on the mean size of the NPs with the smallest ones being obtained at low temperatures. The size of RuNPs is governed by the degree of self-organization of the imidazolium-based ionic liquid. Since the 3D organization of the ionic liquid is better maintained at low temperature, the Ru nuclei are confined inside to a greater degree, giving smaller nanoparticles as a result.

This observation was further confirmed by using a series of imidazolium-derived ionic liquids: [RMIm][NTf$_2$] (R=C$_n$H$_{2n+1}$ with n=2, 4, 6, 8, 10), [R$_2$Im][NTf$_2$] (R=Bu) and [BMMIm][NTf$_2$] (with BMMIm=1-butyl-2,3-dimethylimidazolium) (see Figure 4.6).[26] In all ILs, the size of the RuNPs

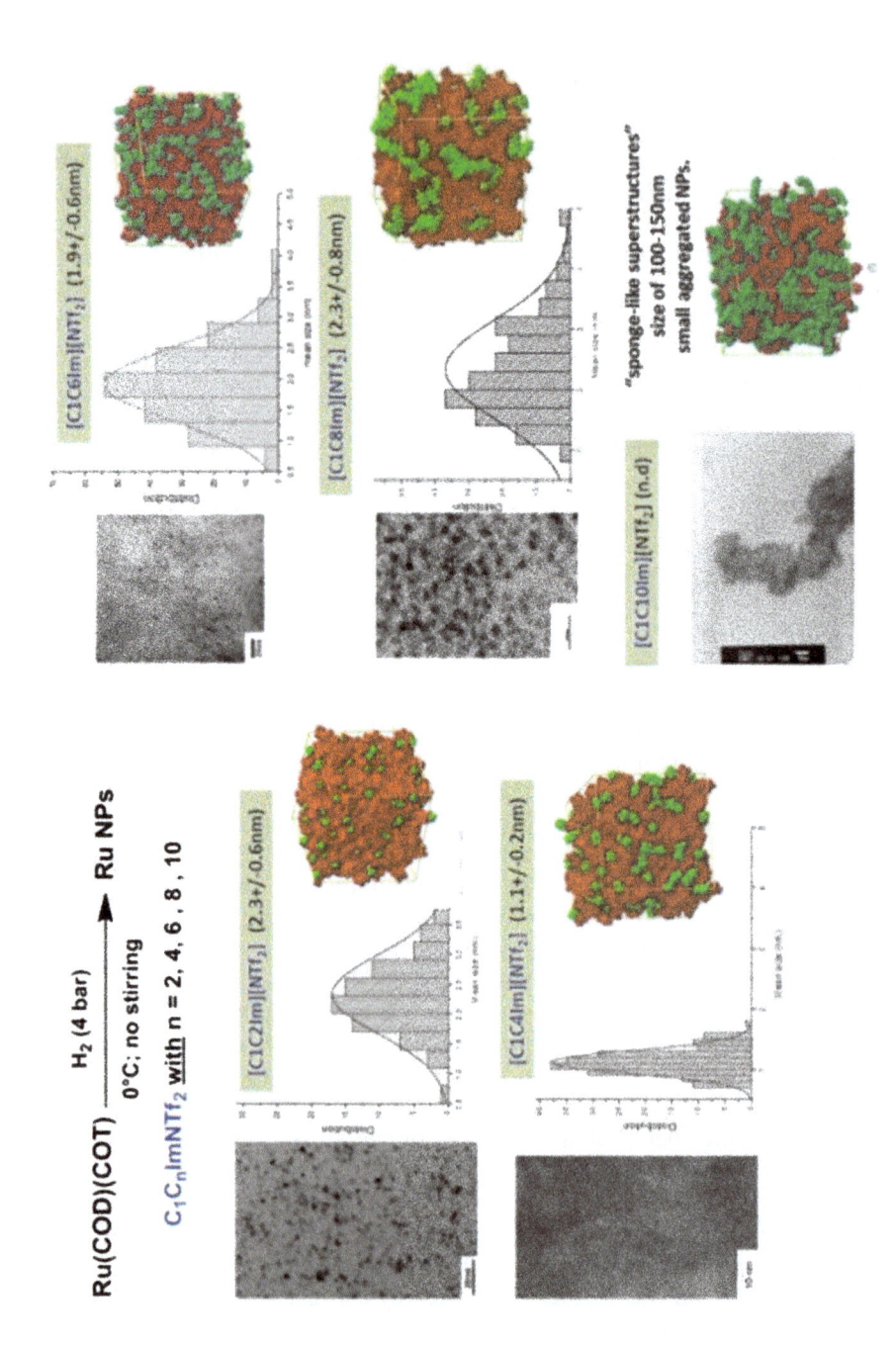

Figure 4.6 Influence of the ionic liquid chain length on the size of the RuNPs.[26]

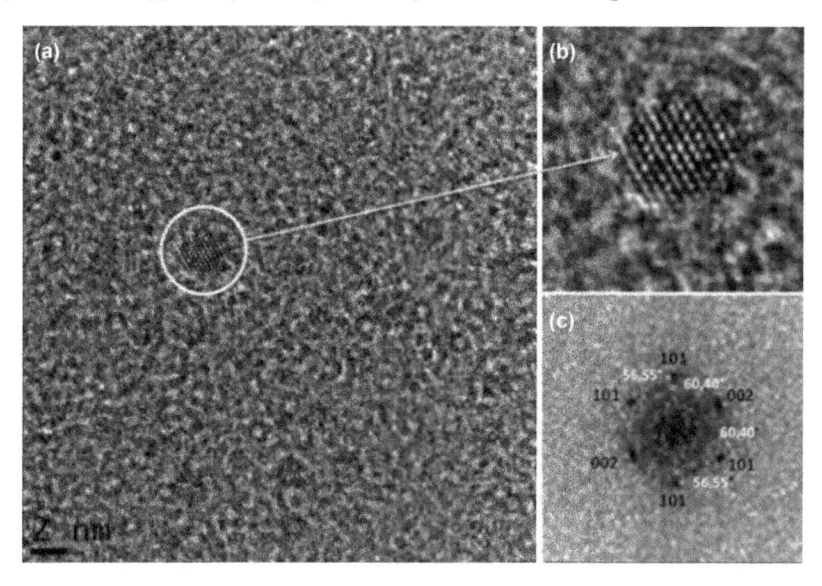

Organometallic Approach for the Synthesis of Noble Metal Nanoparticles 59

Figure 4.7 (a) General HREM image of *C4OA*-stabilized RuNPs in C₁C₄ImNTf₂, (b) HREM image of an isolated and well-crystallised *C4OA*-stabilized RuNP with (c) FFT.[51]

was smaller at 0 °C than at 25 °C, and the stirring induced their agglomeration. Additionally, the increase in alkyl chain lengths was found to be linearly proportional to the mean size of the RuNPs.

The influence of added amines on the size control of IL-prepared RuNPs and on their stability was also studied.[51,52] RuNPs are commonly used as catalysts in hydrogenation reactions but IL-stabilized RuNPs are generally not stable in the necessary catalytic conditions. The ionic liquids [RMIm][NTf₂] (R = C_nH_{2n+1}, n = 2, 4, 6, 8, 10) and two amines, octylamine (OA) and hexadecylamine (HDA), were used as added stabilizers. The synthesis of the NPs was performed following the previously described conditions, but with 0.2 molar equiv. of the chosen ligand. Regardless of the alkyl chain length of the IL, well-dispersed and well-crystallized NPs were obtained with a mean size in the range of 1.1–1.3 nm (see Figure 4.7). These RuNPs displayed a better size dispersity than the corresponding pure amine-stabilized RuNPs. The IL prevents particles from agglomeration through a confinement effect but amines also play an important role in controlling their size and dispersion. This demonstrates the interest in combining the properties of ionic liquids to confine nanoparticles in non-polar domains (*nanoreactors*) with the presence of a ligand that stabilizes the particles at a very small size.

The hydrogenation of toluene was first performed with the RuNPs synthesized in C₁C₄ImNTf₂ in the presence of octylamine, at different temperatures (see Table 4.3). The conversion obtained was low but reproducible. A temperature of 75 °C was further chosen for studying the reaction with the other colloids. Since Ru/HDA and Ru/OA NPs have similar size and shape, the slightly lower conversion obtained by hexadecylamine was explained by

Table 4.3 Hydrogenation of toluene with RuNPs synthesized in $C_1C_4ImNTf_2$ + octylamine.[a]

Nanocatalyst IL/ligand	T (°C)	PhMe/ Ru$_T$	PhMe/ Ru$_S$	Ru$_S$/ Ru$_T$	Conversion (%)	TON[a]	k$_{initial}$[b]
C4·OA	30	38/1	46/1	0.82	4	2	
C4·OA	50	38/1	46/1	0.82	9	4.5	
C4·OA	75	38/1	46/1	0.82	17	8	12±1
C4·OA	100	38/1	46/1	0.82	17	8	
C4·HAD	75	38/1	52/1	0.73	14	6.5	
C4·OA	75	38/1	46/1	0.82	17	8	12±1
C6·OA	75	38/1	49/1	0.77	16	8	8±1
C8·OA	75	38/1	49/1	0.77	11	5.5	5±1
C10·OA	75	38/1	52/1	0.73	14	7	7±1

[a]Reaction conditions: $P(H_2) = 1.2$ bars; reaction time = 5 h; Ru$_T$ = total amount of Ru atoms; Ru$_S$ = amount of surface Ru atoms. [a]Turn over number (moles of product converted per mol of Ru$_S$).
[b]Estimated initial rates in 10^{-2} mol L^{-1} h^{-1}.

the viscosity changes in the reaction mixtures and/or by a higher steric hindrance of HDA compared to that of OA once coordinated on the RuNP surface.

These systems were also tested in the hydrogenation of 1,3-cyclohexadiene (CYD), styrene (STY) and R-(+)-limonene (LIM). The results evidenced that for the hydrogenation of 1,3-CYD, STY and LIM, the activity of the RuNPs increased with σ-donor ligands such as $C_8H_{17}NH_2$ and H_2O, but decreased with the bulkier and π-acceptor ligands, PPhH$_2$ PPh$_2$H and CO. This underlined the pseudo-molecular nature of NPs with sizes <3 nm since their activity can be tuned by the σ-π-ligand character as in homogeneous catalysis.

In summary, the confinement properties of ionic liquids based on imidazolium cations help to control the growth of RuNPs, and their catalytic performance can be further improved by the addition of a stabilizing ligand.

4.3.2 Dehydrogenation Reactions of Amine-borane

Among the hydrogen storage materials, amine-boranes are interesting candidates due to their high hydrogen content (19.6 wt %).[54] Recent studies have shown that the catalytic dehydrogenation of dimethylamine-borane (Me$_2$NHBH$_3$; DMAB) can be achieved under mild conditions. In a collaborative work with the team of Özkar, we studied the catalytic performances of 3-aminopropyltriethoxysilane-stabilized Ru nanoclusters (Ru/APTS) with different sizes in this reaction (see Figure 4.8).[55,56] The mean particle size of Ru/APTS NPs decreased with increasing APTS ligand concentration (see Table 4.4).

After the addition of DMAB into a THF solution of Ru/APTS 3, the hydrogen evolution started immediately with an initial turnover frequency (TOF) of 53 h^{-1} and continued until 1 equiv. H$_2$ per mol DMAB was released. In order to differentiate between homogeneous and heterogeneous catalysis, Hg(0), a poison for heterogeneous metal(0) catalysts, was added into the reaction mixture. The suppression of the catalysis by Hg(0) was evidence of heterogeneous catalysis (see Figure 4.9). The initial TOF value of 53 h^{-1}

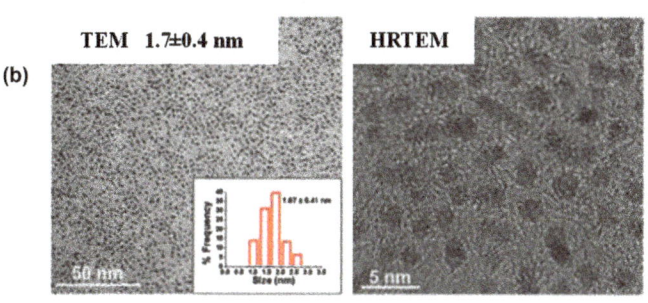

Organometallic Approach for the Synthesis of Noble Metal Nanoparticles 61

(a) Ru(COD)(COT) + 3-APTS $\xrightarrow[\text{THF, 25 °C}]{\text{3 bar H}_2}$ Ru/APTS + 2 ⬡

(b) TEM 1.7±0.4 nm   HRTEM

Figure 4.8 (a) Synthesis of Ru/APTS NPs, (b) TEM and HRTEM images of Ru/APTS NPs (~1.7 nm) immediately after the synthesis.[55]

Table 4.4 Size of Ru/APTS NPs with different [APTS]–[Ru] ratios.[56]

Nanocatalyst	[APTS]–[Ru]	Mean size (nm)[a]
Ru/APTS 1	0.25	2.36 ± 0.61
Ru/APTS 2	0.50	1.96 ± 0.44
Ru/APTS 3	1.0	1.67 ± 0.41
Ru/APTS 4	2.0	1.29 ± 0.31
Ru/APTS 5	4.0	1.16 ± 0.30

[a]The no. of atoms was estimated by using the equation $N = N_0 pV/101.7$, where $N_0 = 6.022 \times 10^{23}$, $p = 12.2$ g cm^{-3} and $V = (4/3)\pi(D/2)^3$.

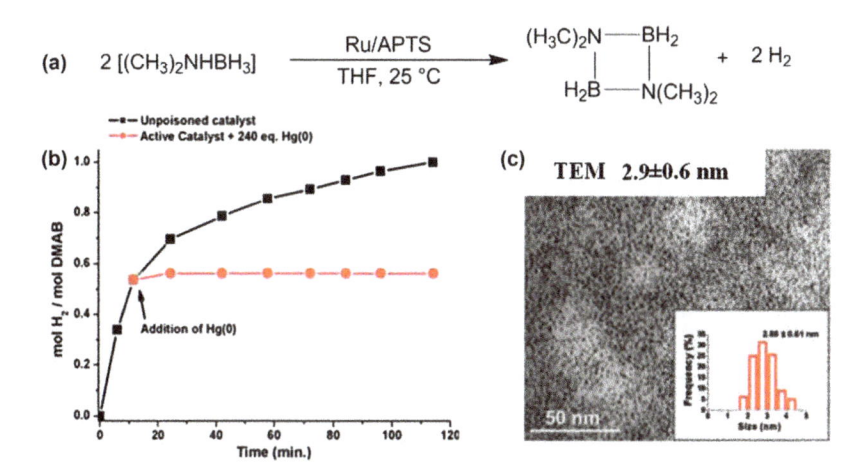

(a) $2\,[(CH_3)_2NHBH_3]$ $\xrightarrow[\text{THF, 25 °C}]{\text{Ru/APTS}}$ $(H_3C)_2N\text{—}BH_2$ $\;\;+\;\; 2\,H_2$
   $\;\;||$
   $H_2B\text{—}N(CH_3)_2$

(b) — Unpoisoned catalyst
    — Active Catalyst + 240 eq. Hg(0)

(c) TEM 2.9±0.6 nm

Figure 4.9 (a) Dehydrogenation of dimethylamine-borane catalyzed by Ru/APTS NPs in THF at r.t., (b) mol H$_2$/mol DMAB vs. time graph ([Ru] = 2.24 mM; [DMAB] = 54 mM; 240 eq. of Hg(0) after ~50% conversion of DMAB), and (c) TEM image of Ru/APTS NPs (~2.9 nm) after the third catalytic run.[56]

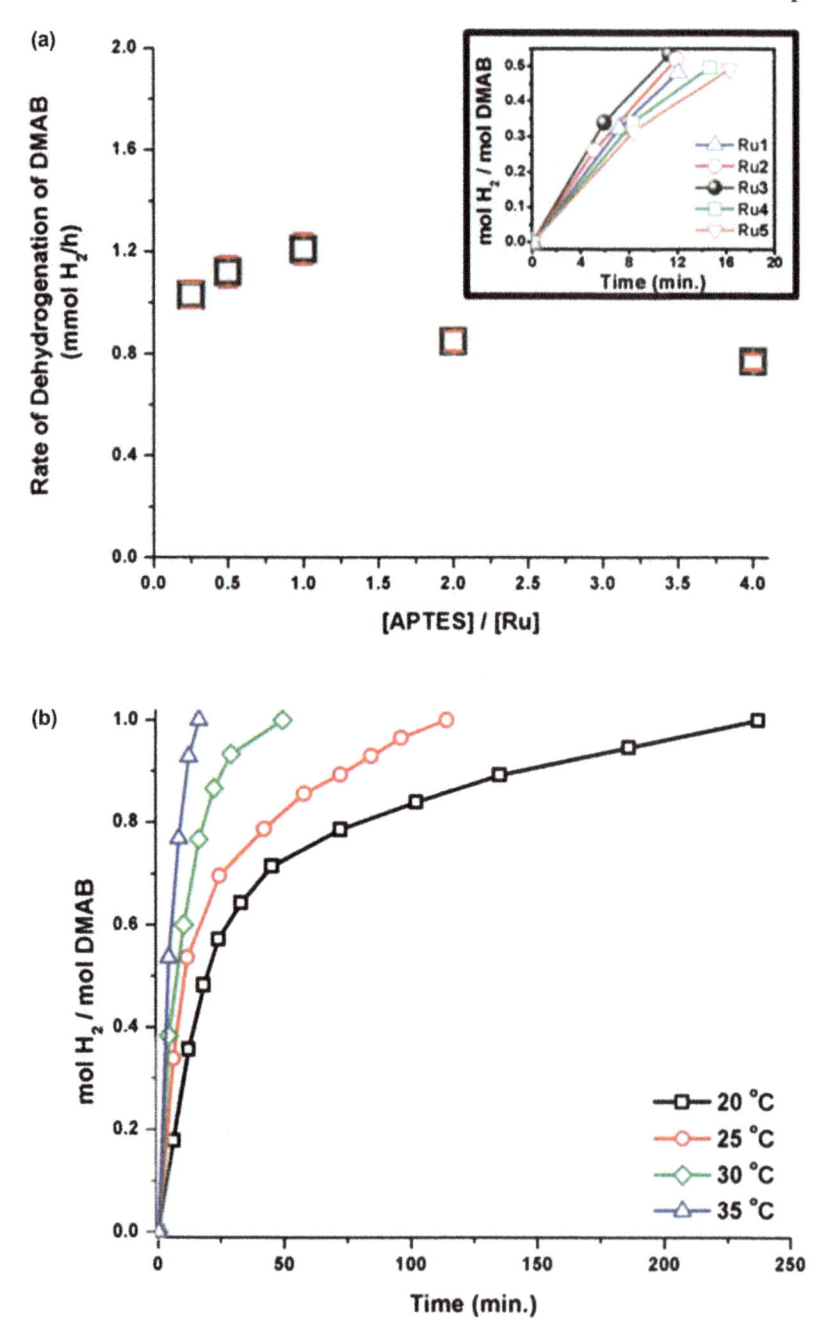

Figure 4.10 (a) The rate of dehydrogenation of 55 mM DMAB vs. [APTS]–[Ru] ratio, using Ru/APTS 1–5 NPs, (b) plot of mol H_2/mol DMAB vs. time for the dehydrogenation rxn (55 mM DMAB; 25 mM Ru/APTS 3).[56]

obtained with this system was comparable to that of the prior best heterogeneous catalyst of rhodium nanoclusters (TOF = 60 h^{-1}). Moreover, it was the first example of an isolable, bottleable and reusable transition metal nanocatalyst for the dehydrogenation of dimethylamine-borane.

An increase of APTS concentration caused a significant decrease in the catalytic activity as a result of a higher coverage of metallic surface (see Figure 4.10 (a)). The maximum catalytic activity was achieved with Ru/APTS 3 NPs, which appeared to have the best compromise between the NP mean size and the surface accessibility. The influence of temperature on the hydrogen generation rate using Ru/APTS 3 NPs gave high activity even at low temperature (see Figure 4.10 (b)).

4.3.3 Carbon–carbon Coupling Reactions

Pd nanoparticles have attracted a great deal of interest as catalysts for C–C coupling reactions, especially in Heck and Suzuki couplings. The first Heck type coupling reaction that was catalyzed by PdNPs stabilized with tetraoctylammonium bromide was reported by Beller *et al.*[57] Such PdNPs were also effective for Suzuki coupling reactions as well as the ones protected with dendrimers in the work of Narayanan *et al.*[58] Despite the progress in asymmetric catalysis, there are few examples of nanocatalysts displaying an enantioselective catalytic activity. Pt(Pd)/cinconidine and Pd/BINAP colloids were used for the hydrogenation of ethyl pyruvates[59,63] and the asymmetric hydrosilylation of styrene,[64] respectively.

4.3.3.1 Pd Nanoparticles Stabilized by Chiral Diphosphite Ligands

In a collaborative project with Gómez, Castillón and Claver *et al.*, chiral diphosphites were used as ligands for the synthesis of PdNPs and corresponding molecular complexes for a comparative study in catalytic allylic alkylation (see Figure 4.11).[65,66]

PdNPs stabilized by the chiral xylofuranoside diphosphite ligand, L1, were achieved by decomposition of [Pd$_2$(dba)$_3$] (THF; P(H$_2$) = 3 bar; r.t.) and were investigated in the allylic alkylation of *rac*-3-acetoxy-1,3-diphenyl-1-propene (*rac*-I) with dimethyl malonate (see Table 4.5). With the Pd/L1 nanocatalyst, this reaction proceeded with only one enantiomer of the racemic substrate (>95% ee), while the corresponding molecular species stabilized with the same ligand yielded both enantiomers. These results demonstrated a very high degree of kinetic resolution with Pd/L1 NPs.

Modifications in the diphosphite ligands changed the selectivity of the reaction, drastically. For instance, Pd/L2 NPs (see Figure 4.11) did not induce any selectivity under the same conditions. When the carbohydrate

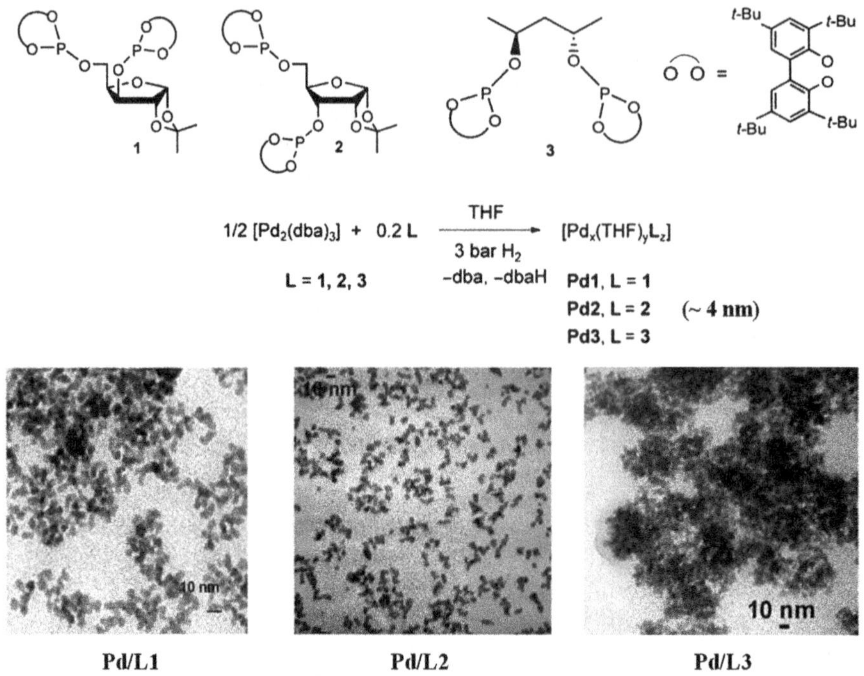

Figure 4.11 Chiral diphosphite ligands and the corresponding organometallic synthesis of PdNPs.[66]

Table 4.5 Asymmetric allylic alkylation of *rac*-3-acetoxy-1,3-diphenyl-1-propene (*rac*-I) catalyzed by molecular and colloidal Pd systems containing chiral diphosphite L1.

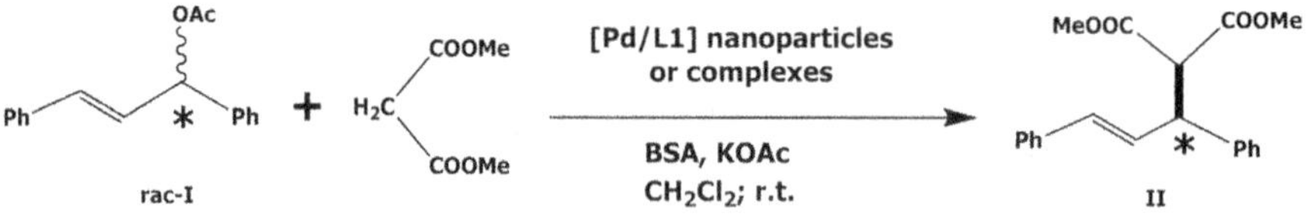

Catalyst	I–Pd–L1	t(h)	Conv. (%)	ee II (%)	ee rac-I (%)
Molecular[a]	100 : 1 : 1.25	1.5	83	90 (S)	0
Colloidal[b]	100 : 1 : 0	24	2	nd	0
Colloidal[b]	100 : 1 : 0.2	24	56	97 (S)	89 (S)
Colloidal[b]	100 : 1 : 1.05	24	56	97 (S)	89 (S)
Colloidal[b]	100 : 1 : 0.2	168	59	97 (S)	89 (S)
Colloidal[b]	100 : 1 : 1.05	168	61	97 (S)	89 (S)

[a]Molecular system generated *in situ* from [Pd(C$_3$H$_5$)Cl]$_2$ and L$_1$*.
[b]20% or 125% of free ligand added into the reaction mixture.

skeleton was substituted by the flexible alkyl 2,4-dimethylpentyl backbone (L3) for less hindrance, the Pd/L3 nanocatalyst appeared highly active and selective towards the allylic alkylation of *rac*-I, similar to the Pd molecular complex synthesized with the same ligand. However, alkyl substituted allyl acetates were not alkylated in contrast to *rac*-I in the presence of stable Pd/L1 and Pd/L3 NPs. This result was consistent with the strong affinity between the metallic surfaces and the aryl groups, most likely due to the π-coordination between the metallic surface and the aromatic group.

These results demonstrated that: (i) catalysis by NPs may be very effective in terms of kinetic resolution; (ii) the system is much more sensitive to the adjustment between the metal, the ligand and the substrate than the corresponding molecular catalysts, in a way which may resemble enzymatic systems. This may explain the very limited number of enantioselective catalytic systems that involved MNPs. Further studies in this field need tuning on these features.

4.3.3.2 Pd Nanoparticles Stabilized by Pyrazole Ligands

In collaboration with García-Antón *et al.*, a new family of hybrid pyrazole derived ligands containing alkylether, alkylthioether or alkylamino moieties was used in the preparation of PdNPs and their corresponding molecular complexes in order to compare their catalytic performance in Suzuki–Miyaura and homocoupling reactions (see Figure 4.12).[67]

Since the best conversions for both iodo- and bromo-toluene were achieved with Pd complexes with ligands, L2 (Y=S) and L3 (Y=NH), most likely due to their strong coordination. PdNPs stabilized with these ligands (Pd/L2 and Pd/L3 NPs) were fully investigated in the Suzuki–Miyaura cross-coupling reaction, using Pd/C as a reference nanocatalyst (see Table 4.6). When 4-iodotoluene was used as a substrate, complete conversion was achieved for both nanocatalysts Pd/L2 and Pd/L3 (0.4 mol% Pd) and 4-methylbiphenyl (BT) was the main product, together with a significant amount of toluene. Pd/C yielded similar chemoselectivity with a lower conversion. A higher load of substrate (0.04 mol% Pd) with Pd/L3 NPs improved the conversion and the chemoselectivity, significantly. 4-Bromotoluene as the substrate yielded only 4,4′-dimethylbiphenyl (TT) due to the homocoupling of 4-bromotoluene and some toluene as the side product. Homocoupling reactions were also investigated in the absence of phenylboronic acid using Pd/pyrazole colloidal solutions. Pd/L3 nanocatalyst gave a remarkable increase in the yield of homocoupled product with a full and 70% conversion to toluene for 4-iodotoluene and 4-bromotoluene as a substrate, respectively. Homocoupling reactions in the absence of the phenylboronic acid using Pd/L3 NPs led to a 70% and full conversion to toluene for 4-bromotoluene and 4-iodotoluene, respectively (see Table 4.6).

For all Pd/pyrazole nanocatalysts, catalytic activity was observed both for the C–C coupling reaction and the dehalogenation of the substrate. The dependence of the chemoselectivity of the reaction on the nature of the catalyst strongly suggested that the homocoupling reaction takes place on the surface of the heterogeneous systems.

Table 4.6 C–C coupling reaction using Pd/pyrazole colloidal solutions.[a]

Catalyst [PhB(OH)]/[Pd]	X = I Conv[b] (%)	BT[c]	TT[c]	PhMe[c]	Yield BT[e] (%)	X = Br Conv[b] (%)	BT[c]	TT[c]	PhMe[c]	Yield TT[f] (%)
Pd/L2 315	100	69	—	31	69	12	—	14	86	2
Pd/L3 315	100	61	—	39	61	37	—	53	47	20
Pd/C 315	31	65	—	35	20	84	17	46	37	39
Pd/L3[d] 3125	100	90	—	10	90	—	—	—	—	—
Pd/L2 —	100	—	—	100	0	18	—	24	76	4
Pd/L3 —	100	—	—	100	0	64	—	70	30	45
Pd/C —	100	—	1	99	0	83	—	41	59	34

[a]Reaction conditions: 1×10^{-2} mmol Pd, 2.5 mmol 4-halogenotoluene, 5.0 mmol ^{t}BuOK, 0.5 mmol naphthalene as internal standard, 8.0 ml DMF–2.0 mL H$_2$O, 100 °C.
[b]Conversion after 6 h.
[c]Product distribution after 6 h.
[d]4-Halogenotoluene–^{t}BuOK–Pd = 2500 : 5000 : 1.
[e]Yield of 4-methylbiphenyl (BT).
[f]Yield of 4,4′-dimethylbiphenyl (TT).

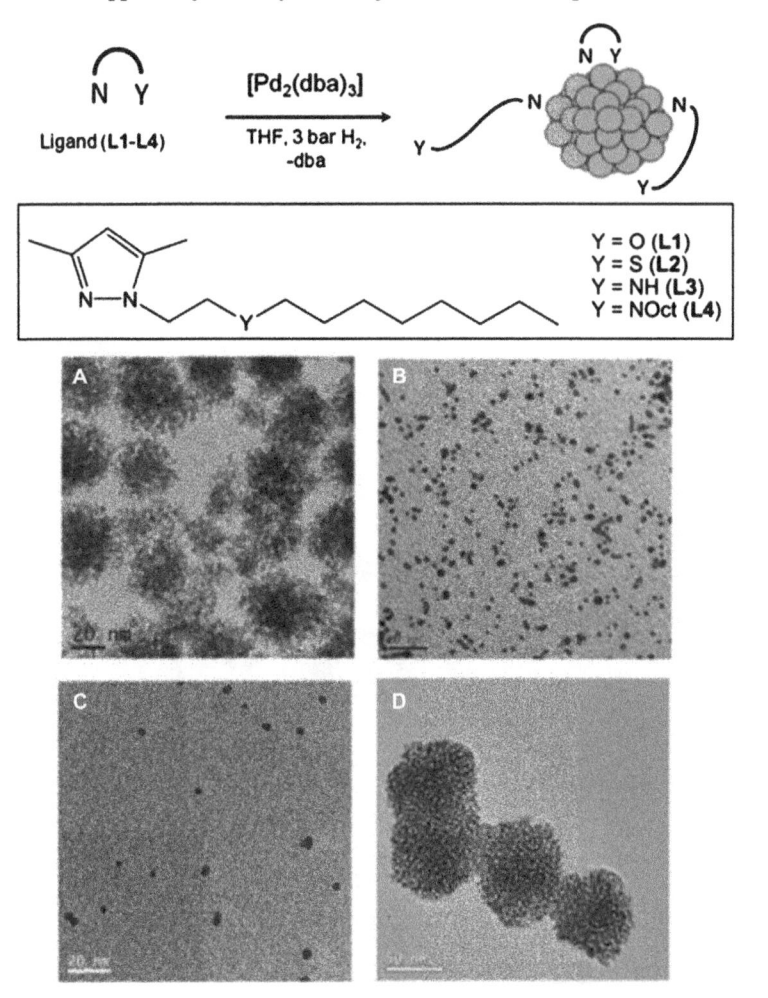

Figure 4.12 Synthesis of PdNPs stabilized by pyrazole derived ligands with $[L_x]$–Pd = 1.0.[67]

4.3.4 Hydroformylation Reactions

In collaboration with Castillón and Claver *et al.*, RhNPs were prepared by decomposition of [Rh(η^3-C$_3$H$_5$)$_3$] and [Rh(μ-OMe)(COD)]$_2$, in the presence of diphosphite ligands (1 or 2; Rh–L = 1 : 0.2; THF; P(H$_2$) = 3 bars; r.t.) (see Figure 4.13).[68] For both, TEM micrographs of Rh1 and Rh2 NP samples derived from [Rh(η^3-C$_3$H$_5$)$_3$] revealed good dispersion of spherical NPs with similar small sizes (Rh1: ~3 nm and Rh2: ~2 nm). Rh3 and Rh4 NPs synthesized from [Rh(μ-OMe)(COD)]$_2$ showed large and sponge-like spherical superstructures (Rh3 NPs: ~50 nm and Rh4 NPs: ~35 nm) with a tendency to agglomerate. Images at higher magnification revealed small individual NPs in the superstructures (~4 nm).

68   *Chapter 4*

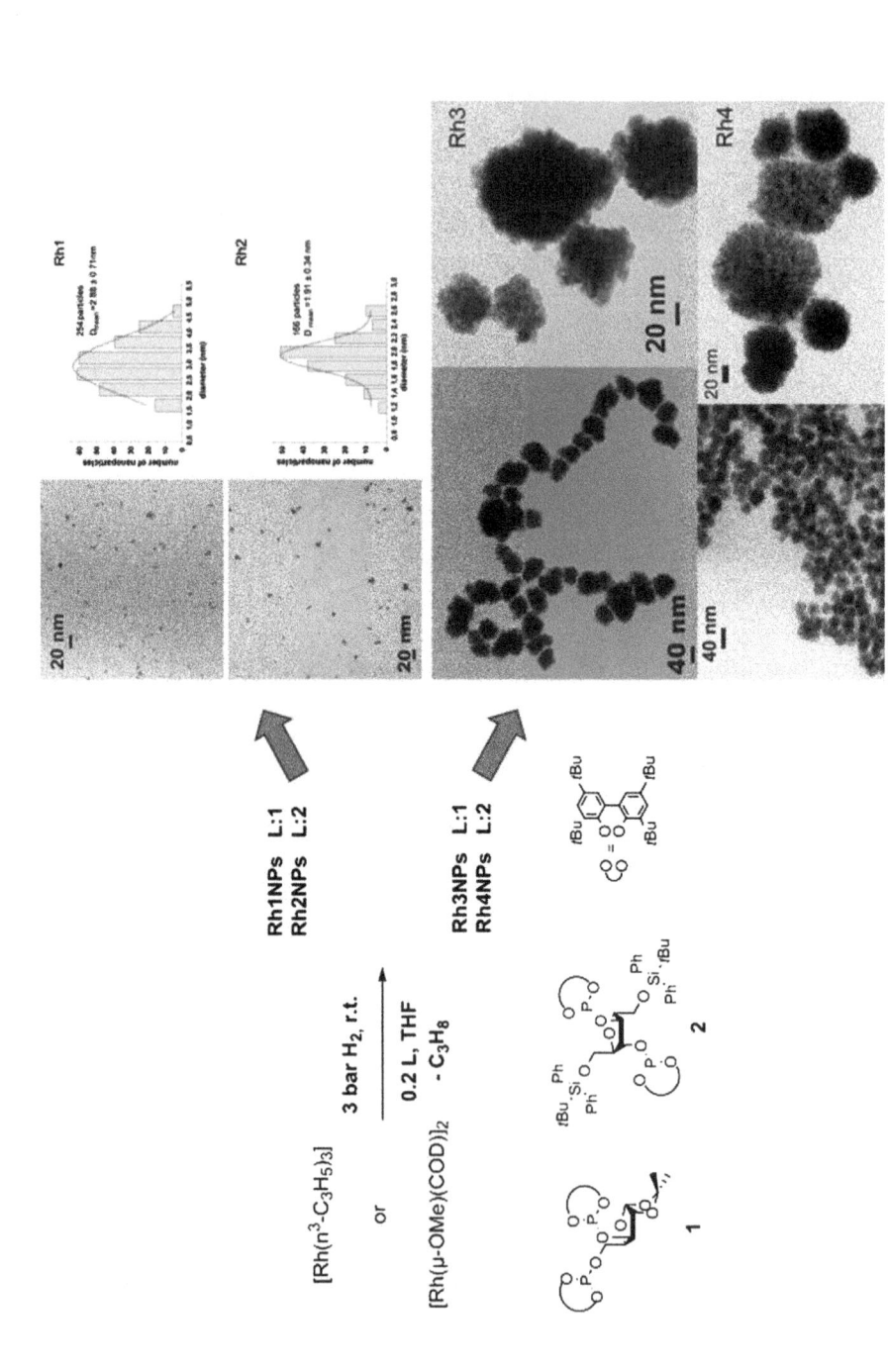

Figure 4.13 Synthesis of RhNPs from $[Rh(\eta^3-C_3H_5)_3]$ and $[Rh(\mu-OMe)(COD)]_2$ in the presence of chiral diphosphite ligands 1 and 2.[68]

Table 4.7 Styrene hydroformylation with diphosphite-stabilized RhNPs.[a,68]

Catalyst	Rh–L–S[b]	t(h)	% Conversion[c]	% Regioselectivity[d]	% ee
Rh3	1 : — : 200	24	28	57	0
Rh3	1 : 0.2 : 200	24	11	>99	40 (S)
Rh4	1 : — : 200	24	95	54	13 (S)
Rh4	1 : 0.2 : 200	24	80	90	24 (S)

[a]Reaction conditions: 3 mg RhNPs, 80 °C, $p = 20$ bar $pCO–H_2 = 1$, Rh–S 1 : 200 styrene 5.8 mmol, 10 ml toluene.
[b]Molar ratio between rhodium, excess ligand added to the catalysis and substrate.
[c]% Conversion styrene G.C.
[d]% 2-Phenylpropanal.

Rh3 and Rh4 NPs were investigated in the asymmetric hydroformylation of styrene (see Table 4.7). Rh3 NPs showed low conversions while Rh4 ones were more active. The addition of free ligand in the reaction media led to less active but more regioselective catalytic systems. The enantioselectivity was also affected by the addition of the free ligand: from 0% to 40% (S) for Rh3 NPs and from 13% to 24% (S) for Rh4 NPs.

In order to shed some light on the exact role of these nanocatalysts, catalytic reactions were performed in similar conditions with molecular complexes prepared with the same ligands. The corresponding molecular catalysts displayed higher activity than the colloidal ones but slightly lower enantioselectivities. Since poisoning tests (addition of CS_2 or Hg) did not induce significant change in the activity and selectivity for both molecular and colloidal catalytic systems, the formation of molecular complexes was envisaged. This was confirmed by high-pressure NMR studies under catalytic conditions which revealed the presence of $[RhH(CO)_4]$ in the nanosystem. So, molecular species formed during the reaction were most likely responsible for the observed catalytic activity, even if some activity coming from the RhNPs could not be excluded. The higher enantioselectivities reached with Rh nanocatalysts could be explained by RhNPs acting as a reservoir for the formation of molecular species. The difference in enantioselectivity from pure molecular complexes is then attributed to the close proximity of such molecular species to the RhNP surface.

4.4 Nanoparticles for Supported Catalysis

In the last fifteen years, the application of supported metal nanoparticles as catalysts in organic synthesis has received a renewed interest. This interest derives from the expected synergistic effect—when a metal is associated to a support, it can help to orientate catalytic reactions. Using supported gold

NPs in carbon monoxide oxidation reactions is a good example of this phenomenon.[69] The recyclability and recovery from reaction media are still major barriers to the widespread use of NPs in catalysis. To overcome these problems, the immobilization of MNPs on solid supports appears to be a promising alternative. Synthetic methods are also being developed to achieve the direct synthesis of MNPs in the presence of a support in a controlled manner. Intensive work is carried out on the functionalization of the support to increase the anchorage of the particles, inspired by the nature of the ligands that are used to stabilize MNPs in solution.

The organometallic approach for the synthesis of MNPs can also be applied for the preparation of composite materials. Some results were obtained using alumina membranes, mesoporous silica and carbon materials as templates for the deposition of MNPs, mainly for hydrogenation and oxidation reactions. Metal oxide nanoparticles could be obtained after a calcination step on the preformed nanoparticles under air without change in size and dispersion. The inclusion of MNPs was performed following two different approaches, either by impregnation of the support using a colloidal solution of pre-formed NPs or by direct synthesis of the NPs in the presence of the support, with or without a ligand.

4.4.1 Alumina as a Support for Hydrogenation and Oxidation Reactions

In collaboration with Schmid *et al.*, the filling of nanoporous alumina membranes of various pore widths has been carried out from [Ru(COD)(COT)] decomposed in THF–MeOH mixtures without any stabilizer in two different ways.[70] The first approach involved the impregnation of an alumina support with colloidal solutions of RuNPs in different sizes, which were dependent on the ratio of MeOH–THF in the reaction mixture. Colloidal solutions were transferred into membranes by vacuum induction. Very few agglomerates were observed outside the pores whereas there were dense areas within the membrane channels. The second approach consisted of decomposition

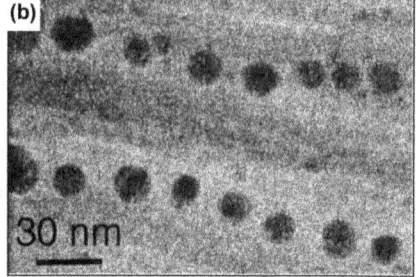

Figure 4.14 TEM images of RuNPs in the pores of alumina membranes by two methods: (a) impregnation and (b) *in situ* decomposition.[70]

under 3 bar of H_2 after the deposition of [Ru(COD)(COT)] inside the pores. In this way, homogeneous materials displaying well dispersed RuNPs were obtained in the pores of alumina membranes (see Figure 4.14). The size of the particles depended on the pore diameter of the template.

These materials were tested in the catalytic hydrogenation of 1,3-butadiene and in CO oxidation in the gas phase and showed an increase in activity with a decrease in particle size.[71]

4.4.2 Silica as a Support for Hydrogenation and Oxidation Reactions

4.4.2.1 Hydrogenation of Olefins (Cyclohexene and Myrcene)

With the aim of improving the stabilization of supported PdNPs in the use of hydrogenation catalysis, a hybrid terpyridine ligand was designed to functionalize a magnetic support in a collaborative project with Rossi et al.[72] The magnetic support consisted of a magnetite core surrounded by a silica shell. Direct decomposition of [Pd$_2$(dba)$_3$] on the terpyridine-modified magnetic support provided well-dispersed Pd NPs (~2.5 nm) (see Figure 4.15). This new composite nanomaterial appeared as a highly active catalyst under mild conditions for the hydrogenation of cyclohexene, reaching turnover frequencies up to ca. 58 000 h^{-1} or 129 000 h^{-1}

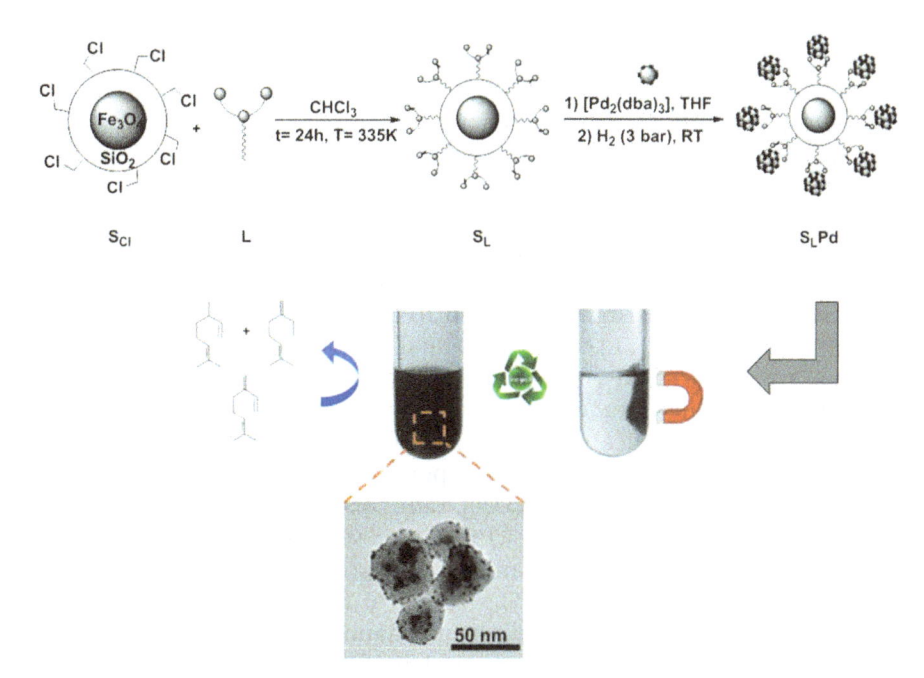

Figure 4.15 Synthesis of PdNPs onto the terpyridine-functionalized magnetic silica support for application in the hydrogenation of myrcene.[72]

when corrected for surface Pd atoms. Furthermore, this nanocatalyst was highly selective for the formation of monohydrogenated compounds in the hydrogenation of β-myrcene. The activity and the selectivity of the nanocatalyst were largely increased in comparison with a similar nanocatalyst that consisted of PdNPs supported on an amino-modified magnetic support or onto Pd/C. This work showed how the design of a hybrid ligand for the functionalization of the support can strongly influence the catalytic properties of the nanomaterial. In terms of *green chemistry*, the magnetic character of the nanocatalyst makes it easily recoverable.

4.4.2.2 Oxidation of Carbon Monoxide and Benzyl Alcohol

In collaboration with Corriu *et al.*, organized mesoporous silica materials containing phosphonate groups were used as a host for controlling the growth of RuNPs.[73] [Ru(COD)(COT)] was first impregnated into the support and then decomposed (3 bar H_2; r.t.). The phosphonate groups present in the pores of the solid acted as a stabilizer for the Ru/SiO$_2$ NPs and allowed their organization in the channels of the host material. A calcination step (air; 400 °C) led to a RuO$_2$/SiO$_2$ nanomaterial (see Figure 4.16).

The use of RuO$_2$/SiO$_2$ composite nanomaterials as catalytic filters for gas sensors revealed successful catalytic behavior for the preferential detection of propane in a propane–CO–NO$_2$–air mixture. The efficiency of the propane sensing was dependent on the metal content of the nanocomposite materials: higher Ru/Si$_w$ induced higher $S_{C_3H_8}$–S_{CO} sensitivity ratios (see Figure 4.17). The RuO$_2$/SiO$_2$ nanomaterials partially removed CO from the gas mixture by its selective oxidation into CO$_2$, leaving the hydrocarbon content unaltered.

To control the size of the NPs inside the silica matrix, the organometallic approach was combined with the sol–gel method for the preparation of RuO$_2$/SiO$_2$ nanocomposite materials (see Figure 4.18).[74] The key point of this new methodology was to use a ligand ((PhCH$_2$)$_2$N(CH$_2$)$_{11}$O(CH$_2$)$_3$Si(OEt)$_3$; benzenemethanamine; L) that has the dual role of being a stabilizing agent for the synthesis of NPs and a precursor for the silica matrix. Hybrid materials with high Ru contents displaying well-dispersed and small sized RuNPs in the silica matrix were obtained by this way. A calcination step (air; 400 °C) gave rise to mesoporous silica materials containing RuO$_2$ NPs. The [L]–[Ru(COD)(COT)] ratio helped to control the mean size of the NPs (see Figure 4.19). Moreover, the addition of tetraethylorthosilicate (TEOS) to the initial [L]–[Ru(COD)(COT)] solution led to materials with higher proportions of silica, also giving higher specific surface areas to the RuNPs. These hybrid nanomaterials displayed a good dispersion inside the silica matrix and interesting surface specific area properties, which made them attractive materials to be used as catalytic filters for gas sensors.

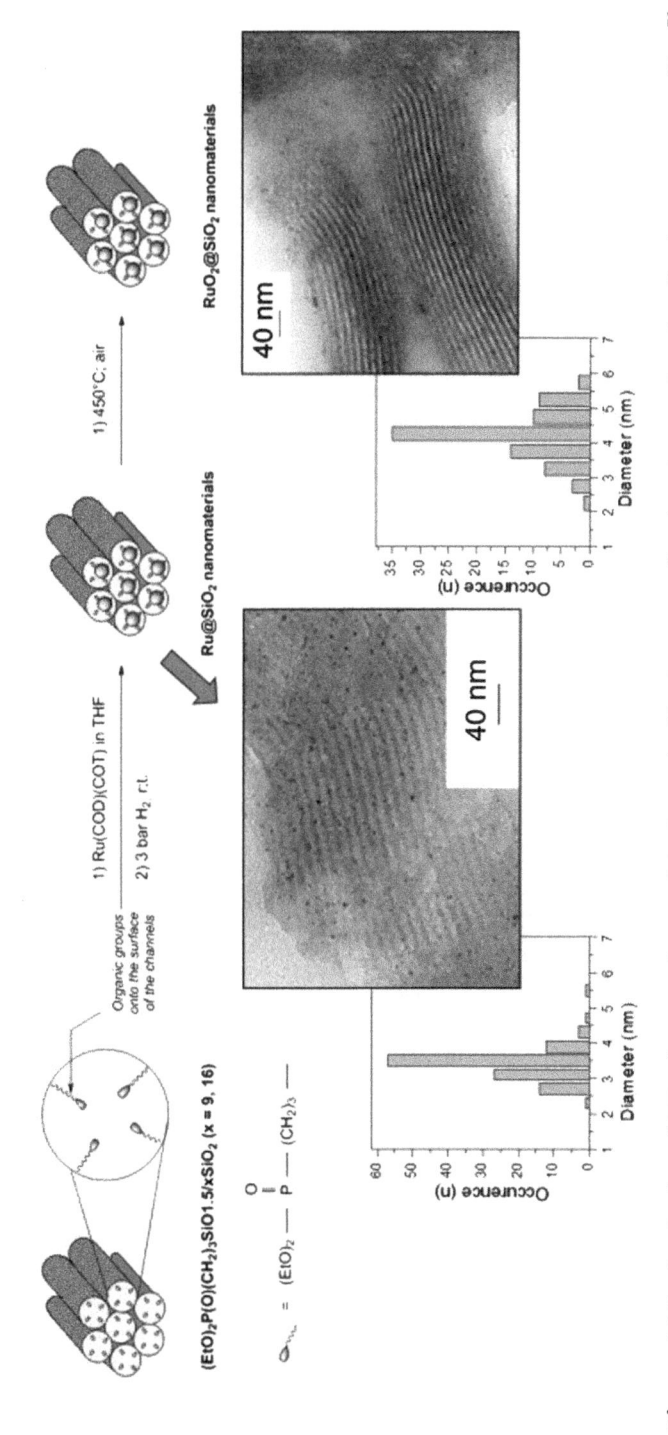

Organometallic Approach for the Synthesis of Noble Metal Nanoparticles 73

Figure 4.16 Synthesis and TEM images of Ru and RuO$_2$NPs inside the pores of a phosphonate-functionalized mesoporous silica.[73]

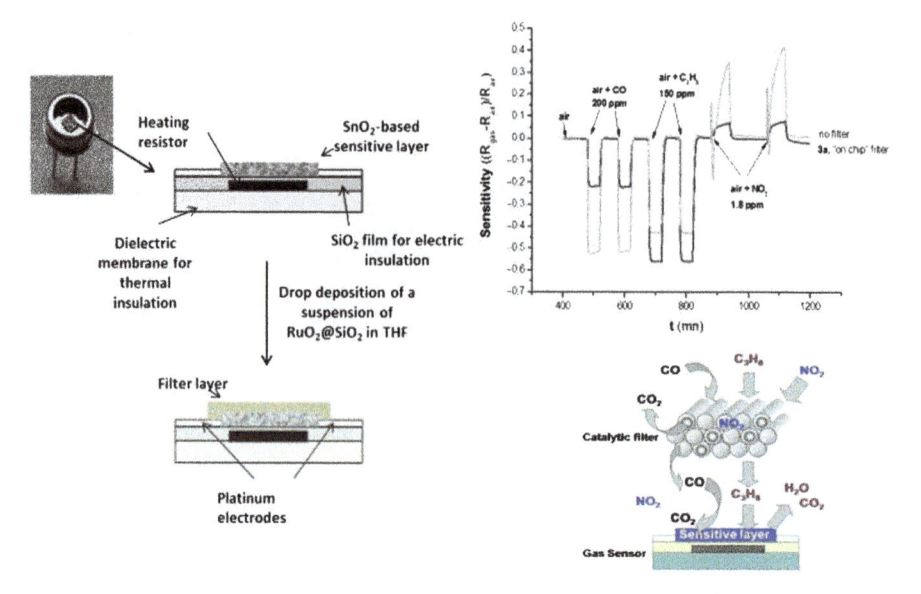

Figure 4.17 (a) Schematic representation of the gas sensors, (b) sensitivity variations of the gas sensor under different gas compositions with (black line) or without (grey line) an "on chip" filter layer, and (c) the role of filter for CO.[73,74]

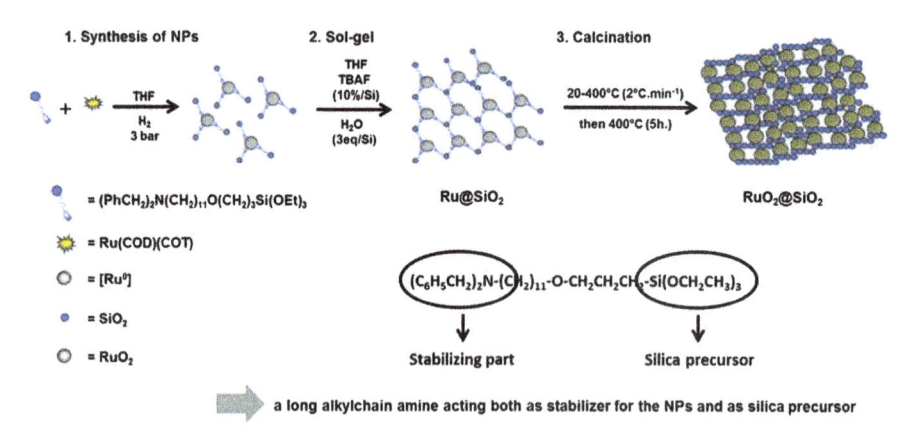

Figure 4.18 Synthesis of Ru and RuO$_2$ NPs embedded in silica using a polycondensable amine.[74]

The extension of this methodology to 3-aminopropyltriethoxysilane and 11-aminoundecyltriethoxysilane, gave rise to very small and reactive RuO$_2$ NPs.[75] Preliminary studies in the aerobic oxidation of benzyl alcohol showed promising results for further development of these types of nanomaterials.

Organometallic Approach for the Synthesis of Noble Metal Nanoparticles 75

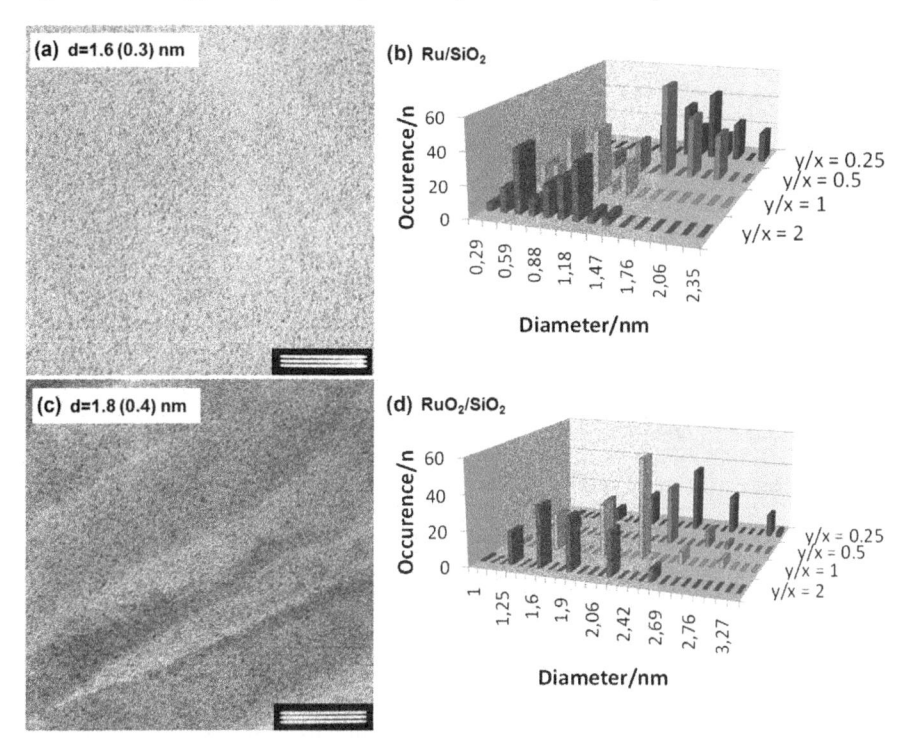

Figure 4.19 (a) TEM image of Ru/SiO$_2$ nanocomposites with L–Ru(COD)(COT) ratio of 2. (b) Distribution histograms of Ru/SiO$_2$ NPs. (c) TEM image of RuO$_2$/SiO$_2$ NPs with L–Ru(COD)(COT) ratio of 0.5. (d) Distribution histograms of RuO$_2$/SiO$_2$ nanocomposites with L–Ru(COD)(COT) = 2, 1, 0.5, 0.25 (scale bars of TEM images = 50 nm).[74]

4.4.3 Carbon Materials as Supports for Hydrogenation and Oxidation Reactions

Today, carbon materials are used often for the immobilization of MNPs since they offer several advantages as solid supports. These include easy availability, relatively low cost, high mechanical strength and chemical stability. In addition, their porous structure makes them attractive for surface chemistry as it enables easy modifications, such as its functionalization in order to suit the immobilization needs of the NPs.

4.4.3.1 Hydrogenation of Cinnamaldehyde

In a collaborative work with Serp *et al.*, the effect of the confinement of MNPs inside carbon nanotubes (CNTs) on their catalytic properties was studied. The synthesis of RuPtL/CNTs (L = 4-(3-phenylpropyl)pyridine) from the organometallic precursors, [Ru(COD)(COT)] and [Pt(CH$_3$)$_2$(COD)], was performed using various functionalized CNTs in different reaction conditions (see Figure 4.20).[76] One option consisted of the preparation of the

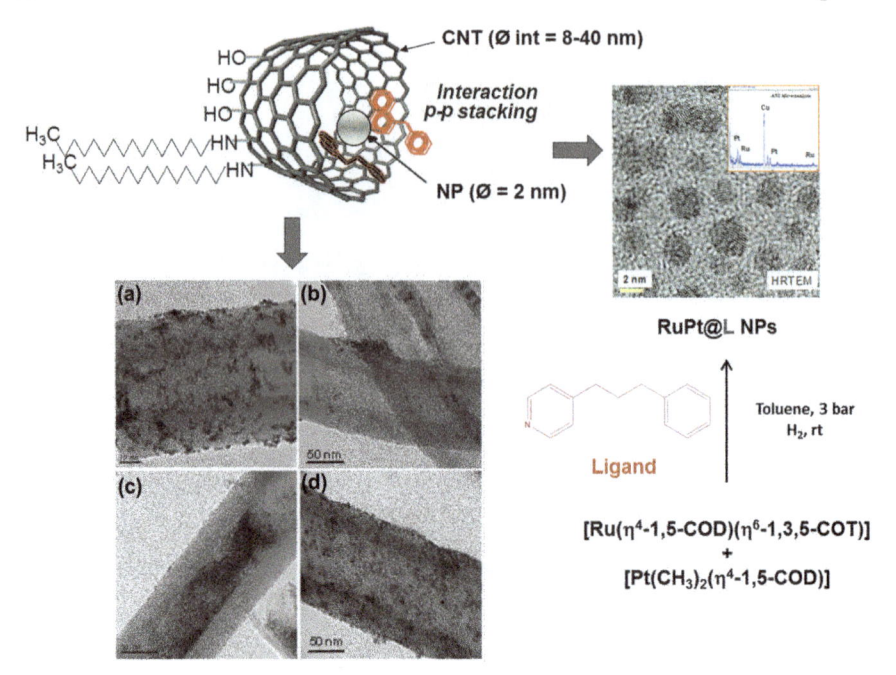

Figure 4.20 Top left: schematic representation of the confinement of the RuPt NPs in the channels of the CNTs. Bottom left: TEM images of the RuPt/L/CNTs nanomaterials with (a) 11% w/w PtRu/L/CNT, (b) 5% w/w PtRu/L/CNT-COOH, (c) 23% w/w PtRu/L/CNT-CO-NH-R (impregnation) and (d) 5% w/w PtRu/L/CNT-COOH (direct decomposition). Right: synthesis and HRTEM image of the RuPt/L NPs.[76]

NPs through stabilization by the ligand first and then their impregnation on the CNTs. An alternative route was the co-decomposition of the two organometallic precursors in the presence of both the ligand L and the CNTs. The best results in terms of confinement of the particles were obtained by the impregnation method employing amide-functionalized CNTs. The resulting hybrid material had ∼2–2.5 nm NPs which were located inside the CNTs (80% of NPs for a 23 wt% of metal). All the prepared systems, in addition to the non-supported RuPt/L NPs, were tested as catalysts in the hydrogenation of cinnamaldehyde (see Table 4.8). The catalytic activity and the selectivity of CNT-supported RuPt/L NPs were higher than those of the non-supported ones. In the case of the NPs mainly located inside the CNTs, a remarkable selectivity towards the formation of cinnamyl alcohol was achieved.

4.4.3.2 Oxidation of Benzyl Alcohol

NP/carbon hybrid systems were obtained by the impregnation of different mesoporous carbons with colloidal solutions of RuNPs or PdNPs prepared previously by the decomposition of [Ru(COD)(COT)] or [Pd$_2$(dba)$_3$], (P(H$_2$) = 3 bar; THF) in the presence of 4(3-phenylpropyl)pyridine or triphenyl- and trioctylphosphine as ligands, respectively.[77] The supported Ru and PdNPs

Table 4.8 Catalytic hydrogenation of cinnamaldehyde with RuPt/CNT nanomaterials.[a,76]

Catalyst	NPs d_{mean} (nm) inside/outside	% NP int.	TOF (h^{-1})	HCAL	HCOL	COL
PtRu/L NPs	2.2	—	30	50	15	35
PtRu/CNT2	2.2/2.2	10	56	33	8	59
PtRu/CNT1	1.6/2.2	30	75	18	12	69
PtRu/CNT3	2/2.5	80	85	0	5	95

[a]Reaction conditions: isopropanol; 20 bar H$_2$; 343 K; 2 h.

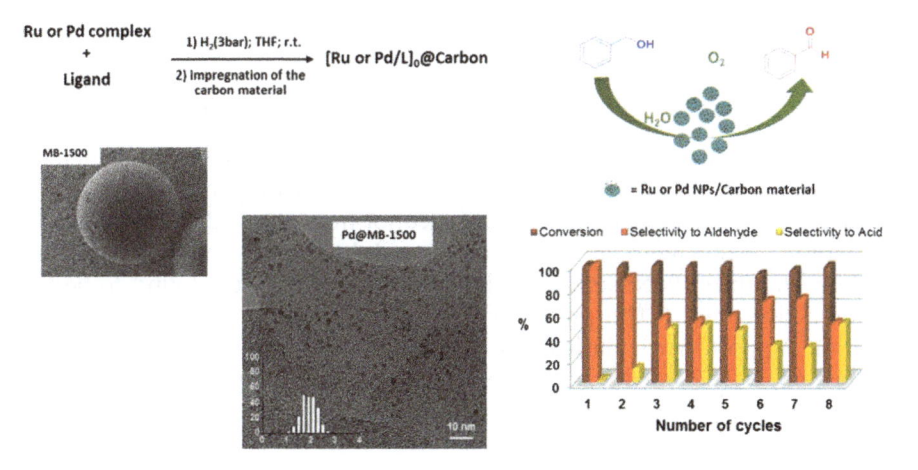

Figure 4.21 Preparation of RuNPs/carbon and PdNPs/carbon nanomaterials for the oxidation of benzyl alcohol in water and recycling of the nanocatalysts.[77]

are well dispersed on the carbon materials and display mean sizes of 1.2–1.3 nm and 1.9–2.2 nm, respectively. These hybrid systems were successfully used as catalysts in the oxidation of benzyl alcohol in water at 80 °C, giving rise to excellent conversion and selectivity (>99%) towards the aldehyde for both metals (see Figure 4.21). The recyclability studies showed that carbon-supported PdNPs can be used in up to eight successive reaction cycles without significant loss in activity. The catalytic activity of these systems was

Table 4.9 Hydrogenation of furfural catalysed by carbon-supported PdNPs/carbon under microwave conditions.[a,78]

Nanocatalyst	Reaction time (min)	Conversion (mol%)	Selectivity in furan (mol%)	Selectivity in THF (mol%)
No catalyst	60	—	—	—
MB-H$_2$O$_2$	60	—	—	—
MB-1500	60	—	—	—
Pd(TOP)/MB-H$_2$O$_2$	30	>90	15	80
Pd(TOP)/MB-1500	30	75	20	72
Pd(TPP)/MB-H$_2$O$_2$	30	69	10	85
Pd(TPP)/MB-1500	30	65	<10	80
Pd/MB-H$_2$O$_2$	45	70	30	70
Pd/MB-1500	45	45	35	65

[a] Reaction conditions: 2 mmol furfural, 1.5 mL formic acid, 1.5 mL water, 0.1 g catalyst, 100 °C, microwave power = 100–150 W, P = 150–200 psi (developed in the systems).

found to be influenced by the hydrophilicity of the carbon support. However, the significance of this influence depended on the hydrophobic character of the ligand used to stabilize the MNPs.

4.4.3.3 Versatile Dual Hydrogenation–oxidation Reactions

The previously described carbon-supported PdNPs appeared as efficient dual hydrogenation–oxidation catalysts, such as in the hydrogenation of furfural and the selective oxidation of benzyl alcohol in water under microwave irradiation. Nanocatalysts based on trioctylphosphine and triphenylphosphine-stabilized PdNPs on oxidized carbon supports were found to be the most active ones.[78] The presence of oxygen groups on the surface of the carbon support, particularly those of acidic character, improved the immobilization of the PdNPs as well as the water affinity, giving rise to high catalytic performances (see Table 4.9).

4.5 Conclusion and Perspective

In this chapter, we presented an overview of the major results obtained in the synthesis of metal nanoparticles from organometallic complexes and their investigation in catalysis for various reactions. Despite the poor handling of metal–organic precursors which are highly sensitive towards oxygen and moisture, the organometallic approach is still a versatile method in the preparation of both soluble and supported MNPs. It allows us to tune the nature of the MNPs for the precise control of size, shape, composition and surface state according to the aim of the catalytic reaction.

In solution, the control of the characteristics of the MNPs is governed by a ligand that is stabilizing by nature and by controlling the key parameters of the syntheses such as the rate of nucleation and the rate of growth. With the experience gained from homogeneous catalysis, the ligands already known to orient catalytic properties, such as phosphine and carbene derivatives, can also be used for very small nanocatalysts. Moreover, ligands can be specifically chosen for their advantageous properties such as solubility or enantioselectivity. For instance, water soluble NPs are prepared using amphiphilic ligands. Enantioselective catalysis could also be achieved by using oxazoline or diphosphite ligands, but keeping in mind the low chiral induction. Concerning the immobilization of MNPs onto supports, the most promising results evidenced the important role of the functionalization of the support to increase the anchorage of the particles and consequently their catalytic performance by increasing the selectivity and the recovery of the catalyst.

The preparation of nanocatalysts displaying desirable catalytic performance is one of the challenges which remain to be met in nanocatalysis. Although numerous efforts with various capping agents have been made over the past few years by researchers all over the world, future progress will probably necessitate the design of molecules more appropriate for the metal surface to induce higher catalytic properties in terms of selectivity. The recycling and the recovery of the nanocatalysts are still to be improved at this point.

References

1. D. Astruc, F. Lu and J. Ruiz Aranzaes, *Angew. Chem., Int. Ed.*, 2005, **44**, 7852–7872.
2. *Nanoparticles and Catalysis*, ed. D. Astruc, Wiley-VCH, Weinheim, 2008.
3. J. M. Thomas, *Chem. Cat. Chem.*, 2010, **2**, 127–132.
4. Z. X. Li, W. Xue, B. T. Guan, F. B. Shi, Z. J. Shi, H. Jiang and C. H. Yan, *Nanoscale*, 2013, **5**, 1213–1220.
5. *Nanoparticles: From Theory to Application*, ed. G. Schmid, Wiley-VCH, Weinheim, 2012.
6. B. R. Cuenya, *Thin Solid Films*, 2010, **518**, 3127–3150.
7. P. Sonström and M. Bäumer, *Phys. Chem. Chem. Phys.*, 2011, **13**, 19270–19284.
8. T. J. Geldbach and P. J. Dyson, *Metal-Catalysed Reactions in Ionic Liquids*, Springer, The Netherlands, 2005.
9. N. Yan, C. Xiao and Y. Kou, *Coord. Chem. Rev.*, 2010, **254**, 1179–1218.
10. B. L. Cushing, V. L. Kolesnichenko and C. J. O'Connor, *Chem. Rev.*, 2004, **104**, 3893–3946.
11. N. Semagina and L. Kiwi-Minsker, *Catal. Rev.: Sci. Eng.*, 2009, **51**, 147–217.
12. *Nanomaterials in Catalysis*, ed. K. Philippot and P. Serp, Wiley-VCH, Weinheim, 2013.
13. D. Uzio and G. Berhault, *Catal. Rev.: Sci. Eng.*, 2010, **52**, 106–131.

14. S. Abbet and U. Heiz, Nanocatalysis, in *The Chemistry of Nanomaterials: Synthesis, Properties and Applications*, ed. C. N. R. Rao, A. Mueller and A. K. Cheetham, Wiley-VCH, Weinheim, 2004, ch. 17, pp. 551–588.
15. K. Philippot and B. Chaudret, *C. R. Chimie*, 2003, **6**, 1019–1034.
16. B. Chaudret and K. Philippot, *Oil & Gas Science and Technology – Revue de l'Institut Français du Pétrole*, 2007, **62**, 799–817.
17. K. Philippot and B. Chaudret, Organometallic Derived Metals, Colloids and Nanoparticles, in *Comprehensive Organometallic Chemistry III*, vol. 12, ed. D. O'Hare, Elsevier, Oxford, 2006, ch. 3, pp. 71–99.
18. P. Lara, M.-J. Casanove, P. Lecante, P.-F. Fazzini, K. Philippot and B. Chaudret, *J. Mater. Chem.*, 2012, **22**, 3578–3584.
19. P. Lara, T. Ayvali, M.-J. Casanove, P. Lecante, P.-F. Fazzini, K. Philippot and B. Chaudret, *Dalton Trans.*, 2013, **42**, 372–382.
20. P. Lara, K. Philippot and B. Chaudret, *ChemCatChem*, 2013, **5**, 28–45.
21. F. Dassenoy, K. Philippot, T. O. Ely, C. Amiens, P. Lecante, E. Snoeck, A. Mosset, M.-J. Casanove and B. Chaudret, *New J. Chem.*, 1998, **22**, 703–712.
22. C. Pan, K. Pelzer, K. Philippot, B. Chaudret, F. Dassenoy, P. Lecante and M.-J. Casanove, *J. Am. Chem. Soc.*, 2001, **123**, 7584–7593.
23. O. Vidoni, K. Philippot, C. Amiens, B. Chaudret, O. Balmes, J.-O. Malm, J.-O. Bovin, F. Senocq and M.-J. Casanove, *Angew. Chem., Int. Ed.*, 1999, **38**, 3736–3738.
24. K. Pelzer, O. Vidoni, K. Philippot, B. Chaudret and V. Collière, *Adv. Funct. Mater.*, 2003, **13**, 118–126.
25. T. Gutel, J. García-Antón, K. Pelzer, K. Philippot, C. C. Santini, Y. Chauvin, B. Chaudret and J.-M. Basset, *J. Mater. Chem.*, 2007, **17**, 3290–3292.
26. T. Gutel, C. Santini, K. Philippot, A. Padua, K. Pelzer, B. Chaudret and J.-M. Basset, *J. Mater. Chem.*, 2009, **19**, 3624–3631.
27. P. S. Campbell, C. C. Santini, D. Bouchu, B. Fenet, K. Philippot, B. Chaudret, A. A. H. Pádua and Y. Chauvin, *Phys. Chem. Chem. Phys.*, 2010, **12**, 4217–4223.
28. E. Ramirez, L. Eradès, K. Philippot, P. Lecante and B. Chaudret, *Adv. Funct. Mater.*, 2007, **17**, 2219–2228.
29. M. Cabié, S. Giorgio, C. Henry, M. R. Axet, K. Philippot and B. Chaudret, *J. Phys. Chem. C*, 2010, **114**, 2160–2163.
30. M. R. Axet, K. Philippot, B. Chaudret, M. Cabié, S. Giorgio and C. R. Henry, *Small*, 2011, **7**, 235–241.
31. V. Hulea, D. Brunel, A. Galarneau, F. Fajula, K. Philippot, B. Chaudret and P. J. Kooyman, *Mesopor. Micropor. Mater.*, 2005, **79**, 185–194.
32. K. Pelzer, K. Philippot and B. Chaudret, *Z. Phys. Chem.*, 2003, **217**, 1539–1547.
33. T. Pery, K. Pelzer, G. Buntkowsky, K. Philippot, H.-H. Limbach and B. Chaudret, *ChemPhysChem*, 2005, **6**, 605–607.
34. A. Adamczyk, Y. Xu, B. Walaszek, F. Roelofs, T. Pery, K. Pelzer, K. Philippot, B. Chaudret, H.-H. Limbach, H. Breitzke and G. Buntkowsky, *Top. Catal.*, 2008, **48**, 75–83.

35. J. García-Antón, M. R. Axet, S. Jansat, K. Philippot, B. Chaudret, T. Pery, G. Buntkowsky and H.-H. Limbach, *Angew. Chem., Int. Ed.*, 2008, **47**, 2074–2078.
36. F. Novio, K. Philippot and B. Chaudret, *Catal. Lett.*, 2010, **140**, 1–7.
37. S. Kinayyigit, P. Lara, P. Lecante, K. Philippot and B. Chaudret, *Nanoscale*, 2014, **6**, 539–546.
38. A. Roucoux and K. Philippot, Homogeneous Hydrogenation: Colloids-Hydrogenation with Noble Metal Nanoparticles in *Handbook of Homogeneous Hydrogenation*, vol. 1, ed. J. G. de Vries and C. J. Elsevier, Wiley-VCH, Weinheim, 2007, ch. 9, pp. 217–253.
39. S. Jansat, D. Picurelli, K. Pelzer, K. Philippot, M. Gómez, G. Muller, P. Lecante and B. Chaudret, *New J. Chem.*, 2006, **1**, 115–122.
40. A. Gual, M. R. Axet, K. Philippot, B. Chaudret, A. Denicourt-Nowicki, A. Roucoux, S. Castillón and C. Claver, *Chem. Commun.*, 2008, 2759–2761.
41. A. Gual, C. Godard, K. Philippot, B. Chaudret, A. Denicourt-Nowicki, A. Roucoux, S. Castillón and C. Claver, *ChemSusChem*, 2009, **2**, 769–779.
42. P. Lara, O. Rivada-Wheelaghan, S. Conejero, R. Poteau, K. Philippot and B. Chaudret, *Angew. Chem., Int. Ed.*, 2011, **50**, 12080–12084.
43. D. Gonzalez-Galvez, P. Lara, O. Rivada-Wheelaghan, S. Conejero, B. Chaudret, K. Philippot and P. W. N. M. van Leeuwen, *Catal. Sci. Technol*, 2013, **3**, 99–105.
44. P.-J. Debouttière, V. Martinez, K. Philippot and B. Chaudret, *Dalton Trans.*, 2009, 10172–10174.
45. P.-J. Debouttière, Y. Coppel, A. Denicourt-Nowicki, A. Roucoux, B. Chaudret and K. Philippot, *Eur. J. Inorg. Chem.*, 2012, 1229–1236.
46. M. Guerrero, A. Roucoux, A. Denicourt-Nowicki, H. Bricout, E. Monflier, V. Collière, K. Fajerwerg and K. Philippot, *Catal. Today*, 2012, **183**, 34–41.
47. J. P. Hallett and T. Welton, *Chem. Rev.*, 2011, **111**, 3508–3576.
48. V. I. Pârvulescu and C. Hardacre, *Chem. Rev.*, 2007, **107**, 2615–2665.
49. A. A. H. Pádua, M. F. Costa Gomes and J. N. A. Canongia Lopes, *Acc. Chem. Res.*, 2007, **40**, 1087–1096.
50. A. S. Pensado and A. A. H. Pádua, *Angew. Chem., Int. Ed.*, 2011, **50**, 8683–8687.
51. G. Salas, C. C. Santini, K. Philippot, V. Collière, B. Chaudret, B. Fenet and P.-F. Fazzini, *Dalton Trans.*, 2011, **40**, 4660–4668.
52. G. Salas, A. Podgorsek, P. S. Campbell, C. C. Santini, A. A. H. Padua, M. F. C. Gomes, K. Philippot, B. Chaudret and M. Turmine, *Phys. Chem. Chem. Phys.*, 2011, **13**, 13527–13536.
53. G. Salas, P. S. Campbell, C. C. Santini, K. Philippot, M. F. Costa Gomes and A. A. H. Padua, *Dalton Trans.*, 2012, **41**, 13919–13926.
54. F. H. Stephens, V. Pons and R. T. Baker, *Dalton. Trans.*, 2007, **25**, 2613–2626.
55. M. Zahmakiran, M. Tristany, K. Philippot, K. Fajerwerg, S. Özkar and B. Chaudret, *Chem. Commun.*, 2010, **46**, 2938–29540.
56. M. Zahmakıran, K. Philippot, S. Özkar and B. Chaudret, *Dalton Trans.*, 2012, **41**, 590–598.

57. M. Beller, H. Fischer, K. Kuhlein, C. P. Reisinger and W. A. Hermann, *J. Organometal. Chem.*, 1996, **520**, 257–259.
58. R. Narayanan and M. A. El-Sayed, *J. Phys. Chem. B*, 2005, **109**, 12663–12676.
59. M. Studer, H.-U. Blaser and C. Exner, *Adv. Synth. Catal.*, 2003, **345**, 45–65.
60. H. Bönnemann and G. A. Braun, *Angew. Chem., Int. Ed.*, 1996, **35**, 1992–1995.
61. X. Zuo, H. Liu, D. Guo and X. Yang, *Tetrahedron*, 1999, **55**, 7787–7804.
62. J. U. Köhler and J. S. Bradley, *Catal. Lett.*, 1997, **45**, 203–208.
63. J. U. Köhler and J. S. Bradley, *Langmuir*, 1998, **14**, 2730–2735.
64. M. Tamura and H. Fujihara, *J. Am. Chem. Soc.*, 2003, **125**, 15742–15743.
65. S. Jansat, M. Gómez, K. Philippot, G. Muller, E. Guiu, C. Claver, S. Castillón and B. Chaudret, *J. Am. Chem. Soc.*, 2004, **126**, 1592–1593.
66. I. Favier, M. Gómez, G. Muller, M. R. Axet, S. Castillón, C. Claver, S. Jansat, B. Chaudret and K. Philippot, *Adv. Synth. Catal.*, 2007, **349**, 2459–2469.
67. D. Peral, F. Gómez-Villarraga, X. Sala, J. Pons, J. Carles Bayón, J. Ros, M. Guerrero, L. Vendier, P. Lecante, J. García-Antón and K. Philippot, *Catal. Sci. Technol.*, 2013, **3**, 475–489.
68. M. R. Axet, S. Castillón, C. Claver, K. Philippot, P. Lecante and B. Chaudret, *Eur. J. Inorg. Chem.*, 2008, 3460–3466.
69. G. C. Bond, C. Louis and D. T. Thompson, *Catalysis by Gold*, Imperial College Press, London, 2006.
70. K. Pelzer, K. Philippot, B. Chaudret, W. Meyer-Zaika and G. Schmid, *Z. Anorg. Allg. Chem.*, 2003, **629**, 1217–1222.
71. H.-P. Kormann, G. Schmid, K. Pelzer, K. Philippot and B. Chaudret, *Z. Anorg. Allg. Chem.*, 2004, **630**, 1913–1918.
72. M. Guerrero, N. J. S. Costa, L. L. R. Vono, L. M. Rossi, E. V. Gusevskaya and K. Philippot, *J. Mater. Chem. A*, 2013, **1**, 1441–1449.
73. S. Jansat, K. Pelzer, J. García-Antón, R. Raucoules, K. Philippot, A. Maisonnat, B. Chaudret, Y. Guari, A. Medhi, C. Reyé and R. J. P. Corriu, *Adv. Funct. Mater.*, 2007, **17**, 3339–3347.
74. V. Matsura, Y. Guari, C. Reyé, R. J. P. Corriu, M. Tristany, S. Jansat, K. Philippot, A. Maisonnat and B. Chaudret, *Adv. Funct. Mater.*, 2009, **19**, 3781–3787.
75. M. Tristany, P. Philippot, Y. Guari, V. Collière, P. Lecante and B. Chaudret, *J. Mater. Chem.*, 2010, **20**, 9523–9530.
76. E. Castillejos, P.-J. Debouttière, L. Roiban, A. Solhy, V. Martinez, Y. Kihn, O. Ersen, K. Philippot, B. Chaudret and P. Serp, *Angew. Chem., Int. Ed.*, 2009, **48**, 2529–2533.
77. E. J. García-Suárez, M. Tristany, A. B. García, V. Collière and K. Philippot, *Micropor. Mesopor. Mater.*, 2012, **153**, 155–162.
78. E. J. García-Suárez, A. M. Balu, M. Tristany, A. B. García, K. Philippot and R. Luque, *Green Chem.*, 2012, **14**, 1434–1439.

CHAPTER 5

Nickel Nanoparticles in the Transfer Hydrogenation of Functional Groups

FRANCISCO ALONSO

Departamento de Química Orgánica, Facultad de Ciencias and Instituto de Síntesis Orgánica (ISO), Universidad de Alicante, Apdo. 99 E-03080 Alicante, Spain
Email: falonso@ua.es

5.1 Introduction

The catalytic transfer hydrogenation of organic compounds[1,2] is an advantageous methodology with respect to other reduction methods for several reasons: (a) the hydrogen source is easy to handle (no gas containment or pressure vessels are necessary), (b) possible hazards are minimized, (c) the mild reaction conditions used can afford enhanced selectivity and, (d) catalytic asymmetric transfer hydrogenation can be applied in the presence of chiral ligands. Among the different types of hydrogen donors, 2-propanol is probably the most popular since it is cheap, non-toxic, volatile, possesses good solvent properties, and it is transformed into acetone, which is environmentally friendly and easy to remove from the reaction medium. Some other sources of hydrogen, such as formic acid derivatives or hydrazine, are in less common use.

Transition metal catalysis is dominated by the elements of the three triads, where ruthenium, rhodium, iridium and, above all, palladium, generally stand out from the others because of their incomparable catalytic activity and

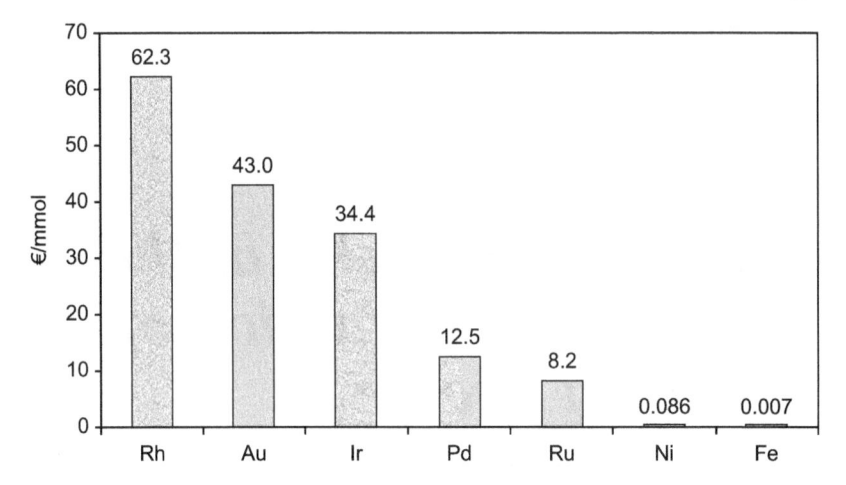

Figure 5.1 Relative prices for some transition metals (Sigma-Aldrich, 2012).

efficiency. These noble metals have obscured the role of the first triad metals in catalysis until the recent revival of iron.[3] Indeed, methodologies based on "modest" transition metals that can promote organic transformations typically reserved for noble metals are challenging and praiseworthy. This assertion gains points when we compare the relative prices of some transition metals compared to the corresponding anhydrous chlorides. Thus, rhodium, gold, and iridium are extraordinarily expensive; palladium and ruthenium are somewhat cheaper, whereas nickel and iron are about 100-fold and 1000-fold cheaper than ruthenium, respectively (Figure 5.1).

It is noteworthy that the combination of inexpensive transition metals with green and safe sources of hydrogen has been practically unexplored until recently. In this sense, the potential of nickel has been undervalued in favor of more "exotic" and attractive catalytic systems containing noble metals. On the other hand, in recent years, nano-catalysis has emerged as a sustainable and competitive alternative to conventional catalysis since the metal nanoparticles possess a high surface-to-volume ratio, which enhances their activity and selectivity, while at the same time maintaining the intrinsic features of a heterogeneous catalyst.[4–7]

This chapter summarizes the application of catalytic systems based on nickel nanoparticles as an alternative to noble-metal catalysts in the hydrogen-transfer reduction of functional groups. In particular, we will deal with the transfer hydrogenation of olefins, carbonyl compounds, and the reductive amination of aldehydes.

5.2 Antecedents

In 1996,[8,9] due to our incipient interest in active metals,[10,11] we discovered the $NiCl_2 \cdot 2H_2O$-Li-arene(cat.) combination as a useful and versatile mixture

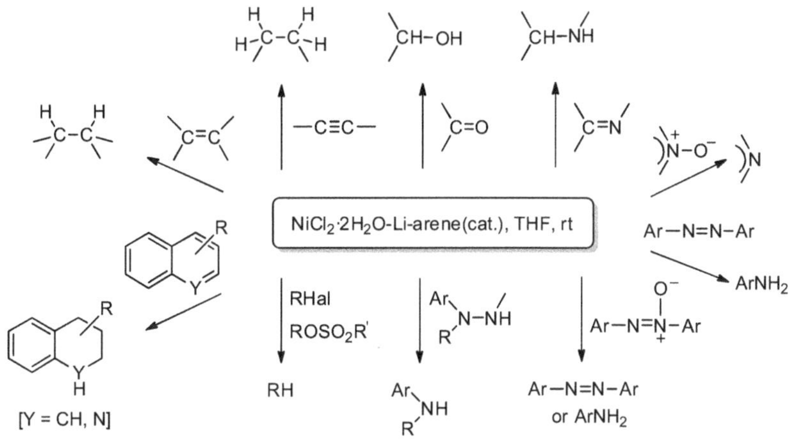

Scheme 5.1 Reduction of organic functionalities with the NiCl$_2$·2H$_2$O-Li-arene (cat.) system.

able to reduce a broad range of functionalities bearing carbon–carbon multiple bonds, as well as carbon–heteroatom and heteroatom–heteroatom single and multiple bonds (Scheme 5.1).[12–14] Two different and independent processes were postulated to take part in the reduction pathway:[13] (a) the reduction of nickel(II) to nickel(0) by the action of the radical anion and/or the dianion derived from the activation of metal lithium by the arene (which acts as an electron carrier), and (b) the generation of molecular hydrogen by reaction of excess lithium and the crystallization water present in the nickel salt. The combination of this molecular hydrogen with the highly reactive nickel(0) would lead to a sort of nickel hydride species, which was considered to be the real reducing agent. Interestingly, the analogous deuterated combination, NiCl$_2$·2D$_2$O-Li-arene(cat.), allowed the easy incorporation of deuterium in the reaction products. In addition, a polymer-supported arene could be used as an electron transfer agent instead of the free arene; this methodology facilitates the recovery and reuse of the arene component. Alternatively, almost the same type of functionalities as those mentioned could be reduced by catalytic hydrogenation utilizing active nickel, generated from anhydrous nickel(II) chloride, and external molecular hydrogen at atmospheric pressure [NiCl$_2$-Li-arene(cat.)-H$_2$ or NiCl$_2$-Li-polymer-supported arene)(cat.)-H$_2$].[15,16]

Some years later, we demonstrated that the highly reactive nickel, obtained by reduction of anhydrous nickel(II) chloride with lithium powder and a catalytic amount of DTBB (4,4'-di-*tert*-butylbiphenyl) in THF at room temperature, was in the form of nanoparticles (2.5 ± 1.5 nm).[17] Similar systems to NiCl$_2$-Li-DTBB(cat.), utilized in reduction reactions, such as NiCl$_2$-Li-copolymer(cat.)-H$_2$, NiCl$_2$·2H$_2$O-Li-DTBB(cat.), NiCl$_2$-Li-DTBB(cat.)-EtOH, or NiCl$_2$-Li-copolymer(cat.)-EtOH, were also seen to generate nano-sized metallic nickel in the absence of any anti-agglomeration additive or nucleation catalyst at room temperature.[18] The introduction in the reducing system of an alcohol as a source of molecular hydrogen, by reaction with Li, was a very convenient method for the highly selective semihydrogenation of alkynes and dienes,[19,20] and conjugate reduction of α,β-unsaturated carbonyl compounds[21] under mild reaction conditions. We observed, however, that a hydrogen-transfer process might be operating, at least in part, in the reactions involving an alcohol. From this finding, we envisaged the possibility of applying the nickel nanoparticles (NiNPs) in hydrogen-transfer reactions.[22]

5.3 Hydrogen-transfer Reduction of Alkenes

The transfer hydrogenation of olefins has been studied little in comparison with that of carbonyl compounds, using alcohols, formic acid or hydroaromatic compounds, mainly in the presence of noble-metal catalysts.[23–25] To the best of our knowledge, only two reports describe the application of nickel and 2-propanol to the transfer hydrogenation of olefins. In the first, Raney nickel (10–50 wt% of total substrate) was used under reflux, showing high conversions for cinnamates and cyclic olefins and low conversions for acyclic olefins.[26] In the second report, activated metallic nickel, prepared by thermal decomposition of *in situ* generated nickel diisopropoxide in boiling 2-propanol, was more effective in the reduction of non-functionalized and non-activated olefins (10–30 mol% Ni, 95–100 °C).[27]

We demonstrated that NiNPs [prepared from anhydrous nickel(II) chloride, metal lithium, and a catalytic amount of DTBB (5 mol%), as an electron carrier] can effectively catalyze the heterogeneous transfer hydrogenation of olefins using 2-propanol as the hydrogen donor.[28] Raney nickel behaved in a similar way to the NiNPs in the reduction of 1-octene but a longer reaction time was needed in order to achieve the same conversion at 76 °C, while the activity of the former drastically dropped at room temperature. Other commercially available nickel catalysts, such as Ni-Al, Ni/SiO$_2$-Al$_2$O$_3$, and NiO, were totally inactive under these conditions. A variety of non-functionalized and functionalized olefins were successfully reduced under the optimized reaction conditions (20 mol% NiNPs, *i*-PrOH, 76 °C) (Scheme 5.2). The process was shown to be highly chemoselective for substrates that are prone to undergoing isomerization (allylic alcohols) or hydrogenolysis (allyl benzyl ether or allylcyclohexylamine), where a clear superiority of the NiNPs was demonstrated with respect to other nickel catalysts. Moreover, the monoterpene (±)-linalool, which contains both a mono- and a tri-substituted carbon–carbon double bond, was transformed into tetrahydrolinalool in a moderate isolated yield after prolonged heating. Unfortunately, an important decrease in the conversion of the olefins was observed in the second cycle when reutilization of the catalyst was attempted, due to progressive deactivation.

The transfer hydrogenation of styrene derivatives was reported by Pitchumani *et al.* using hydrazine as the hydrogen source and a catalyst consisting of

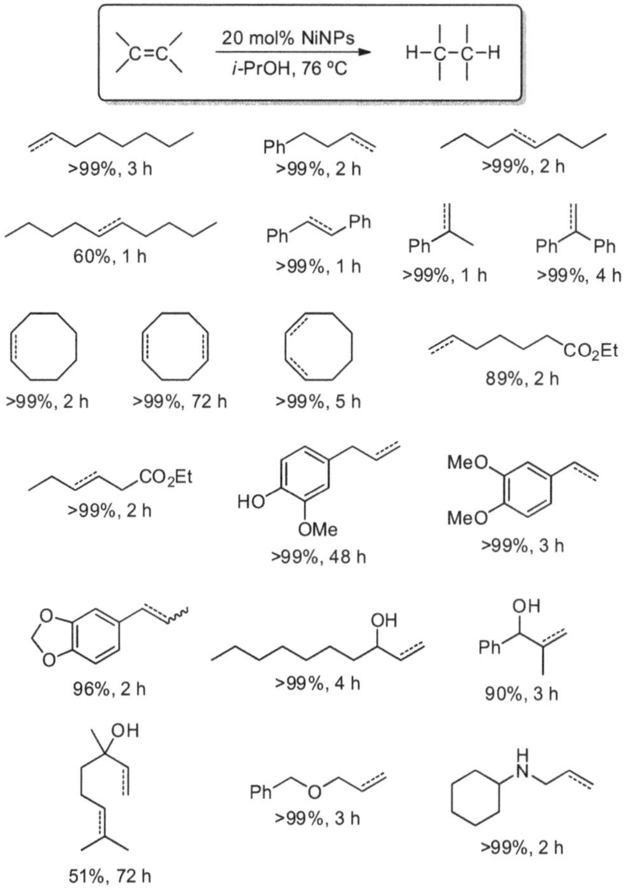

Scheme 5.2 Transfer hydrogenation of olefins catalyzed by NiNPs in 2-propanol. The dashed bond indicates the original position of the carbon–carbon double bond.

clay-entrapped NiNPs.[29] The NiNPs were entrapped in montmorillonite K10 by reducing Ni^{+2}-exchanged montmorillonite K10 with hydrazine in an aqueous alkaline medium. Characterization of the NiNPs by various means revealed that they consisted of pure nickel with face-centered-cubic structure and an average particle size of 15–20 nm. Apparently, the nitrogen gas evolved from hydrazine during the reaction created an inert atmosphere that prevented the oxidation of the NiNPs. Besides styrenes, other substrates, including cyclohexene, α,β-unsaturated carbonyl compounds (conjugate reduction), an imine, phenylacetylene (to ethyl benzene), and allyl phenyl thioether (to phenyl propyl thioether), were also reduced in moderate-to-high yields using 5 mol% NiNPs and hydrazine hydrate in ethanol at 70 °C for 8 h (Scheme 5.3). Interestingly, the catalyst could be reused for three consecutive runs without any appreciable loss of activity.

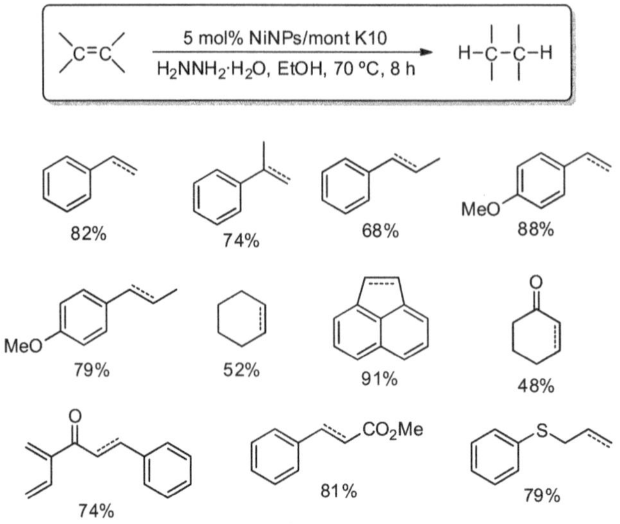

Scheme 5.3 Transfer hydrogenation of olefins with hydrazine hydrate catalyzed by NiNPs entrapped in montmorillonite K10.

5.4 Hydrogen-transfer Reduction of Carbonyl Compounds

The transfer hydrogenation of carbonyl compounds has been mostly accomplished using 2-propanol as a hydrogen donor, under homogeneous conditions in the presence of noble-metal complexes (such as those of Ru, Rh, Ir).[30] In this field, ruthenium complexes have been by far the most studied catalysts, especially for the asymmetric transfer hydrogenation of aromatic ketones[31] and from a mechanistic point of view.[32,33] Nickel has been used either under homogeneous[34,35] or heterogeneous conditions,[27,35,36] mainly for aromatic substrates. In all these cases, the addition of an external base was mandatory for the reaction to take place. More recently, an excess of Raney nickel in refluxing 2-propanol, containing a trace of HCl, reduced a series of aliphatic aldehydes[37] and ketones[38] to the corresponding primary and secondary alcohols, respectively.

We reported on the application of NiNPs (*ca.* 1.75 ± 1.00 nm) to the transfer hydrogenation of an array of carbonyl compounds with 2-propanol in the absence of added base at 76 °C. Both a stoichiometric[39] and a sub-stoichiometric (20 mol% Ni)[40] version were studied and compared. Modest-to-good yields of the corresponding alcohols were obtained in relatively short reaction times. The reduction of substituted acetophenones was shown to be dependent on the electronic character and position of the substituent (Scheme 5.4), whereas, in general, better yields were attained for aliphatic ketones (Scheme 5.5). The reducing system was shown to be diastereoselective for most of the cyclic ketones studied, with the stereoselectivity

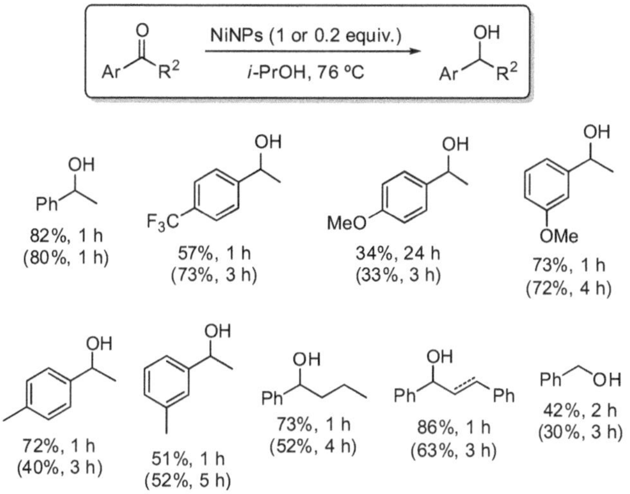

Nickel Nanoparticles in the Transfer Hydrogenation of Functional Groups

Scheme 5.4 Transfer hydrogenation of aromatic carbonyl compounds catalyzed by NiNPs in 2-propanol. Data in parentheses refer to 20 mol% NiNPs. The dashed bond indicates the original position of the carbon–carbon double bond.

achieved with 20 mol% NiNPs being somewhat lower than that achieved with stoichiometric NiNPs. The NiNPs exhibited a superior performance in comparison with other forms of nickel under the same reaction conditions. Moreover, the NiNPs could be easily separated by decantation with an extra amount of 2-propanol and reutilized up to four times (with 1 equiv. NiNPs, 1 h, 94–95%) or five times (20 mol% NiNPs, 1 h, 87–77%) in a very simple reaction medium composed of the NiNPs, 2-propanol and the substrate, with no base. Deactivation of the catalyst with further repeated reuse was ascribed to surface oxidation.[41] According to some deuteration experiments, the reaction seemed to proceed through a dihydride-type mechanism, in which the two hydrogens of the donor are transferred to the metal, thus losing their original identity and becoming equivalent (Scheme 5.6). After our work, several reports emerged describing the application of supported NiNPs, prepared by different methods, to the transfer hydrogenation of carbonyl compounds with isopropanol (see below).

Laboratory-made magnetic nano-ferrites were surface-modified with dopamine followed by the addition of NiCl$_2$ at basic pH and reduction with NaBH$_4$. The resulting Ni-coated nano-ferrite (10–13 nm) was used in the transfer hydrogenation of aromatic ketones in the presence of KOH and isopropanol at 100 °C under microwave irradiation (45 min (Scheme 5.7).[42] Although high yields were generally attained, brominated acetophenones (either on the ring or α position) underwent hydrodebromination whereas nitroacetophenones were transformed into the corresponding anilines. The catalyst could be recovered magnetically and reused, after simple treatment (washing with MeOH and drying at 60 °C for 30 min), five times with no apparent loss of activity (98–95%).

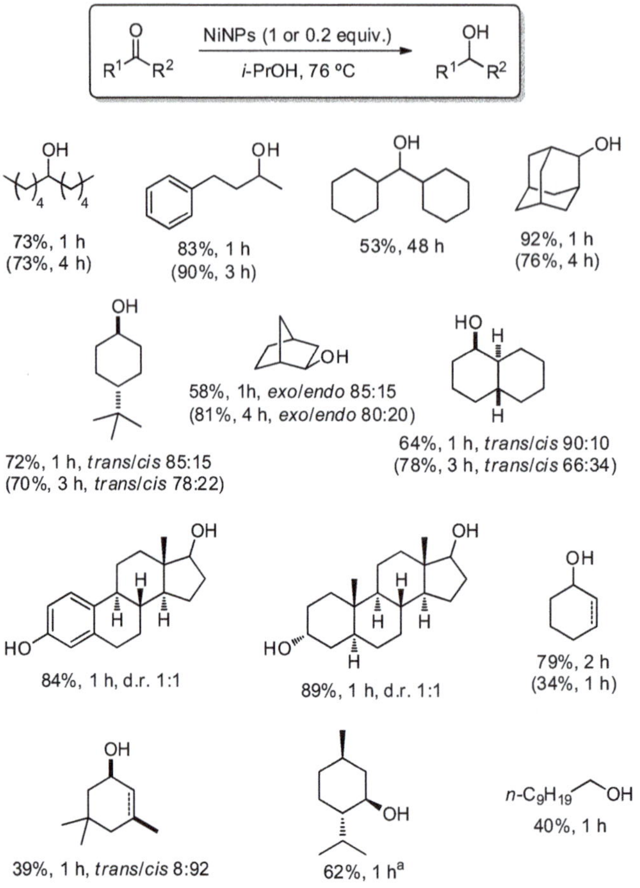

Scheme 5.5 Transfer hydrogenation of aliphatic carbonyl compounds catalyzed by NiNPs in 2-propanol. Data in parentheses correspond to 20 mol% NiNPs. [a]Diastereomeric ratio: (−)-menthol–neoisomenthol–neomenthol 77 : 9 : 14.

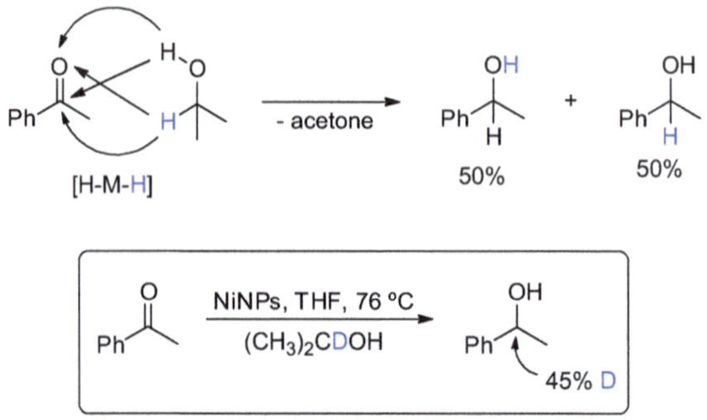

Scheme 5.6 The dihydride-type mechanism.

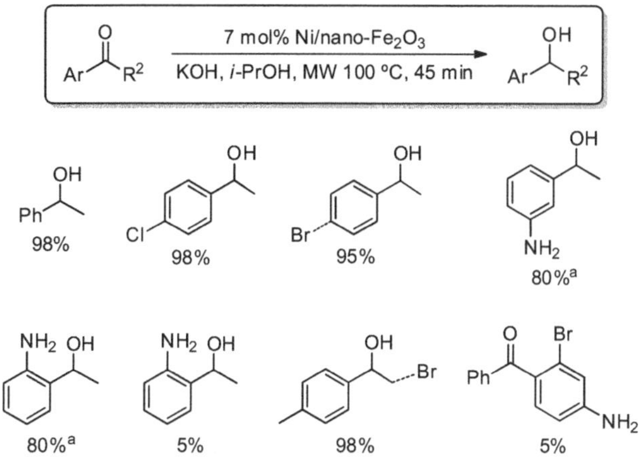

Scheme 5.7 Transfer hydrogenation of aromatic ketones catalyzed by nanoferrite-Ni in 2-propanol. [a]Starting from the nitrocompound.

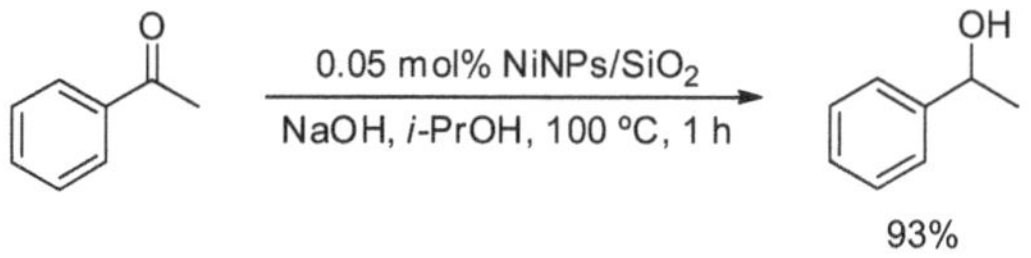

Scheme 5.8 Transfer hydrogenation of acetophenone catalyzed by nickel-silica hybrid nanostructures.

The group of Song prepared NiNPs by thermal decomposition of nickel-oleylamine complexes, which were successfully coated with silica using the microemulsion method. The nickel cores inside silica shells formed nickel phyllosilicate with an urchin-like morphology under hydrothermal conditions, and NiNPs (24 ± 1 nm) were regenerated on the silica spheres by hydrogen reduction.[43] The resulting morphology of the Ni/SiO$_2$ nanostructure can be regarded as active NiNPs embedded in the silica support. Acetophenone was converted into 1-phenylethanol in a high yield (93%), using only 0.05 mol% of the catalyst and NaOH in isopropanol at 100 °C for 1 h (Scheme 5.8). The catalyst could be easily separated from the reaction medium by centrifugation and reused five times without any loss of catalytic activity (97–92%).

The same group reported the synthesis of Ni@SiO$_2$ yolk–shell nanocatalysts with nickel cores of *ca.* 3 nm. The NiNPs were coated with silica through the microemulsion method, with the resulting Ni@SiO$_2$ yolk–shell NPs being subjected to partial etching by acid treatment. Uniform yolk–shell nanocatalysts were obtained after calcination of the NPs.[44] The transfer hydrogenation of cyclohexanone and a series of aromatic ketones was effected with low catalyst loading (0.03 mol%) and NaOH (20 mol%) at 150 °C for 30 min (Scheme 5.9). Although the yields and TOF (6000 h^{-1}) were high, the reaction conditions are rather harsh and might be incompatible with the

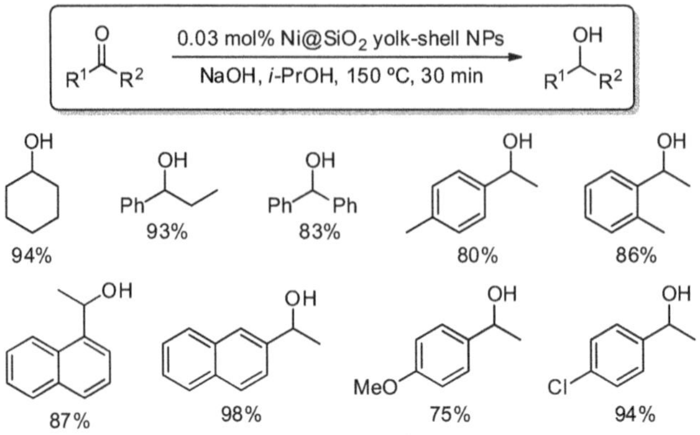

Scheme 5.9 Transfer hydrogenation of ketones catalyzed by a Ni@SiO$_2$ yolk–shell nanocatalyst.

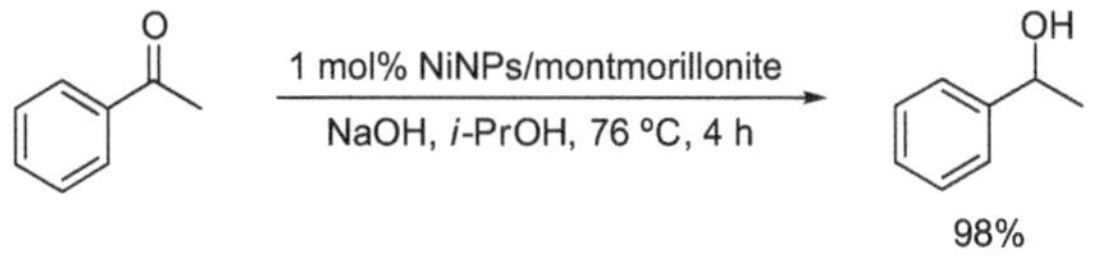

Scheme 5.10 Transfer hydrogenation of acetophenone catalyzed by NiNPs on acid-activated montmorillonite.

presence of other functional groups. Nonetheless, the catalyst could be reused for six cycles maintaining an excellent performance (92–90%).

The impregnation of Ni(OAc)$_2$ into the nanopores of two acid-modified montmorillonite supports, followed by polyol reduction (with ethylene glycol) provided the corresponding catalysts of NiNPs (<1–8 nm) differing in the time of activation of the support.[45] The catalyst with the longer activation time (2 h) reached a higher yield and selectivity in the hydrogen-transfer reduction of acetophenone, using low catalyst loading (1 mol% Ni) and NaOH in isopropanol at 76 °C under nitrogen (Scheme 5.10). The higher catalytic activity of the latter was related to the wider specific pore volume of the catalyst matrix. The presence of base and absence of oxygen were mandatory for the reaction to take place. The catalyst could be recovered by filtration and reused (after washing with water and drying) in three runs with a decline in the conversion (98–64%).

More recently, a highly efficient transfer hydrogenation catalyst was prepared by incorporation of NiNPs into a mesoporous aluminosilicate framework.[46] The latter was obtained by the sol-gel method, followed by calcination at 800 °C. Reduction of the impregnated support with hydrazine gave rise to the nickel aluminosilicate nanocomposite. The catalyst was fully characterized, and the powder XRD pattern showed the presence of nickel oxide with a fcc lattice and a crystallite size of *ca.* 5 nm. The TPR profile indicated that the major portion of the original NiNPs is on the surface of the framework and

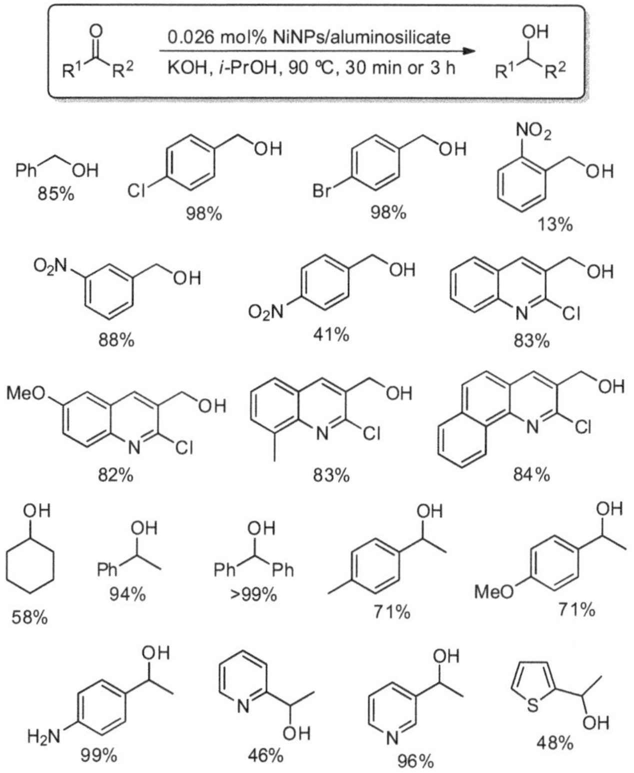

Scheme 5.11 Transfer hydrogenation of aldehydes (30 min) and ketones (3 h) catalyzed by NiNPs on aluminosilicate.

was oxidized to NiO in the atmosphere. The catalyst exhibited a high performance in the transfer hydrogenation of aldehydes and ketones with 0.026 mol% Ni, KOH (2 equiv.) at 90 °C for 30 min (aldehydes) or 3 h (ketones) (Scheme 5.11). GC yields >82% were generally recorded for aromatic and heteroaromatic aldehydes as well as for ketones, though cyclohexanone was shown to be more reluctant to react. The process was highly chemoselective, with the reaction conditions being compatible with the presence of halogens (Cl and Br), nitro, methoxy, or amino groups. Furthermore, the catalyst was recyclable for four cycles with no significant loss in activity (94–92%).

Kidwai *et al.* prepared NiNPs by reduction of Ni(II) to Ni(0) in a reverse micellar system containing an aqueous solution of Ni(NO$_3$)$_2$ and an alkaline solution of NaBH$_4$. The NiNPs (10–85 nm) were efficiently applied to the transfer hydrogenation of aldehydes[47] and ketones,[48] using ammonium formate as a hydrogen source. The maximum reaction rate was recorded for an average particle diameter of about 20 nm. All reactions proceeded in relatively short reaction times at room temperature (for aldehydes, Scheme 5.12) or 45–50 °C (for ketones, Scheme 5.13) in THF under nitrogen and were high yielding. Moreover, the NiNPs could be recycled, with a

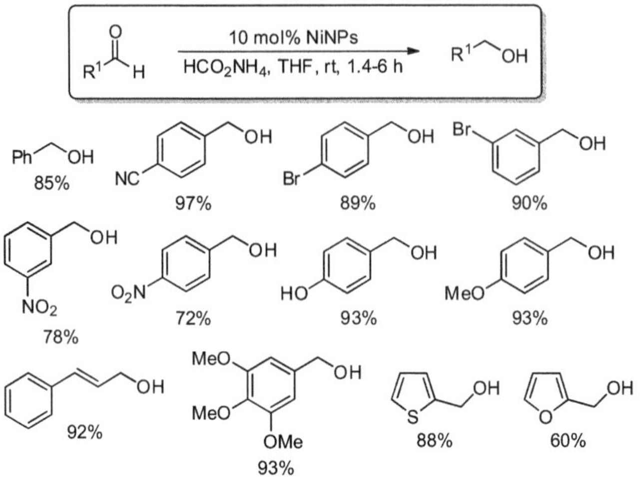

Scheme 5.12 Hydrogen-transfer reduction of aldehydes with ammonium formate catalyzed by NiNPs.

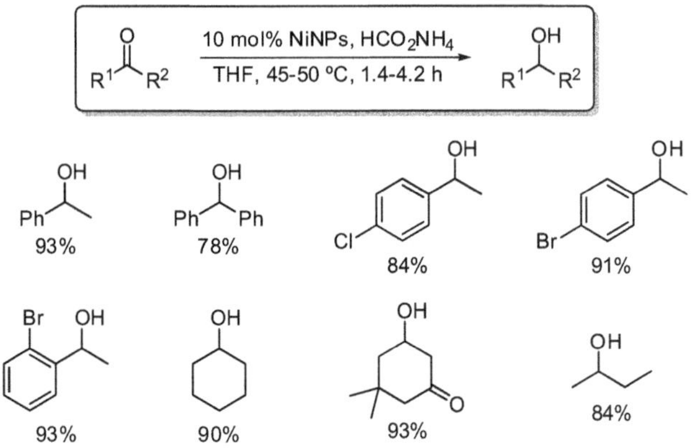

Scheme 5.13 Hydrogen-transfer reduction of ketones with ammonium formate catalyzed by NiNPs.

decrease in the catalytic activity of up to 76% after the sixth cycle, which was attributed to partial particle agglomeration.

5.5 Hydrogen-transfer Reductive Amination of Aldehydes

The reductive amination of aldehydes and ketones, an important transformation in biological systems, is a direct and convenient route to amines.[49,50] Reductive amination with metal hydrides[51,52] or by catalytic hydrogenation[53,54] are widespread methods, whereas transfer hydrogenation is more restricted to the reduction of pre-formed imines.[55] It is worth noting that, in

Nickel Nanoparticles in the Transfer Hydrogenation of Functional Groups 95

contrast with other transition metals, nickel has been scarcely used in reductive aminations or in the transfer hydrogenation of imines. Reductive aminations involving nickel were performed with hydrazine and borohydride exchange resin–nickel acetate[56] or with stoichiometric nickel boride in methanol.[57] On the other hand, the transfer hydrogenation of pre-formed imines was carried out with 2-propanol and an excess of aluminium isopropoxide in the presence of Raney nickel (the reaction failed in the absence of aluminium isopropoxide),[58] or with sodium isopropoxide catalyzed by a nickel(0)–N-heterocyclic carbene complex.[59]

To the best of our knowledge, however, nickel-mediated reductive-amination reactions by transfer hydrogenation had not been reported previous to our work. We proved that NiNPs can catalyze the reductive amination of aldehydes by transfer hydrogenation, using 2-propanol as the hydrogen source in the absence of any added base (Scheme 5.14).[60] The

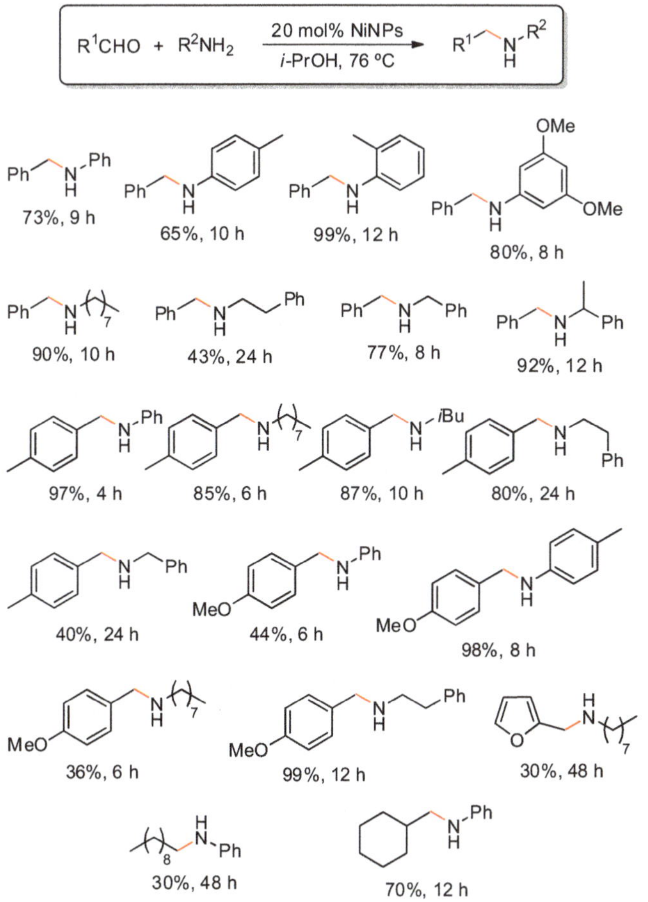

Scheme 5.14 Hydrogen-transfer reductive amination of aldehydes catalyzed by NiNPs.

process was especially effective in the reductive amination of aromatic aldehydes, most of the corresponding secondary amines being obtained in good-to-excellent yields. The application to aliphatic aldehydes was, however, more limited. Although any attempt to reuse the NiNPs failed, this methodology can be considered advantageous because (a) the preparation step of the imine is avoided, aldehydes and amines being used directly as starting materials, (b) the source of hydrogen is 2-propanol, a cheap and environmentally friendly solvent, and (c) the NiNPs were shown to be superior to other nickel catalysts in this reaction.

5.6 Conclusions

In view of the above results, we can state that NiNPs are a real alternative to noble-metal based catalysts in hydrogen-transfer reactions. Simple catalytic systems, such as that composed of NiNPs and 2-propanol, exhibited a high performance in the transfer hydrogenation of olefins, carbonyl compounds, and reductive amination of aldehydes. This represents the first transition-metal catalyzed transfer hydrogenation in 2-propanol performed in the absence of any base or acid. Other unsupported NiNPs also succeeded in the transfer hydrogenation of carbonyl compounds with ammonium formate at room temperature. However, NiNPs immobilized on inorganic supports are more stable and allow the reactions to proceed with very low metal loadings, albeit the presence of a base is mandatory. It must be underlined that, due to the heterogeneous nature of all the catalytic systems shown herein, they can be easily recovered and reused in 3–6 cycles maintaining high catalytic activities and displaying a reactivity superior to other nickel catalysts, including Raney nickel. The results presented herein open an array of possibilities for further research in reactions where the more expensive, and sometimes sophisticated, noble-metal catalysts might be replaced by the cheaper and, generally, easy-to-prepare NiNP-based catalysts.

References

1. R. A. W. Johnstone and A. H. Wilby, *Chem. Rev.*, 1985, **85**, 129–170.
2. R. M. Kellogg, Reduction of C=X to CHXH by hydride delivery from carbon in *Comprehensive Organic Synthesis*, ed. B. M. Trost and I. Fleming, Pergamon, Oxford, 1st edn, 1991, vol. 8, ch. 1.3, pp. 79–106.
3. *Iron Catalysis in Organic Chemistry. Reactions and Applications*, ed. B. Plietker, Wiley-VCH, Weinheim, 2008.
4. *Nanoparticles and Catalysis*, ed. D. Astruc, Wiley-VCH, Weinheim, 2008.
5. M. Kidwai, Nanoparticles in Green Catalysis in *Handbook of Green Chemistry*, ed. P. T. Anastas and R. H. Crabtree, Wiley-VCH, Weinheim, 2009, vol. 2, ch. 4, pp. 81–92.
6. N. Yan, C. Xiao and Y. Kou, *Coord. Chem. Rev.*, 2010, **254**, 179–1218.
7. V. Polshettiwar and R. S. Varma, *Green Chem.*, 2010, **12**, 743–754.
8. F. Alonso and M. Yus, *Tetrahedron Lett.*, 1996, 37, 6925–6928.

9. F. Alonso and M. Yus, *Tetrahedron Lett.*, 1997, **38**, 149–152.
10. R. D. Rieke, *Top. Curr. Chem.*, 1975, **59**, 1–31.
11. A. Fürstner, *Active Metals*, Wiley-VCH, Weinheim, 1996.
12. F. Alonso, G. Radivoy and M. Yus, *Russ. Chem. Bull., Int. Ed.*, 2003, **52**, 2563–2576.
13. F. Alonso and M. Yus, *Chem. Soc. Rev.*, 2004, **33**, 284–293.
14. F. Alonso and M. Yus, *Pure Appl. Chem.*, 2008, **80**, 1005–1012.
15. F. Alonso and M. Yus, *Adv. Synth. Catal.*, 2001, **343**, 188–191.
16. F. Alonso, P. Candela, C. Gómez and M. Yus, *Adv. Synth. Catal.*, 2003, **345**, 275–279.
17. F. Alonso, J. J. Calvino, I. Osante and M. Yus, *Chem. Lett.*, 2005, **34**, 1262–1263.
18. F. Alonso, J. J. Calvino, I. Osante and M. Yus, *J. Exp. Nanosci.*, 2006, **1**, 419–433.
19. F. Alonso, I. Osante and M. Yus, *Adv. Synth. Catal.*, 2006, **348**, 305–308.
20. F. Alonso, I. Osante and M. Yus, *Tetrahedron*, 2007, **63**, 93–102.
21. F. Alonso, I. Osante and M. Yus, *Synlett*, 2006, 3017–3020.
22. F. Alonso, P. Riente and M. Yus, *Acc. Chem. Res.*, 2011, **44**, 379–391.
23. M. Kitamura and R. Noyori, Hydrogenation and transfer hydrogenation in *Ruthenium in Organic Synthesis*, ed. S.-I. Murahashi, Wiley-VCH, Weinheim, 2004, pp. 31–32.
24. J. M. Brunel, *Tetrahedron*, 2007, **63**, 3899–3906.
25. S. Horn and M. Albrecht, *Chem. Commun.*, 2011, **47**, 8802–8804.
26. M. J. Andrews and C. N. Pillai, *Indian J. Chem.*, 1978, **16B**, 465–468.
27. G. P. Boldrini, D. Savoia, E. Tagliavini, C. Trombini and A. A. Umani-Ronchi, *J. Org. Chem.*, 1985, **50**, 3082–3086.
28. F. Alonso, P. Riente and M. Yus, *Tetrahedron*, 2009, **65**, 10637–10643.
29. A. Dhakshinamoorthy and K. Pitchumani, *Tetrahedron Lett.*, 2008, **49**, 1818–1823.
30. D. Klomp, U. Hanefeld and J. A. Peters, Transfer hydrogenation including the Meerwein–Ponndorf–Verley reduction in *The Handbook of Homogeneous Hydrogenation*, ed. J. G. de Vries C. J. and Elsevier, Wiley-VCH, Weinheim, 2007.
31. C. Wang, X. F. Wu and J. L. Xiao, *Chem.–Asian J.*, 2008, **3**, 1750–1770.
32. J. S. M. Samec, J.-E. Bäckwall, P. G. Andersson and P. Brandt, *Chem. Soc. Rev.*, 2006, **35**, 237–248.
33. M. D. Le Page and B. R. James, *Chem. Commun.*, 2000, 1647–1648.
34. P. Phukan and A. Sudalai, *Synth. Commun.*, 2000, **30**, 2401–2405.
35. T. T. Upadhya, S. P. Katdare, D. P. Sabde, V. Ramaswamy and A. Sudalai, *Chem. Commun.*, 1997, 1119–1120.
36. S. K. Mohapatra, S. U. Sonavane, R. V. Jayaram and P. Selvam, *Org. Lett.*, 2002, **24**, 4297–4300.
37. R. C. Mebane and A. J. Mansfield, *Synth. Commun.*, 2005, **35**, 3083–3086.
38. R. C. Mebane, K. L. Holte and B. H. Gross, *Synth. Commun.*, 2007, **37**, 2787–2791.
39. F. Alonso, P. Riente and M. Yus, *Tetrahedron*, 2008, **64**, 1847–1852.

40. F. Alonso, P. Riente and M. Yus, *Tetrahedron Lett.*, 2008, **49**, 1939–1942.
41. F. Alonso, P. Riente, J. A. Sirvent and M. Yus, *Appl. Catal., A: Gen.*, 2010, **378**, 42–51.
42. V. Polshettiwar, B. Baruwati and R. S. Varma, *Green Chem.*, 2009, **11**, 127–131.
43. J. C. Park, H. J. Lee, J. U. Bang, K. H. Park and H. Song, *Chem. Commun.*, 2009, 7345–7347.
44. J. C. Park, H. J. Lee, H. Y. Kim, K. H. Park and H. Song, *J. Phys. Chem. C*, 2010, **114**, 6381–6388.
45. D. Dutta, B. J. Borah, L. Saikia, M. G. Pathak, P. Sengupta and D. K. Dutta, *Appl. Clay Sci.*, 2011, **53**, 650–656.
46. N. Neelakandeswari, G. Sangami, P. Emayavaramban, S. G. Babu, R. Karvembu and N. Dharmaraj, *J. Mol. Catal. A: Chem.*, 2012, **356**, 90–99.
47. M. Kidwai, V. Bansal, A. Saxena, R. Shankar and S. Mozumdar, *Tetrahedron Lett.*, 2006, **47**, 4161–4165.
48. M. Kidwai, N. Kumar Mishra, V. Bansal, A. Kumar and S. Mozumdar, *Catal. Commun.*, 2008, **9**, 612–617.
49. V. A. Tarasevich and N. G. Kozlov, *Russ. Chem. Rev.*, 1999, **68**, 55–72.
50. S. Gomez, J. A. Peters and T. Maschmeyer, *Adv. Synth. Catal.*, 2002, **344**, 1037–1057.
51. E. W. Baxter and A. B. Reitz, *Org. React.*, 2002, **59**, 1–714.
52. A. F. Abdel Magid and S. J. Mehrman, *Org. Process Res. Dev.*, 2006, **10**, 971–1031.
53. V. I. Tararov, R. Kadyrov, T. H. Riermeier, U. Dingerdissen and A. Boerner, *Org. Prep. Proced. Int.*, 2004, **36**, 99–120.
54. V. I. Tararov and A. Boerner, *Synlett*, 2005, 203–211.
55. P. Roszkowski and Z. Czarnocki, *Mini-Rev. Org. Chem.*, 2007, **4**, 190–200.
56. J. H. Nah, S. Y. Kim and N. M. Yoon, *Bull. Korean Chem. Soc.*, 1998, **19**, 269–270.
57. I. Saxena, R. Borah and J. C. Sarma, *J. Chem. Soc., Perkin Trans.*, 2000, **1**, 503–504.
58. M. Botta, F. De Angelis, A. Gambacorta, L. Labbiento and R. Nicoletti, *J. Org. Chem.*, 1985, **50**, 1916–1919.
59. S. Khul, R. Schneider and Y. Fort, *Organometallics*, 2003, **22**, 4184–4186.
60. F. Alonso, P. Riente and M. Yus, *Synlett*, 2008, 1289–1292.

CHAPTER 6

Ammonium Surfactant-capped Rh(0) Nanoparticles for Biphasic Hydrogenation

AUDREY DENICOURT-NOWICKI*[a,b] AND ALAIN ROUCOUX*[a,b]

[a] Ecole Nationale Supérieure de Chimie de Rennes, CNRS, UMR 6226, 11 allée de Beaulieu, CS 50837, F-35708 Rennes cedex 7, France;
[b] Université Européenne de Bretagne, France
*Email: Audrey.Denicourt@ensc-rennes.fr; Alain.Roucoux@ensc-rennes.fr

6.1 Introduction

In the drive towards the development of environmentally-friendly and economically viable processes, nanometre-sized metallic species, finely dispersed in water, have appeared as an unavoidable family of active and reusable catalysts under biphasic conditions.[1] Thus, soluble metal nanoparticles have emerged as sustainable alternatives to conventional molecular catalysts, potentially being easily recycled as heterogeneous systems.[2-4] These nanocatalysts have proved to be efficient and selective for various reactions, owing to a great number of potential active sites and an original surface reactivity.[5-7] In particular, they have showed great promise in arene hydrogenation reactions,[8,9] presenting high activities in mild conditions, especially with Rh(0) nanoparticles. Recently, the preparation of metal particles in green solvents, such as water, ionic liquids, fluorous medium or supercritical fluids, has received great attention.[10] Among them, water has appeared as a relevant alternative to conventional organic media for economic and environmental reasons,[11,12] as well as the easy catalyst recovery through a biphasic approach thanks to its absence of miscibility with the usual organic phases.[13,14] Since 'naked' nanoparticles tend to aggregate, protective agents are usually used and chosen according to the reaction media. Besides the use of ligands (polymers, cyclodextrins, dendrimers, *etc.*) as stabilizers,[1,15] surfactants, such as ammonium salts, are widely known as efficient protective agents for electrosteric stabilization of particles within the aqueous phase, thanks to their amphiphilic character. Generally, nanospecies with well-controlled size were obtained, providing pertinent reactivity in catalysis.

In this chapter, we will mainly focus on the use of Rh(0) nanoparticles in hydrogenation reactions in pure biphasic liquid–liquid conditions, namely when a liquid substrate constitutes the only organic phase (no solvent added).

6.2 Nanoparticles as Relevant Catalysts for Biphasic Hydrogenation

Surfactants, and particularly quaternary ammonium salts,[16,17] proved to be relevant protective agents of various transition metal nanospecies for catalytic applications in pure biphasic conditions.

In that field, Roucoux and co-workers have largely contributed to the efficient stabilization of Rh(0) nanocatalysts through the design of a library of easily tunable *N*-(hydroxyalkyl)ammonium salts (HAAnX). These ionic compounds (Figure 6.1) could be easily synthesized by quaternarization of *N*,*N*-dimethylethanolamine with the corresponding functionalized halogenoalkanes for the HEA series[18,19] or of hexadecylamine with chloroethanol for THEA16Cl.[20] Based on cetyl-ammonium chloride salt as a common skeleton, *N*,*N*-dimethyl-*N*-cetyl-*N*-(3-hydroxypropyl) ammonium chloride (HPA16Cl) containing a hydroxypropyl head group was also reported.[21] Various structural parameters of these ionic compounds could be modulated such as: (i) the length of the lipophilic chain (C12–C18),[22] (ii) the nature of the polar head (mono- or poly-hydroxylated with various lengths)[20,21] and (iii) the nature of the counter-ion thanks to anionic metathesis.[18,23,24] Surfactants

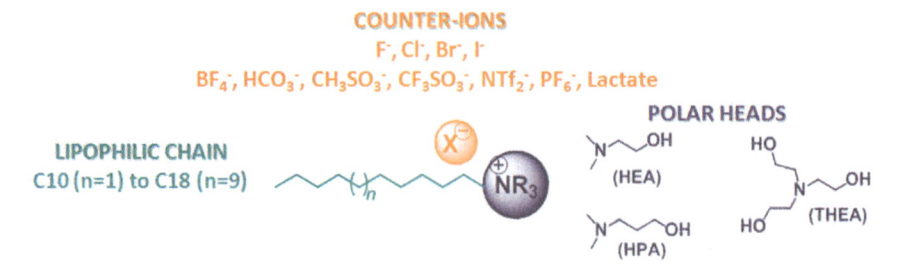

Figure 6.1 Library of *N*,*N*-dimethyl-*N*-alkyl-*N*-(hydroxyalkyl)ammonium salts as efficient protective agents of Rh(0) nanoparticles (HEA = hydroxyethylammonium, HPA = hydroxypropylammonium, THEA = trishydroxyethylammonium).

possessing halogens (F⁻, Cl⁻, Br⁻, I⁻), mesylate (CH$_3$SO$_3^-$), tetrafluoroborate (BF$_4^-$), hydrogen carbonate (HCO$_3^-$), triflate (CF$_3$SO$_3^-$), bis(trifluoromethane)sulfonimidate (NTf$_2^-$) or hexafluorophosphate (PF$_6^-$) as counter-ions were reported by our group.

These ammonium compounds bearing a lipophilic alkyl chain of 16 carbons display surfactant behaviour (Table 6.1) and self-aggregate into micelles above the critical micellar concentration (cmc). The cmc values decrease according to the increase in the length of the lipophilic chain and to the nature of the counter-ion in the following order: F⁻ > Cl⁻ > Br⁻ > CH$_3$SO$_3^-$ > BF$_4^-$ ≈ HCO$_3^-$ > CF$_3$SO$_3^-$ > I⁻ in the HEA16 series.[18,22,23] The larger and more polarisable anions tend to decrease the electrostatic repulsions between the polar groups, thus promoting micelle formation. Compared to HEA16Cl and HPA16Cl,[21] micelles are formed at lower concentrations with THEA16Cl,[20] probably owing to its more sterically hindered polar head. The decrease in the surface tension of these aqueous solutions of surfactants below the cmc is related to their adsorption at the water–air interface according to Gibbs law.

Aqueous suspensions of metallic rhodium(0) colloids were easily synthesized at room temperature by chemical reduction of rhodium trichloride with sodium borohydride in dilute aqueous solutions of quaternary hydroxylated ammonium salts (Figure 6.2). From optimization studies, the stabilization of Rh(0) nanoparticles depends on the alkyl chain length of the protective groups. Indeed, effective electrosteric stabilization was achieved owing to a sufficiently lipophilic C16 chain, combined with a highly hydrophilic hydroxylated polar head. Moreover, based on the molar ratio [surfactant]/[metal] investigation,[22] a value of 2 proves to be an optimal choice to prevent aggregation of nanoparticles within the aqueous phase and to provide a pertinent activity. Thanks to thermogravimetric analyses,[18] the

Table 6.1 Physico-chemical properties of some N-cetyl-ammonium derivatives and their predicted aggregate geometry (adapted from ref. 18, 21, 23).

Entry	Surfactant	CMC^a (mmol L^{-1})	γCMC^b (mN m^{-1})	A^c (Å^2)	CPP^d	Geometry
1	HEA16Cl	1.23	40.5	33.4	0.63	Bilayers
2	HEA16F	2.12	43.2	87.1	0.24	Spheres
3	HEA16HCO$_3$	0.6	40.3	46.6	0.45	Rods
4	HEA16BF$_4$	0.62^e	30.3	37.3	0.55	Rods
5	HEA16CF$_3$SO$_3$	0.25	32.0	49.6	0.42	Rods
6	HPA16Cl	2.00	36.9	91	0.23	Spheres
7	THEA16Cl	0.63	33.5	128	0.16	Spheres

^{a}CMC = critical micellar concentration measured at 296K.
^{b}γCMC = surface tension at CMC.
^{c}A = area of polar head per molecule.
^{d}CPP = critical packing parameter, spherical micelles for CPP < $\frac{1}{3}$, cylindrical or rod-like micelles for $\frac{1}{3}$ < CPP < $\frac{1}{2}$, bilayers with CPP > $\frac{1}{2}$.
^{e}CMC measured at 303 K.

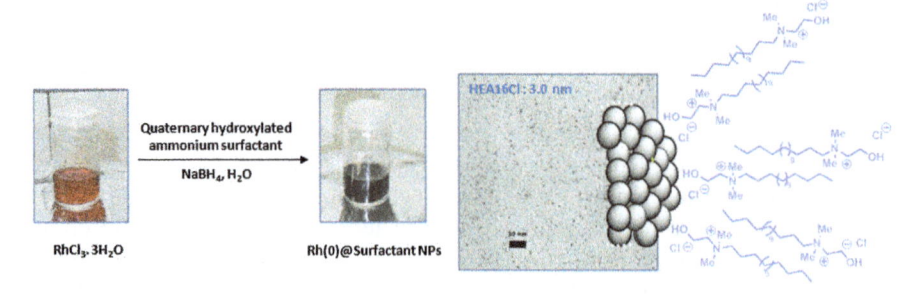

Figure 6.2 Preparation of surfactant-capped Rh(0) colloids and a representation of the bilayer structure on the particle's surface.

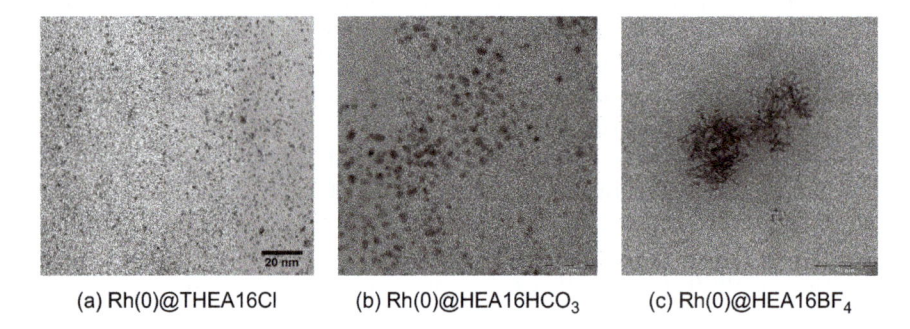

Figure 6.3 TEM pictures of Rh(0) nanocatalysts capped with various quaternary hydroxylated ammonium surfactants (adapted from ref. 18, 20).

authors presume that the surfactants self-organize in a bilayer structure at the nanoparticle's surface, as already described in the case of ammonium salt-stabilized Cu and Ni nanoparticles or Au nanorods by Chen[25] and El Sayed.[26] The inner layer of the surfactants is bound to the particle's surface *via* the charged groups and is connected to the outer layer through hydrophobic interactions while the polar groups of the last one are directed towards the aqueous solution. Finally, the potential zeta ζ, measured from dynamic light scattering (DLS) analyses, possesses an apparently positive charge in the range 40–50 mV for the colloids coated with various surfactants in solution.[18] These values are in agreement with cationic protective types and show the primordial role of electrostatic repulsion (coulombic interaction) in nanocluster stabilization.

TEM studies have revealed the influence of the protective agent (nature of the counter-ion and of the polar head) on the average particle size and morphology of various Rh(0)@HEA16X nanosystems (Figure 6.3). Usually, a mean diameter in the range of 2–3 nm was obtained depending on the counter-ion's nature. While spherical geometries were observed with THEA16Cl or HEA16X ($X^- = Cl^-$, Br^-, HCO_3^-) as protective agents, the use of HEA16BF$_4$ and HEA16CF$_3$SO$_3$ led to worm-like structures. In the case of Rh(0)@HEA16F NPs, spherical and elongated NPs were obtained, probably owing to a weaker stabilization of the spherical particles inducing a sintering of the nanospecies.[18]

The surfactant-capped Rh(0) NPs proved to be very active in the hydrogenation of various arene derivatives, at atmospheric hydrogen pressure and room temperature, under biphasic water/substrate conditions. Optimization studies showed that an optimal turnover frequency (TOF) was achieved when the surfactant concentration was proximate to the cmc value.[27] A comparison of various protective agents of Rh(0) colloids in benzene reduction is presented in Table 6.2.

Ammonium-capped Rh(0) colloids, acting as micellar nanoreactors, proved to be reference catalysts for this reaction with quite high TOF values. The catalytic activity was mostly influenced by a pertinent combination of a cetyl lipophilic chain (C16) with the counter-ion and a mono- or polyhydroxylated polar head. In the case of HEA16I, no reduction occurred due to the simultaneous formation of iodine, well-known as a poison for nanospecies.[23] In most cases, an increase in hydrogen pressure gave rise to an activation of the catalytic suspension.[18] The use of these nanocatalysts was extended to various aromatic derivatives (alkylated, functionalized and disubstituted arenes),[21] heteroaromatics[28] and halogenoarenes.[29] Some of the results are gathered in Table 6.3, considering the usual Rh(0)@HEA16Cl system under mild biphasic conditions.

The ammonium-capped Rh(0) nanoparticles showed remarkable activities towards a large variety of alkylated or functionalized aromatics (entries 1–3) and oxygen- or nitrogen-containing heteroaromatics (entries 4–8). The selective reduction of disubstituted arenes, such as xylene isomers (entries 9–11), leads to the major formation of the thermodynamically less stable cis-product, as usually observed with heterogeneous catalysts.[30] This selectivity could be attributed to a "continuous" coordination of the substrate to the catalyst's surface during the reaction, thus favouring the addition of

Table 6.2 Biphasic benzene hydrogenation under 1 bar of H_2 and 20 °C using various ammonium-protected Rh(0) colloids[a] (adapted from ref. 18, 21–23).

Entry	Surfactant	t (h)	TOF [b] (h^{-1})
1	HEA18Br	9.1	33
2	HEA16Br	5.3	57
3	HEA16Cl	3.6	83
4	HEA16F	6	50
5	HEA16BF$_4$	3.7	81
6	HEA16HCO$_3$	6	50
7	HEA16CH$_3$SO$_3$	3.7	81
8	HEA16CF$_3$SO$_3$	3	100
9	HPA16Cl	5	60
10	THEA16Cl	1	300

[a]Reaction conditions: Rh (3.8×10^{-5} mol), [substrate]/[metal]/[surfactant] = 100 : 1 : 2, 10 mL H_2O, 1 bar H_2, 20 °C.
[b]Turnover frequency determined by number of moles of consumed H_2 per mole of Rh per hour.

Table 6.3 Biphasic hydrogenation of arene derivatives with HEA16Cl-capped Rh(0) colloids[a] (adapted from ref. 21, 28, 29).

Entry	Substrate	Product yield[b] (%)	t (h)	TOF[c] (h^{-1})
1	Toluene	Methylcyclohexane (100)	3.6	83
2	Anisole	Methoxycyclohexane (70)–cyclohexanone (30)	4	75
3	Ethyl benzoate	Ethylcyclohexanoate (100)	4.7	64
4	Aniline	Cyclohexylamine (100)	10	30
5	Furan	1,2,3,4-Tetrahydrofuran (100)	2.1	95
6	Pyridine	Piperidine (100)	3.9	77
7	Quinoline	1,2,3,4-Tetrahydroquinoline (100)	9.1	22
8	1,3,5-Triazine	1,3,5-Hexahydrotriazine (100)	1.7	176
9	o-Xylene	1,2-Dimethylcyclohexane (cis-trans: 91 : 9)	5.3	57
10	m-Xylene	1,3-Dimethylcyclohexane (cis-trans: 80 : 20)	6.8	44
11	p-Xylene	1,4-Dimethylcyclohexane (cis-trans: 65 : 35)	4.1	73
12	Chorobenzene	Cyclohexane (100)	1.7	—
13	2-Chloroanisole	Methoxycyclohexane (53)–cyclohexanone (47)	11	34

[a]Reaction conditions: Rh (3.8 × 10^{-5} mol), [substrate]/[metal]/[surfactant] = 100 : 1 : 2, 10 mL H$_2$O, 1 bar H$_2$, 20 °C.
[b]Determined by gas chromatography.
[c]Turnover frequency determined by number of moles of consumed H$_2$ per mole of Rh per hour.

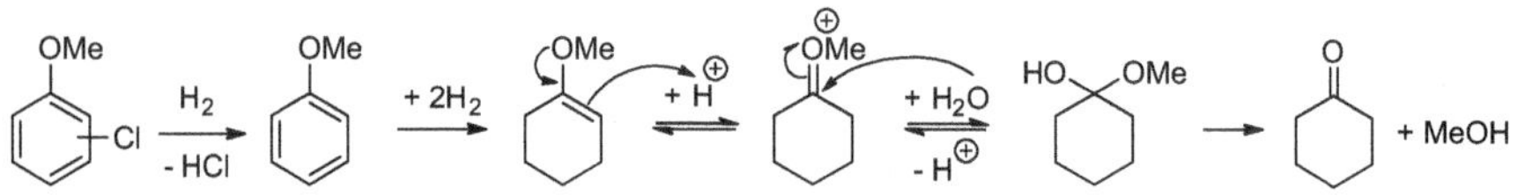

Figure 6.4 Proposed mechanism of cyclohexanone formation during the chloroanisole hydrogenation (adapted from ref. 29).

hydrogen atoms to only one face of the arene.[31,32] The formation of the minor *trans* diastereoisomers could be explained through a roll-over mechanism, in which the partially hydrogenated intermediate dissociates from the catalyst's surface and then re-associates through the opposite face before further reduction occurs.[30,33] In all cases, the nanocatalysts proved to be easily recycled by simple extraction of the product with the adequate solvent (alkanes or ether), during several successive reduction reactions with similar TOFs and no metal leaching. In the case of halogenoarenes (entries 12–13), a tandem dehalogenation–hydrogenation reaction occurred with Rh(0) colloids.[34] Moreover, the hydrodechlorination of halogenoanisoles as model substrates of endocrine disruptors (entry 13) yielded the relatively non-toxic saturated product and added-value cyclohexanone in a significant amount (nearly 50% GC yield).[29] The formation of the ketone could be attributed to the decomposition of the hemiacetal intermediate, generated from the partially hydrogenated product in acidic conditions due to the release of HCl during the dehalogenation step (Figure 6.4). However, in some cases, a deactivation and an aggregation of the catalyst were observed owing to a chlorine poisoning effect of the catalyst surface, which could be circumvented by a heterogeneization of the colloidal suspensions onto inorganic supports such as SiO_2 or TiO_2.[29,35,36]

To conclude, quaternary hydroxylated ammonium surfactants, used as protective agents of Rh(0) nanospecies, provide pertinent micellar nanoreactors for benzene derivative hydrogenation under mild pressure and temperature conditions in biphasic liquid–liquid media, with high activities and a good recyclability.

6.3 Asymmetric Nanocatalysis: a Great Challenge

In the last decades, soluble metal nanoparticles have showed great potential in many catalytic reactions owing to their unique surface properties. However, their use as effective catalytically active species in asymmetric processes remains a challenge in nanocatalysis.[37] In this approach, the catalytic reaction takes place on the particle's surface and the enantiocontrol will depend on the way that the capping agents will transmit their chiral information to the substrate in the vicinity of the chirally modified particles. Generally, this methodology has no equivalent in homogeneous asymmetric catalysis, thus rendering these systems challenging but also valuable.[38] Although some examples of metallic nanospecies modified by chiral ligands were reported for asymmetric allylic alkylations[39,40] and arene hydrogenations,[41,42] the use of finely dispersed metallic nanoparticles in neat water, stabilized by water-soluble protective agents, has been far less described.

Following their approach based on the electrosteric stabilization of nanospecies through the use of quaternary hydroxylated ammonium salts, Roucoux and co-workers developed a new library of water-soluble and optically active ammonium salts possessing various polar heads, derived from N-methylephedrine, N-methylprolinol or cinchona derivatives, in combination with different counter-ions, such as Br^-, HCO_3^-, (S)- or (R)-lactate (Figure 6.5).[43,44] The bromide surfactants were easily synthesized from commercially available precursors by quaternarization with the corresponding bromoalkane with high yields (70–95%). The other ammonium

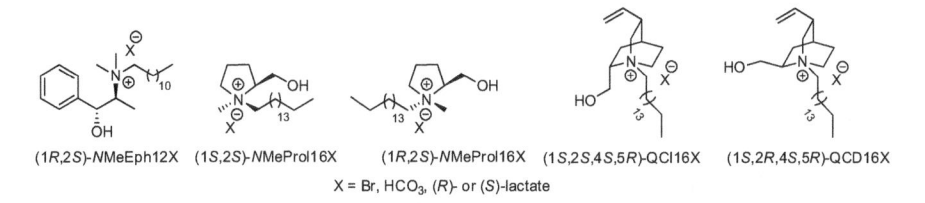

(1R,2S)-NMeEph12X   (1S,2S)-NMeProl16X   (1R,2S)-NMeProl16X   (1S,2S,4S,5R)-QCl16X   (1S,2R,4S,5R)-QCD16X

X = Br, HCO₃, (R)- or (S)-lactate

Figure 6.5 Optically active ammonium surfactants as protective agents of Rh(0) NPs. N-Dodecyl-N-methylephedrium (NMeEph12)⁺ salt, N-hexadecyl-N-methyl-L-prolinolinium (NMeProl16)⁺ salt, N-hexadecyl-quicorinium (QCI16)⁺ salt, N-hexadecyl-quicoridinium (QCD16)⁺ salt (adapted from ref. 43, 44).

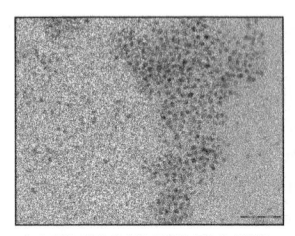

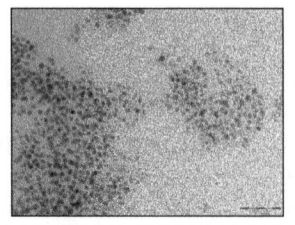

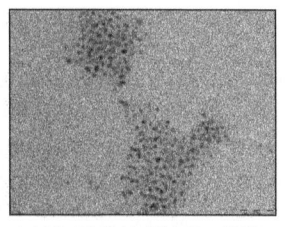

(a) Rh(0)@(-)-NMeEph12Br   (b) Rh(0)@(-)-NMeEph12HCO$_3$ (c) Rh(0)@(-)-NMeProl16Br

Figure 6.6 TEM pictures of chirally modified Rh(0) nanocatalysts (adapted from ref. 43, 44).

derivatives, presenting a hydrogenocarbonate (HCO$_3^-$) or a lactate (CH$_3$CH(OH)COO$^-$) as a counter-ion, were obtained by an anionic metathesis reaction starting from the bromide salt. Depending on the polar head, the length of the lipophilic chain was modulated to provide efficient water-solubility.

These optically active surfactants proved to be efficient protective agents for the stabilization of spherical rhodium(0) nanospecies with mean sizes in the range of 0.8–2.5 nm and narrow size distributions, according to the surfactant's nature (Figure 6.6). The chirally modified colloidal suspensions showed a positive apparent charge in solution, ranging from 38 to 73 mV.

These optically active amphiphilic ammonium salts were tested in the enantioselective reduction of ethyl pyruvate as a model reaction to evaluate the catalytic performances of Rh(0) nanoparticles as well as the diastereoselective hydrogenation of prochiral 3-methylanisole as a challenging reaction.

6.3.1 Ethylpyruvate

In a first set of experiments, the obtained nanocatalysts were investigated in the asymmetric hydrogenation of ethylpyruvate in neat water. Various parameters, such as the nature of the counter-ion in the N-methylephedrine-based surfactants and of the polar head in the bromide series, and the hydrogen pressure, were optimized. The use of cinchonidine, already known as a promising external chiral inducer in the presence of Pt nanoparticles capped with an achiral surfactant, was also investigated.[45] The results are gathered in Table 6.4, considering an optimum 40 bar H$_2$.

First, concerning the optically active polar head, the steric hindrance seems to govern the enantiocontrol, the best results being achieved with the N-methylephedrinium salt. In the Rh(0)@(-)-NMeEph12X series, the bromide and the (S)-lactate counter-ions proved to be the most effective in terms of asymmetric induction up to 13% e.e. (entries 1 and 5). The best value of 15% e.e. was obtained with the dextrorotatory enantiomer (+)-NMeEph12Br (entry 2). The use of (-)-cinchonidine as an external inducer did not significantly improve the enantiomeric excess, with e.e. up to 18% (entry 12).

Table 6.4 Enantioselective hydrogenation of ethylpyruvate with Rh(0) nanoparticles capped with optically active ammonium salts in neat water[a] (adapted from ref. 43, 44).

Entry	Surfactant	t (h)	Conversion[b] (%)	e.e.[b] (%)
1	(1R,2S)-(−)-NMeEph12Br	1	100	12 (R)
2	(1S,2R)-(+)-NMeEph12Br	1	100	15 (S)
3	(1R,2S)-(−)-NMeEph12HCO$_3$	1.5	100	4 (R)
4	(1R,2S)-(−)-NMeEph12(R)-lactate	1	100	9 (R)
5	(1R,2S)-(−)-NMeEph12(S)-lactate	1	100	13 (R)
6	(2S)-(−)-NMeProl16Br	0.5	100	1
7	(1S,2S)-(−)-NMeProl16Br	1	100	3
8	(1R,2S)-(+)-NMeProl16Br	1	100	5
9	(1S,2S,4S,5R)-(+)-QCI16Br	2[c]	100	1
10	(1S,2R,4S,5R)-(+)-QCD16Br	2	60	0
11	(1R,2S)-(−)-NMeEph12Br[d]	1	100	15 (S)
12	(1R,2S)-(−)-NMeEph12(S)-lactate[d]	1	100	18 (S)

[a]Reaction conditions: Rh (3.8 × 10^{-5} mol), [substrate]/[metal]/[surfactant] = 100:1:2, H$_2$O (10mL), 40 bar H$_2$, 20 °C.
[b]Determined by gas chromatography analysis using a chiral column (Chiralsil-Dex CB).
[c]Non-optimized.
[d]Addition of (−)-cinchonidine.

Unsatisfactorily, QCI, QCD and NMeProl derivatives demonstrated poor results with e.e. <5%.

The low asymmetric induction observed could be attributed to a lack of steric hindrance around the rhodium(0) particles and also to the dynamic behavior of the protective agent at the metal's surface.

6.3.2 Prochiral Arenes

The stereoselective hydrogenation of monocyclic polysubstituted arenes provides an elegant synthetic approach to optically active cyclohexyl compounds and remains one of the last challenging reactions in asymmetric catalysts due the lack of efficient catalysts.[46,47] In the drive towards ecocompatible chemistry, heterogeneous catalytic asymmetric processes are of great interest. The first attempts described by Lemaire and co-workers, using either a chiral lipophilic amine or a chiral auxiliary covalently bound to the substrate, gave only low asymmetric induction in the hydrogenation of 2-methylanisole.[48,49]

More recently, Roucoux and co-workers have investigated Rh(0) nanoparticles capped with the previously described quaternary hydroxylated ammonium salts in the reduction of 3-methylanisole as a model substrate, at room temperature under pure biphasic conditions (Figure 6.7).[43,44]

The influence of several parameters, such as the hydrogen pressure and the protective agent, has been studied (Table 6.5). First, reduction of the aromatic ring under atmospheric pressure is kinetically slow (100% conversion in 3 days) and thus was performed under 40 bar H$_2$, without

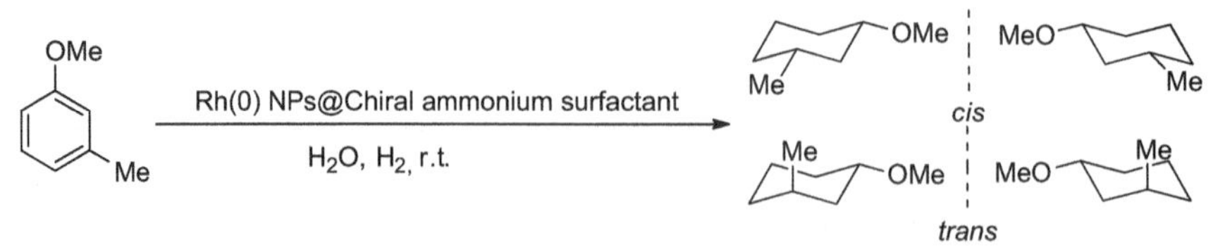

Figure 6.7 Hydrogenation of 3-methylanisole with chirally modified Rh(0) nanoparticles.

Table 6.5 Asymmetric hydrogenation of 3-methylanisole with Rh(0) colloids capped with optically active ammonium salts in neat water[a] (adapted from ref. 43, 44).

Entry	Surfactant	t (h)	Conv.[b] (%)	Yield cis–trans (e.e.) (%)[b]
1	(1R,2S)-(−)-NMeEph12Br	3	100	62 (0) : 38 (0)
2	(1R,2S)-(−)-NMeEph12(R)-lactate	3	100	65 (1) : 35 (0)
3	(1R,2S)-(−)-NMeEph12(S)-lactate	3	100	66 (0) : 34 (0)
4	(2S)-(−)-NMeProl16Br	16[c]	100	60 (1) : 40 (1)
5	(1S,2S,4S,5R)-(+)-QCI16Br	1	82	70 (1) : 30 (2)
6	(1S,2R,4S,5R)-(+)-QCD16Br	1	55	70 (1) : 30 (2)

[a]Reaction conditions: Rh (3.8×10^{-5} mol), [substrate]/[metal]/[surfactant] = 100 : 1 : 2, H$_2$O (10mL), 40 bar H$_2$, 20 °C.
[b]Determined by gas chromatography analysis using a chiral column (Chiralsil-Dex CB).
[c]Non-optimized.

modifying the selectivity in favor of the *cis* isomers (*cis–trans* ratio = 65 ± 5 : 35 ± 5).

These nanocatalysts were also active in the hydrogenation of 3-methylanisole, yielding high diastereoselectivity in the *cis*-product, but without asymmetric induction. This selectivity essentially depended on the chiral skeletons of head groups and varied in the following order: (+) − QCD ≈ (+) − QCI > (−)-NMeEph ≈ (−)-NMeProl. No influence of the (S)- or (R)-lactate counter-ion was observed with a similar *cis–trans* ratio and no asymmetric induction (entries 2–3). We could presume that optically active ammonium surfactants could protect the metal nanospecies through the adsorption of anions at the particle's surface, as described in colloid stabilization,[50,51] and also through the cationic lipophilic tail, providing steric stabilization and hindrance. However, contrary to coordinating ligands, the interaction of ammonium salts within the metallic surface remains weak, allowing the displacement of the protective agents by incoming substrates, and thus lowering the enantiofacial discrimination.

6.4 Conclusions

In summary, hydroxylated ammonium surfactants have been widely applied as protective agents for the efficient electrosteric stabilization of rhodium(0) nanoparticles in neat water. The easy modulation of the skeleton such as the lipophilic chain length, and the nature of the associated counter-ion or the polar head allow tuning of the stability of the aqueous suspensions and finally the hydrogenation catalytic properties. The obtained micelles are able to confine metallic species within their cores, thus acting as promising nanoreactors. In contrast to homogeneous species, these nanocatalysts proved to be active in the reduction of arene derivatives under mild conditions and easily recyclable under biphasic conditions (water/substrate). Undoubtedly, the development of asymmetric catalysis with well-defined nanospecies will constitute an ambitious goal in the field of catalysis, which remains unexplored in neat water.

References

1. A. Denicourt-Nowicki and A. Roucoux, Metallic Nanoparticles in Neat Water for Catalytic Applications in *Nanomaterials in Catalysis*, ed. P. Serp and K. Philippot, Wiley-VCH Verlag GmbH & Co. KGaA, Weinheim, 1st edn, 2013, pp. 55–95.
2. R. Narayanan and M. El-Sayed, *Top. Catal.*, 2008, **47**, 15–21.
3. V. Polshettiwar and R. S. Varma, *Green Chem.*, 2010, **112**, 743–754.
4. D. Astruc, F. Lu and J. R. Aranzaes, *Angew. Chem., Int. Ed.*, 2005, **44**, 7852–7872.
5. J. P. Wilcoxon and B. L. Abrams, *Chem. Soc. Rev.*, 2006, **35**, 1162–1194.
6. *Nanoparticles and Catalysis*, ed. D. Astruc, Wiley-VCH Verlag GmbH & Co. KGaA, Weinheim, 2008.
7. A. Roucoux, A. Nowicki and K. Philippot, Rhodium and Ruthenium nanoparticles in catalysis in *Nanoparticles and Catalysis*, ed. D. Astruc, Wiley-VCH Verlag GmbH & Co. KGaA, Weinheim, 2008, pp. 349–388.
8. A. Roucoux, Stabilized noble metal nanoparticles: An unavoidable family of catalysts for arene derivatives hydrogenation, in *Topics in Organometallic Chemistry: Surface and Interfacial Organometallic Chemistry and Catalysis*, ed. C. Copéret and B. Chaudret, Springer GmbH, Heidelberg, 2005, vol. 16 pp. 261–279.
9. A. Gual, C. Godard, S. Castillon and C. Claver, *Dalton Trans.*, 2010, **39**, 11499–11512.
10. N. Yan, C. X. Xiao and Y. Kou, *Coord. Chem. Rev.*, 2010, **254**, 1179–1218.
11. U. M. Lindstrom, *Chem. Rev. (Washington, DC, U. S.)*, 2002, **102**, 2751–2772.
12. C.-J. Li and L. Chen, *Chem. Soc. Rev.*, 2006, **35**, 68–82.
13. *Comprehensive Organic Reactions in Aqueous Media*, ed. C. J. Li and T. H. Chan, Wiley, New-York, 1997.
14. *Aqueous-Phase Organometallic Catalysis: Concepts and Applications*, ed. B. Cornils and W. A. Hermann, Wiley-VCH Verlag GmbH & Co. KGaA, Weinheim, 2nd edn, 2005.
15. A. Denicourt-Nowicki and A. Roucoux, *Curr. Org. Chem.*, 2010, **14**, 1266–1283.

16. R. W. Albach and M. Jautelat, *Process for the preparation of six-ring carbocyclic compounds*, Ger. Pat., DE19981007995, BAYER AG, 1999.
17. J. Schulz, A. Roucoux and H. Patin, *Chem. Commun.*, 1999, 535–536.
18. E. Guyonnet Bilé, R. Sassine, A. Denicourt-Nowicki, F. Launay and A. Roucoux, *Dalton Trans.*, 2011, **40**, 6524–6531.
19. G. Cerichelli, L. Luchetti, G. Mancini and G. Savelli, *Tetrahedron*, 1995, **51**, 10281–10288.
20. C. Hubert, A. Denicourt-Nowicki, J. P. Guegan and A. Roucoux, *Dalton Trans.*, 2009, 7356–7358.
21. C. H. Pélisson, C. Hubert, A. Denicourt-Nowicki and A. Roucoux, *Top. Catal.*, 2013, **56**, 1220–1227.
22. J. Schulz, A. Roucoux and H. Patin, *Chem.–Eur. J*, 2000, **6**, 618–624.
23. A. Roucoux, J. Schulz and H. Patin, *Adv. Synth. Catal.*, 2003, **345**, 222–229.
24. V. Mevellec, B. Leger, M. Mauduit and A. Roucoux, *Chem. Commun.*, 2005, 2838–2839.
25. S. Chen and K. Kimura, *Langmuir*, 1999, **15**, 1075–1082.
26. B. Nikoobakht and M. A. El-Sayed, *Langmuir*, 2001, **17**, 6368–6374.
27. J. Schulz, S. Levigne, A. Roucoux and H. Patin, *Adv. Synth. Catal.*, 2002, **344**, 266–269.
28. V. Mévellec and A. Roucoux, *Inorg. Chim. Acta*, 2004, **357**, 3099–3103.
29. C. Hubert, E. Guyonnet Bilé, A. Denicourt-Nowicki and A. Roucoux, *Appl. Catal., A*, 2011, **394**, 215–219.
30. M. Boutros, F. Launay, A. Nowicki, T. Onfroy, V. Herledan-Semmer, A. Roucoux and A. Gédéon, *J. Mol. Catal. A: Chem.*, 2006, **259**, 91–98.
31. T. Q. Hu, C.-L. Lee, B. R. James and S. J. Rettig, *Can. J. Chem.*, 1997, **75**, 1234–1239.
32. C. M. Hagen, L. Vieille-Petit, G. Laurenczy, G. Suss-Fink and R. G. Finke, *Organometallics*, 2005, **24**, 1819–1831.
33. A. Kalantar Neyestanaki, P. Mäki-Arvela, H. Backman, H. Karhu, T. Salmi, J. Väyrynen and D. Y. Murzin, *J. Catal.*, 2003, **218**, 267–279.
34. B. Léger, A. Nowicki, A. Roucoux and J.-P. Rolland, *J. Mol. Catal. A: Chem.*, 2007, **266**, 221–225.
35. V. Mévellec, A. Nowicki, A. Roucoux, C. Dujardin, P. Granger, E. Payen and K. Philippot, *New J. Chem.*, 2006, **30**, 1214–1219.
36. L. Barthe, A. Denicourt-Nowicki, A. Roucoux, K. Philippot, B. Chaudret and M. Hemati, *Catal. Commun.*, 2009, **10**, 1235–1239.
37. S. Roy and M. A. Pericas, *Org. Biomol. Chem.*, 2009, 7, 2669–2677.
38. K. V. S. Ranganath and F. Glorius, *Catal. Sci. Technol.*, 2011, **1**, 13–22.
39. S. Jansat, M. Gomez, K. Philippot, G. Muller, E. Guiu, C. Claver, S. Castillon and B. Chaudret, *J. Am. Chem. Soc.*, 2004, **126**, 1592–1593.
40. S. Jansat, D. Picurelli, K. Pelzer, K. Philippot, M. Gomez, G. Muller, P. Lecante and B. Chaudret, *New J. Chem.*, 2006, **30**, 115–122.
41. A. Gual, M. R. Axet, K. Philippot, B. Chaudret, A. Denicourt-Nowicki, A. Roucoux, S. Castillon and C. Claver, *Chem. Commun.*, 2008, 2759–2761.

42. A. Gual, C. Godard, K. Philippot, B. Chaudret, A. Denicourt-Nowicki, A. Roucoux, S. Castillón and C. Claver, *ChemSusChem*, 2009, **2**, 769–779.
43. E. Guyonnet Bilé, A. Denicourt-Nowicki, R. Sassine, P. Beaunier, F. Launay and A. Roucoux, *ChemSusChem*, 2010, **3**, 1276–1279.
44. E. Guyonnet Bilé, E. Cortelazzo-Polisini, A. Denicourt-Nowicki, R. Sassine, F. Launay and A. Roucoux, *ChemSusChem*, 2012, **5**, 91–101.
45. V. Mévellec, C. Mattioda, J. Schulz, J.-P. Rolland and A. Roucoux, *J. Catal.*, 2004, **225**, 1–6.
46. F. Glorius, *Org. Biomol. Chem.*, 2005, **3**, 4171–4175.
47. Y.-G. Zhou, *Acc. Chem. Res.*, 2007, **40**, 1357–1366.
48. K. Nasar, F. Fache, M. Lemaire, J. C. Beziat, M. Besson and P. Gallezot, *J. Mol. Catal.*, 1994, **87**, 107–115.
49. F. Fache, S. Lehuede and M. Lemaire, *Tetrahedron Lett.*, 1995, **36**, 885–888.
50. M. E. Labib, *Colloids Surf.*, 1988, **29**, 293–304.
51. J. D. Aiken, Y. Lin and R. G. Finke, *J. Mol. Catal. A: Chem.*, 1996, **114**, 29–51.

CHAPTER 7

Pd Nanoparticles in C–C Coupling Reactions

DENNIS B. PACARDO AND MARC R. KNECHT*

Department of Chemistry, University of Miami, 1301 Memorial Drive, Coral Gables, Florida 33146
*Email: knecht@miami.edu

7.1 Introduction

Transition metals have played an important role in synthetic chemistry as they have the unique capacity of activating various organic compounds towards chemical transformation and/or generation of new bonds. As such, organic ligand-stabilized metal catalysts based on Mg,[1–4] Fe,[5–9] Ni,[10–16] Au,[17–23] Ag,[24–30] Pd,[31–42] and Pt[43–49] have been synthesized to afford industrially important transformations such as carbon–carbon (C–C) coupling reactions, alkylation, hydrogenation, oxygen and nitro-group reduction, and carbon dioxide oxidation. These reactions, however, typically involve the use of organic solvents, high temperatures, and high metal loadings. Furthermore, these processes also produce toxic by-products and are becoming economically unsustainable and environmentally unviable.[50,51] As such, new processes are being developed to reduce the energy consumption and ecological impact of these reactions that are essential to maintain current technological advances.[52]

One reaction that has garnered significant attention over the past decade is the C–C coupling reaction using Pd-based catalysts due to its applications in pharmaceutical, natural products, and materials synthesis.[53–56] The coupling reaction, shown in Scheme 7.1, typically involves the formation of a

RSC Catalysis Series No. 17
Metal Nanoparticles for Catalysis: Advances and Applications
Edited by Franklin Tao
© The Royal Society of Chemistry 2014
Published by the Royal Society of Chemistry, www.rsc.org

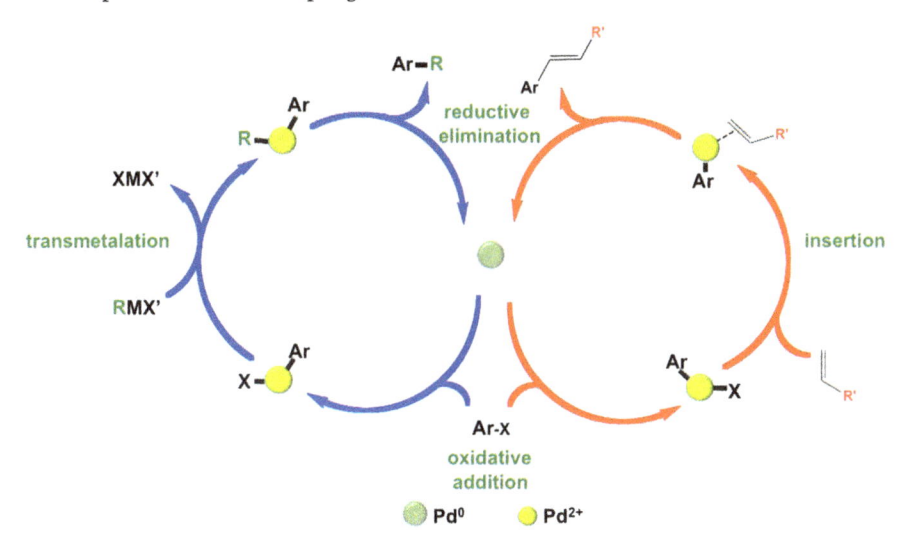

Scheme 7.1 General mechanism of Pd-catalyzed C–C coupling.

new C–C bond between an electrophile (usually an aryl or vinyl halide) and a nucleophile (either an olefin or organometallic reagent), using zerovalent Pd metal to drive the reaction. The first step in the coupling is oxidative addition wherein Pd^0 reacts with the electrophile by insertion between the C–X bond to form an organopalladate complex, oxidizing the Pd^0 to Pd^{2+}. The nucleophilic component will then attack the organopalladate complex either through transmetalation or alkene insertion, positioning the two carbon groups closer to each other on the metal surface. After reductive elimination, the new C–C bond is formed, the product is released, and the Pd^{2+} is reduced back to Pd^0.

The modern use of Pd catalysts for C–C coupling reactions was pioneered by Heck with the publication of his work in 1972 on Pd-catalyzed arylation of olefins to produce substituted olefinic compounds.[57] In this reaction, catalytic quantities of Pd were used to form new C–C bonds between aromatic and olefinic compounds producing a standard protocol for metal-catalyzed coupling reactions.[57] For many years thereafter, several Pd-catalyzed coupling reactions (Scheme 7.2)[52,58] were developed by different groups by modifying the nucleophilic component of the reaction. Sonogashira used alkynes instead of alkenes for cross-coupling reactions with aryl halides.[59] Stille, on the other hand, initiated the use of organometallic reagents as the nucleophile by using organostannane compounds for the Pd-catalyzed coupling reaction with aryl halides;[60] however, the toxic nature of tin compounds limited their large-scale use. As an alternative to tin reagents, less toxic organometallic compounds such as organozinc was used by Negishi in the coupling reaction generating a mild and highly selective transmetalation agent that can tolerate a variety of functional groups.[61] Kumada used organomagnesium reagents for transmetalation,[62] although the highly reactive

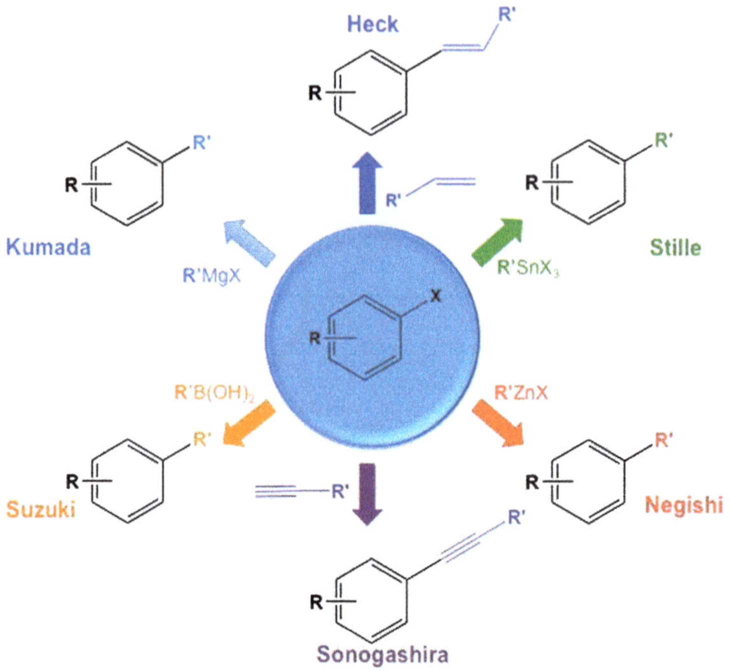

Scheme 7.2 Select Pd-catalyzed C–C coupling reactions.

Grignard reagent also restricted their applications to a selection of functionalities. As an alternative to highly toxic tin and highly reactive Grignard transmetalation agents, Suzuki developed boron-based organometallic reagents for coupling reactions with aryl halides,[63,64] generating a non-toxic and practical alternative that can be used under mild reaction conditions. In this reaction, activation of organoboron by a base is required to produce the boronate intermediate that initiates transmetalation, where a wide variety of functional groups can be employed.

The importance of Pd-catalyzed C–C coupling reactions in organic and material synthesis, as well as industrial applications, was recognized by the 2010 Nobel prize in chemistry for the pioneering works of Heck, Suzuki, and Negishi.[65,66] In more recent times, technological advancements have led to novel synthetic approaches for producing highly active Pd catalysts by using different types of organic-based stabilizers that can control the reactivity and selectivity of the metal catalyst. Furthermore, these new Pd materials have unique characteristics and properties that enable their use in catalytic transformations for a wide array of applications in material,[67–69] natural products,[70–75] and pharmaceutical synthesis.[54,76,77]

With the advent of nanotechnology, new synthetic strategies for the generation of Pd nanocatalysts have been developed with particular emphasis on the control of the particle size, shape, and functionality.[78–83] Since nanomaterials encompass the gap between the atomic level and bulk materials, interesting physical, chemical, electronic, and optical properties can be obtained,[81,84,85] thereby generating a new class of structures for specific applications. Modern synthetic methods allow for the production of nano-sized catalysts with optimized surface-to-volume ratios, allowing for more efficient catalytic reactivity as compared to traditional small molecule materials. The use of Pd nanoparticles for C–C catalysis marks a transition from using traditional organic solvent-based and high temperature reactions toward more ambient conditions such as aqueous solutions, mild temperatures, and low catalyst loadings.[86-88] This is achieved by employing highly functionalized stabilizers such as dendrimers,[33,86,89-93] proteins,[31] peptides,[87,88,94] and other small molecules.[95-98] These materials allow for size-specific interactions, as well as control over the nanoparticle shape, thereby generating solution stability, but at the same time exposing active sites for interaction with substrates.

This chapter will cover select recent advances regarding the use of Pd nanoparticles for C–C coupling reactions. Due to the large volume of work done in this area, this report will mostly focus on homogeneous Pd nanoparticles; however, it will not discuss bimetallic catalysts for coupling reactions. Emphasis will be given to Pd nanoparticles used in energy-efficient and environmentally benign catalytic reactions, as well as information regarding their catalytic mechanisms.

7.2 Synthetic Scheme for the Fabrication of Pd Nanoparticles

There are several common methods employed for the synthesis of Pd nanoparticles such as chemical reduction,[37,38,86,94] vapor phase deposition,[99-101] laser ablation,[102-105] sonochemical reduction,[106-110] electrochemical methods,[39,111-113] seed-mediated approaches,[35,114] and microwave reduction.[40,115-117] For this review, however, we will focus on the chemical reduction approach, which typically involves the use of metal salts as a precursor, organic and/or biological ligands as surface passivants, and a reductant to reduce the metal ions to the zerovalent state. Using this method, the metal precursor, which is usually K_2PdCl_4[88] or K_2PdCl_6,[118] is dissolved in aqueous solution and added to a separate ligand solution. In this process, the interaction of metal ions and ligands can lead to a ligand-to-metal charge transfer band in UV-vis analysis.[88,119,120] Upon addition of a reducing agent, typically $NaBH_4$, the metal ions are reduced to form metallic Pd nanoparticles as evidenced by the formation of a dark brown solution with a broad absorbance spectrum.[86,88,94] During the reduction process, several nucleation sites are generated, thus resulting in rapidly growing nanoparticles that can ripen and precipitate as bulk metal in the absence of stabilizing agents. As such, the ligands present in the reaction prevent particle agglomeration and control the size and shape of the resulting nanoparticles. By design, these stabilizing agents do not completely coat the nanoparticles, which, therefore, allows for surface accessibility by the substrates in the reaction; however, the porous network of ligands can also be designed to control access to the metal surface to develop catalytic selectivity. Commonly used stabilizers in Pd nanoparticle synthesis are amine-,[121–123] thiol-,[124–127] and phosphorus-based ligands,[128–130] polymers,[34,38,96,97] dendrimers,[86,89,90,92,119] surfactants,[131–133] ionic liquids,[134–137] and biomacromolecules such as peptides.[87,88,94]

While this method has produced a variety of homogeneous nanocatalysts for different C–C coupling reactions, their wide scale application in industry has not yet been realized due to the difficulty in recovering and recycling the catalyst. In this regard, attaching the nanoparticles to a solid support offers an alternative process for easier catalyst recovery and recyclability. Although there are many advantages for homogeneous catalysts such as efficiency and selectivity,[138–140] supported nanocatalysts can be more practical, easily separated, and cost-effective in large-scale use. They also greatly reduce the amount of metal impurities in the isolated products, thus lowering the degree of separation required post reaction. Solid supports such as carbon nanotubes,[141–144] metal oxides,[145–148] silica,[149–151] alumina,[152–154] zeolites,[155–157] clays,[158] polymer coatings,[159–161] molecular sieves,[162] aerogels,[163,164] and naturally occurring porous diatomite[165] have been used to immobilize Pd nanoparticles. The synthesis of a solid-supported nanocatalyst generally involves the adsorption, grafting, or attachment of the stabilizing agent to the solid material, after which Pd ions are introduced to interact with the ligands, followed by reduction. The solid materials can be easily separated *via* filtration, washed to remove excess ligands, and redispersed in various solvents for catalytic application.

7.3 C–C Coupling Reaction Mechanism for Pd Nanocatalysts

Nanoparticles have gained significant attention due to their enhanced surface-to-volume ratio generating more efficient exposure of the catalytic materials. In C–C coupling, the nature of the nanocatalyst and the actual catalytic mechanism remain a subject of intense debate. The colloidal nature of Pd nanoparticles suggests that they can act as homogeneous or heterogeneous catalysts in solution and, as such, two catalytic mechanisms are possible. On one side, researchers suggest that the reaction occurs directly at the surface of the nanoparticles without inducing changes in the material structure, size, and shape, indicating a purely heterogeneous nature of the catalyst.[52,166] On the other hand, substrates could interact with the active metal surface and abstract atoms from the nanoparticle surface, from which the catalytic cycle could then occur in solution. This homogeneous process, often referred to as the atom-leaching mechanism, results in the nanoparticles serving as a reservoir of active catalytic species.[52,167] In this process,

Pd Nanoparticles in C–C Coupling Reactions 117

Pd atoms are abstracted from the surface of the nanoparticles during oxidative addition.

A thorough study of the leaching mechanism of Pd nanoparticles in coupling reactions was investigated using a 2-compartment reactor, shown in Figure 7.1(a).[168,169] This system separates the two compartments with an

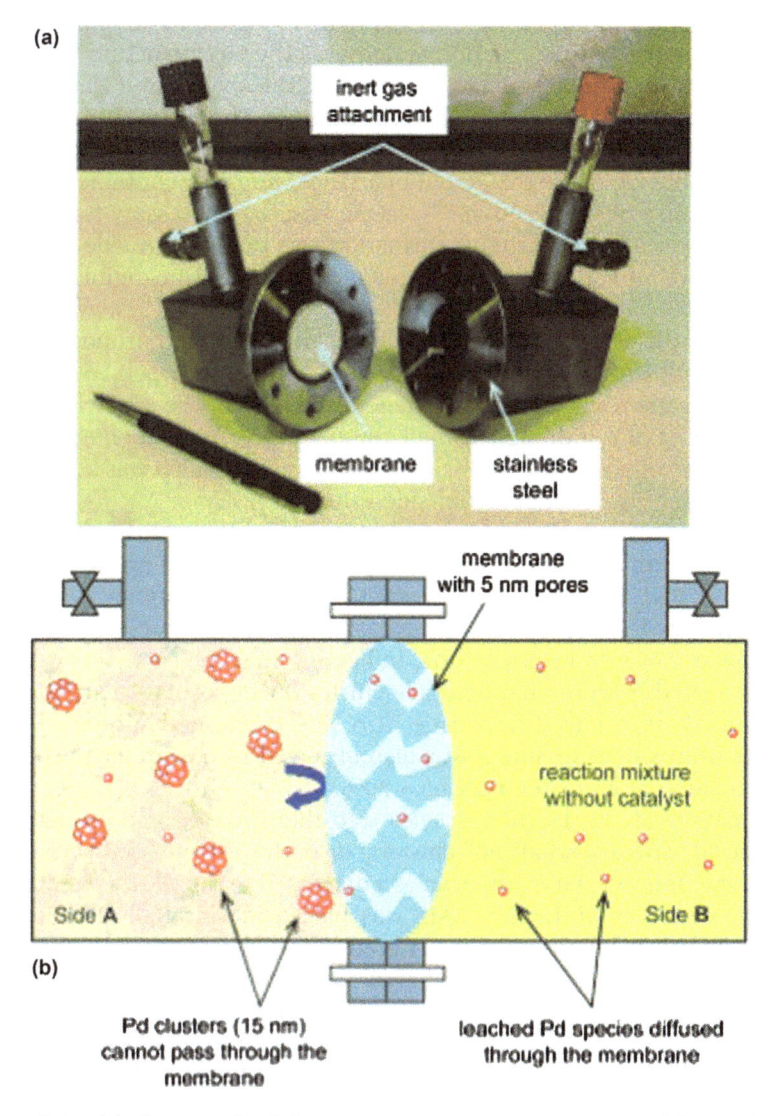

Figure 7.1 (a) Photograph of the two-component reactor separated by an alumina membrane and (b) schematic diagram of the reactor showing the cluster-exclusion process.[168]
Reproduced with permission from M. B. Thathagar, J. E. ten Elshof and G. Rothenberg, *Angew. Chem., Int. Ed.*, 2006, **45**, 2886–2890. Copyright © 2006 WILEY-VCH Verlag GmbH & Co. KGaA, Weinheim.

alumina membrane that only allows for passage of metal species of <5 nm.[169] As such, the nanocluster exclusion process was achieved in which only small Pd clusters, Pd⁰ atoms, and/or Pd²⁺ ions can traverse the alumina membrane, thereby monitoring and analysis of the leached species was possible. Pd nanoparticles were prepared by mixing the metal precursor, Pd(OAc)$_2$, with tetraoctylammonium glycolate (TOAG) acting as both a stabilizer and a reducing agent to produce ~14 nm particles.[169] The C–C coupling initially studied using the unique reaction system was the Heck reaction between *n*-butyl acrylate and iodobenzene using DMF as the solvent and NaOAc as the base to generate *n*-butylcinnamate as the product. The two starting materials were added to both sides of the reaction chamber, except the solid base was placed only in side B and the Pd nanoparticles were fully contained in side A at the initiation of the reaction.[169] NaOAc base was necessary for the Heck coupling to occur so no product formation was expected in side A. Control studies showed that the solubility of NaOAc in DMF was negligible thus base diffusion from side B to side A was not observed.[168] In this system, however, diffusion of the active Pd species to side B through the mesoporous membrane will allow the reaction to proceed. To this end, Heck coupling was observed for the reaction at side B generating 88% of the *n*-butylcinnamate product after 120 h, although a small amount of the product (4.9%) was also formed in the opposite side of the reaction chamber, which was attributed to the diffusion of the product from side B to side A.[169]

To further verify the leaching of the Pd species through the membrane, another reaction was performed using the Suzuki coupling between phenylboronic acid and *p*-iodotoluene with the same conditions as the Heck coupling. Approximately 50% of the total anticipated product was formed on side B after 250 h of reaction time, indicating that the catalytic Pd species was abstracted from the nanoparticle cluster during oxidative addition and subsequently transferred from side A to side B; however, diffusion of the product from side B to side A was also observed to be similar to the Heck reaction albeit at a greater yield of ~20%.[169]

The nature of the leached species was then investigated to determine whether Pd⁰ species, small Pd⁰ clusters, or both were released through the membrane. Initially, Pd nanoparticles were added to the DMF solvent in side A of the chamber while only DMF was placed in side B.[169] The leaching process was then monitored for 144 h at 100 °C by taking small aliquots of the reaction mixture in side B and analyzing the particles using inductively coupled plasma (ICP) and transmission electron microscopy (TEM).[169] TEM results indicated that small Pd clusters do not diffuse through the membrane, whereas only 20% of the total Pd was found on side B after 144 h, as determined by ICP analysis.[169] Only the Pd⁰ species was observed in side B, forming irregular shaped aggregates stabilized by the presence of TOAG ions that were released from the nanoparticle surface and diffused through the membrane due to concentration gradient.[169] This was verified by a UV-vis experiment using samples from both sides of the reaction chamber, wherein only a broad spectrum was observed for side B indicating the presence of Pd aggregates, while an absorbance shoulder at 390 nm was obtained from samples in side A suggesting the presence of Pd clusters in solution.[169]

Another experiment was performed to probe the leaching of Pd^{2+} as a result of oxidative addition in the presence of aryl halides. To accomplish this study, iodobenzene was dissolved in DMF and placed on both sides of the reaction chamber with Pd nanoparticles added only to side A.[169] After 24 h, aliquots were taken from both reactions wherein it was observed that the black solution in side A turned into a dark reddish color while the reaction in side B turned dark red after another 24 h indicating the migration of Pd species from side A to side B. As such, leaching of catalytically active Pd^0 atoms from the nanoparticles was observed in the presence of an oxidizing aryl iodide substrate, which supports an atom-leaching mechanism during oxidative addition for Pd nanoparticles.[169]

From these studies, three different scenarios could occur for the transfer of Pd species from one side of the reactor to the other through an alumina membrane (Figure 7.2).[169] The first mechanism involves the fragmentation of Pd nanoparticles into smaller nanoclusters, which can then pass through the porous membrane; however, the results indicate that this process was not occurring as there were no small Pd nanoclusters observed in side B of the reactor. Although the fragmentation of Pd nanoparticles results in the formation of small Pd clusters, these do not appear to be able to pass through the alumina membrane. In the second mechanism, Pd^0 atoms were leached from the nanoparticles under non-oxidizing conditions and then transferred to side B through the porous membrane as indicated by the experimental results. In contrast, the third mechanism involved the leaching of Pd species from the nanoparticles' surface under oxidizing conditions due to the presence of aryl halides in solution. In this process, the oxidative attack of the aryl halides resulted in leaching of Pd species and formation of Pd^{2+} complexes that were then transferred to side B. These Pd^{2+} complexes can then pass through the porous membrane and continue with the catalytic cycle in side B if the coupling components were present. To this end, transfer of Pd species from side A to side B and their corresponding catalytic activity can be attributed to the leached Pd^0 atoms and/or the Pd^{2+} complex.

In another study by Niu *et al.*, the design of Pd nanoparticles was employed to probe the catalytic mechanism promoted by oxidative addition in the presence of aryl halide substrates.[170] Observations of tunable characteristics of the nanocatalysts, such as crystallinity, composition, and surface structure, were used for qualitative scrutiny of the reaction mechanism by employing two differently synthesized particles: 1. poorly crystalline (disordered) Pd nanoparticles and 2. bimetallic PdAu nanocrystals.[170] Using highly disordered nanoparticles in the coupling reaction, the overall change in the structure and crystallinity of the particles after the reaction was monitored from which information regarding the mechanism could be obtained. Meanwhile, using bimetallic PdAu nanoparticles provided evidence concerning the catalytic mechanism by the formation of Pd nanostructures *via* phase segregation in the reaction as a result of leaching during oxidative addition.[170]

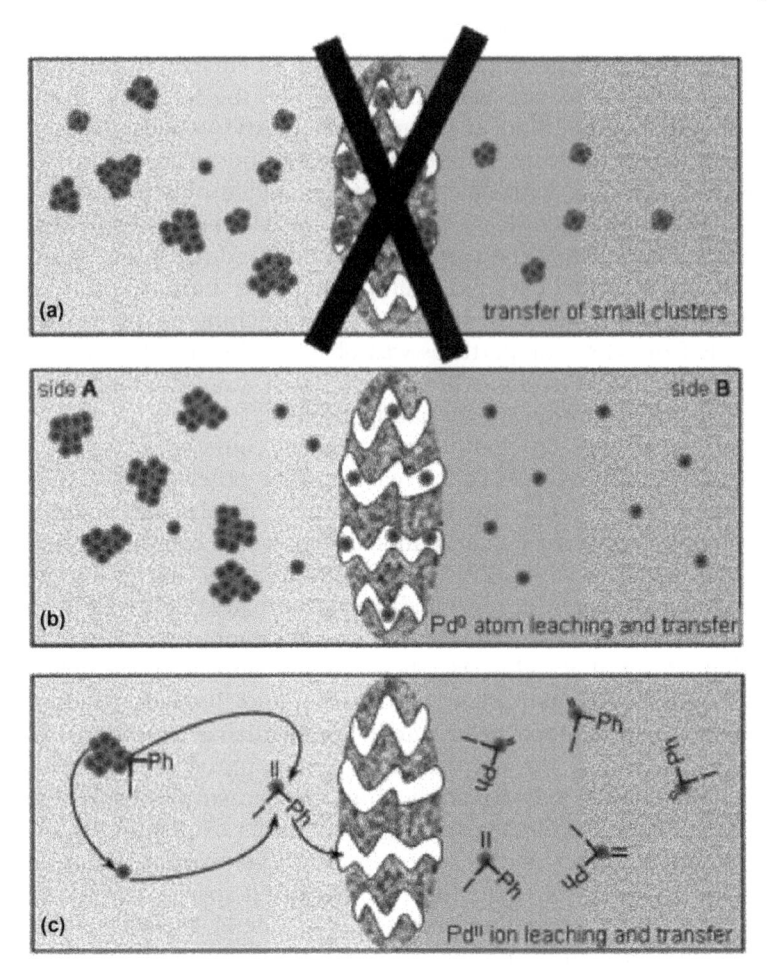

Figure 7.2 Possible transfer mechanisms of leached Pd species in a two-component reactor. (a) Nanoparticle fragmentation and transfer of small clusters does not occur, (b) leaching of Pd^0 atoms and ensuing transfer, and (c) formation of Pd^{2+} complexes during oxidative addition and subsequent leaching and transfer.[169]
Reproduced with permission from A. V. Gaikwad, A. Holuigue, M. B. Thathagar, J. E. ten Elshof and G. Rothenberg, *Chem.-Eur. J.*, 2007, **13**, 6908–6913. Copyright © 2007 WILEY-VCH Verlag GmbH & Co. KGaA, Weinheim.

As illustrated in Scheme 7.3, these nanoparticles were employed in the Suzuki coupling reaction between iodoanisole and phenylboronic acid using Na$_2$CO$_3$ as a base. In the case of the poorly crystalline Pd nanoparticles, HRTEM analysis, shown in Figure 7.3(g) and (h), after the reaction illustrated sintering and crystallinity enhancements of the particles, which were very different from the original nanocatalysts (Figure 7.3(d)).[170] High resolution

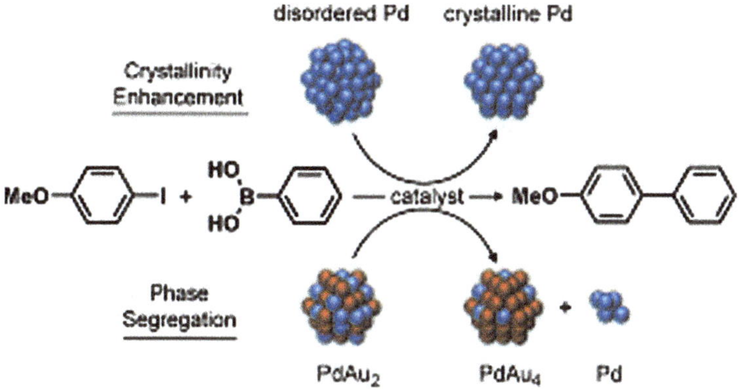

Pd Nanoparticles in C–C Coupling Reactions 121

Scheme 7.3 Representation of changes in crystallinity and morphology of Pd and Au-Pd nanoparticles as a result of the atom-leaching mechanism in Suzuki coupling.[170]
Reproduced with permission from Z. Niu, Q. Peng, Z. Zhuang, W. He and Y. Li, *Chem.–Eur. J.*, 2012, **18**, 9813–9817. Copyright © 2012 WILEY-VCH Verlag GmbH & Co. KGaA, Weinheim.

TEM (HRTEM) and selected area electron diffraction (SAED) experiments still did show some poorly crystalline nanoparticles, suggesting that not all of the materials were involved in the catalytic cycle. The crystalline nanoparticles ranged from spherical to irregularly shaped structures with sizes between 3–22 nm, where SAED and energy-dispersive X-ray (EDX) analyses confirmed the formation of pure Pd nanocrystals as shown in Figure 7.3(f) and (j), respectively.[170]

To further probe the improved crystal structure of the Pd nanoparticles in the reaction, several control experiments were performed wherein the results indicated that the presence of aryl halides was responsible for crystallinity enhancement. In this process, the aryl halides abstract Pd atoms from the surface of the disordered nanoparticles during oxidative addition, generating nanoparticles with an enhanced crystalline structure. To this end, when aryl halides were removed from the reaction system, retention of poor crystallinity was observed for the Pd nanoparticles.[170] In a separate study, in a reaction system where only the aryl halide was present in the absence of the transmetalation reagent, nanoparticle crystallinity enhancement was again observed due to Pd leaching during oxidative addition.[170] These results suggest that the catalytic mechanism driven by the Pd nanoparticles follows an atom-leaching process wherein the aryl halides oxidatively abstract Pd^0 atoms. Following this, the leached complexes undergo transmetalation and reductive elimination, releasing the coupling product and free Pd^0 atoms in solution. The Pd^0 atoms can then undergo nucleation and growth to produce crystalline nanoparticles with no size and shape control due to the absence of stabilizing agents in the reaction system.[170] On the other hand, if the catalytic mechanism was exclusively a surface-based reaction, only minimal to no changes in nanoparticle morphology should be

122   Chapter 7

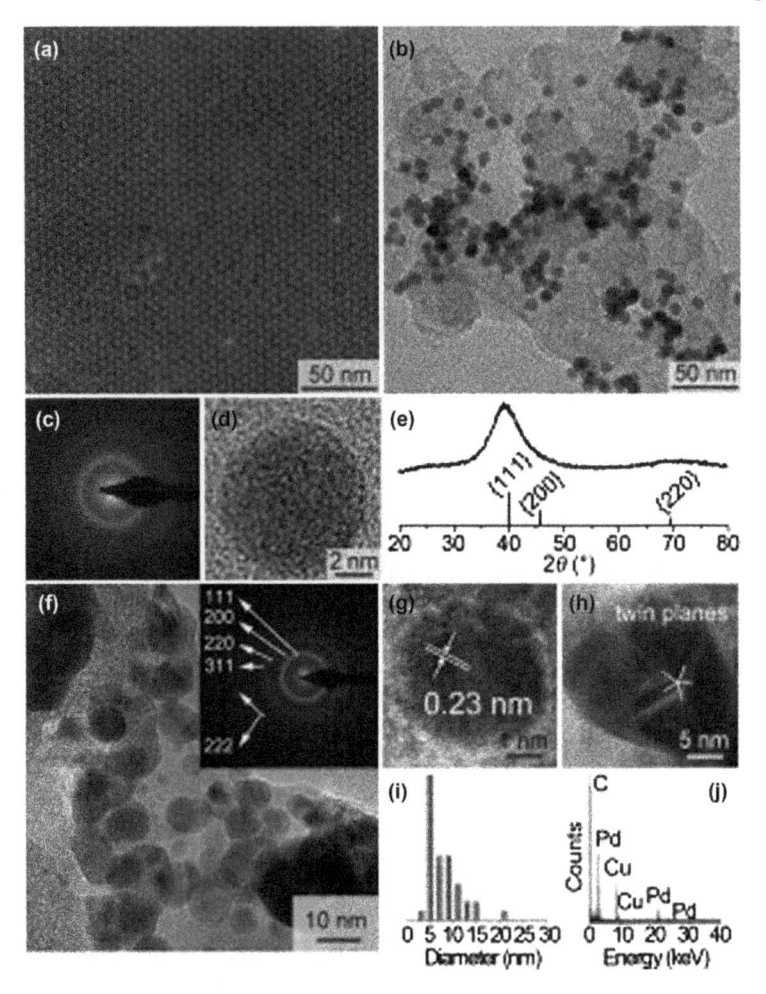

Figure 7.3 (a, b) TEM images, (c) SAED pattern, (d) HRTEM image, and (e) XRD analysis of the Pd nanoparticles before catalysis. (f) TEM image and SAED pattern (inset), (g and h) HRTEM images, (i) size distribution histogram, and (j) EDX spectrum of Pd nanoparticles after catalysis.[170] Reproduced with permission from Z. Niu, Q. Peng, Z. Zhuang, W. He and Y. Li, *Chem.–Eur. J.*, 2012, **18**, 9813–9817. Copyright © 2012 WILEY-VCH Verlag GmbH & Co. KGaA, Weinheim.

observed, with no formation of irregular shaped, crystalline Pd nanoparticles.[170]

Inductively coupled plasma-mass spectrometry (ICP-MS) analysis was also performed to measure the concentration of soluble Pd species in the reaction solution with and without the aryl halide. Using the poorly crystalline Pd nanoparticles as the catalyst, 60 ppb Pd species were determined to be in solution for the reaction containing 4-iodoanisole; however, only 9 ppb Pd species were found in the reaction without the aryl halide substrate.[170] These results suggest that oxidative addition in the presence of aryl halides causes abstraction of Pd species from the nanoparticle surface, thereby increasing the amount of Pd released to solution. On the other hand, only minimal leaching was observed without aryl halides, which can be attributed to the inherent instability of the nanoparticles.

While the crystallinity results suggest leaching of Pd atoms during oxidative addition, further evidence was obtained using a bimetallic Pd-Au nanoparticle system. In this case, co-reduction of Pd and Au metal precursors was achieved by adding oleylamine at 230 °C to produce spherical nanoparticles with a ~9 nm average diameter size.[170] HRTEM and EDX mapping analysis revealed that the particles were composed of a 31 : 69 Pd-Au ratio with lattice spacings between the expected value for face-centered cubic (fcc) Pd and Au. Catalytic reactions between 4-iodoanisole and phenylboronic acid were used to probe the leaching process driven by the Pd-Au nanoparticles. TEM analysis of the bimetallic nanocatalyst after the reaction revealed that the shape and size of the particles remained the same; however, EDX results showed the Pd-Au atomic ratio to be 19 : 81, indicating a decreased amount of Pd atoms in the nanoparticles. Interestingly, highly crystalline Pd nanoparticles ~5 nm in size were formed in this system, as confirmed by TEM and EDX, suggesting that the Pd atoms were selectively abstracted from the bimetallic nanoparticle during the coupling reaction.[170] To verify the role of oxidative addition by the aryl halide in the leaching process, control reactions were done without 4-iodoanisole, wherein the EDX analysis of the materials after the process displayed a Pd-Au ratio of 27 : 73, which was similar to the original bimetallic ratio.[170] These results indicate that the aryl halides leached the Pd^0 species during the oxidative addition process in the catalytic reaction.

Overall, the results of these studies indicated that the atom-leaching mechanism promoted by oxidative addition of the aryl halide was the likely catalytic process for Pd nanoparticles in the Suzuki coupling reaction. These results offer direct evidence using various analytical techniques and different types of Pd nanoparticles wherein atom abstraction during the catalytic cycle was both qualitatively and quantitatively analyzed. Aside from indirect processes such as monitoring the TOF values, these studies provided physical validation of an atom-leaching process for Pd nanoparticles.

The other side of the debate regarding the catalytic mechanism of Pd nanoparticles centers on the argument that the entire reaction occurs directly on the surface of the particles in a straightforward heterogeneous reaction. To this end, evidence of surface-based Suzuki coupling catalyzed by Pd nanoparticles was examined by monitoring the local coordination environment of the metal in real time using a recirculating reactor set up for operando fluorescence X-ray absorption spectroscopy (XAS) illustrated in Figure 7.4(a).[171] Polymer-stabilized Pd nanoparticles were prepared by refluxing H_2PdCl_4 and polyvinylpyrrolidone (PVP) with water and ethanol to generate 1.8-4.0 nm Pd^0 particles.[171] Suzuki coupling between iodoanisole and phenylboronic acid was monitored for these studies wherein the

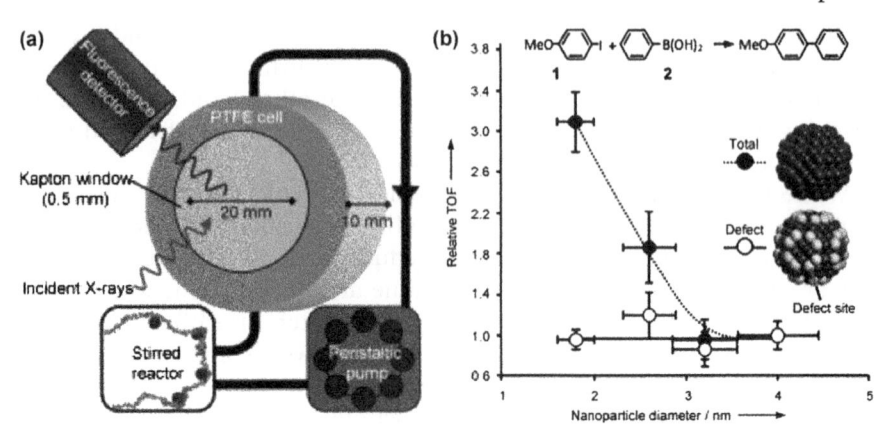

Figure 7.4 (a) Schematic diagram of the recirculating reactor setup for XAS, (b) normalized TOF data for Suzuki coupling reaction showing rate dependence on correct active sites and not on particle size.[171] Reproduced with permission from P. J. Ellis, I. J. S. Fairlamb, S. F. J. Hackett K. Wilson and A. F. Lee, *Angew. Chem., Int. Ed.,* 2010, 49, 1820–1824. Copyright © 2010 WILEY-VCH Verlag GmbH & Co. KGaA, Weinheim.

catalytic activity was expressed as turnover frequency (TOF) values. When the reactivity of the nanocatalyst was monitored with respect to particle diameter, the TOF values decreased as the size of the nanoparticles increased as expected; however, when the TOF values were normalized with respect to the edge and vertex atoms on the surface of the nanoparticles, statistically similar TOF values were obtained regardless of particle size (Figure 7.4(b)).[171] This suggests that the reactivity of Pd nanoparticles is associated with the defect atoms on the particle surface.

As such, monitoring of the possible leached species was performed using XAS, which can sensitively measure the average Pd–Pd coordination number (CN), as well as provide information concerning the particle size and morphology. From the extended X-ray absorption fine structure (EXAFS) intensities for simulated 1.8 nm Pd nanoparticles, a Pd–Pd CN value of 9.46 was calculated, which is in agreement with the observed value of 9.58.[171] Leached Pd species, however, were not observed as there was neither sintering nor dissolution of Pd atoms since there is no change in the CN, which supported the stability of the nanoparticles in the reaction. HRTEM analysis before and after the reaction also illustrated no change in the size and shape of the nanoparticles.[171] Simulation of analogous 1.8 nm particles with 48 Pd defect atoms removed into the solution would lead to a drop of CN to 7.23; however, this phenomenon was not observed, which indicated that there was no solubilization and homogeneous contribution of catalytically active Pd species.[171]

To further probe the system with regards to leached Pd atoms, trace amounts of dissolved Pd(OAc)$_2$ were added to the reaction, which was expected to enhance the catalytic activity of the particles; however, no such enhancement was observed.[171] On the other hand, adding Hg as a selective catalyst poison immediately inhibits the catalytic activity, suggesting that a surface-based process was occurring for the PVP-Pd nanoparticles. Analysis of the nanocatalyst using *ex situ* X-ray photoelectron spectroscopy (XPS) showed a 1 : 1 ratio between surface Pd and Hg atoms indicating full coverage of the active catalytic sites.[171] When the reaction was deliberately spiked with Pd(OAc)$_2$ atoms after addition of Hg, no additional catalytic activity was observed. These results indicated that the catalytic activity for PVP-Pd nanoparticles happens at the metal surface, while the additional Pd(OAc)$_2$ species were not catalytically reactive.

One final method was employed to study this leaching process. For this, thiolated porous silica was used to probe the nature of the active species wherein the sulfur-containing substrate should interact with soluble Pd species abstracted from the nanoparticles and prevent catalysis if the leached atoms were responsible for the reaction. Using these functionalized silica materials, the catalytic reaction of the nanocatalyst decreased; however, this was due to nanoparticle entrapment inside the pores of the silica beads and not due to the sequestration of Pd species by thiols.[171] To this end, employing unfunctionalized silica with identical pore sizes over the thiolated silica resulted in the same catalytic rate decrease, indicating that the coupling reaction was inhibited by the nanocatalyst being trapped within the porous material, as confirmed *via* TEM, XAS, and XPS analyses.[171] Kinetic studies comparing the reactivities of PVP-Pd nanoparticles and Pd(OAc)$_2$ showed a first-order dependence on Pd concentration for the former and negative-order dependence on the latter, suggesting that a surface-based catalytic reaction was at play.[171] Furthermore, if the reaction was based on the dissolved or leached Pd species, a decrease in coupling rate would be observed as the amount of Pd loading was increased; however, the opposite results were obtained wherein increases in initial rates were directly proportional to the increased Pd concentration. Overall, the results of these studies offered a strong argument that the catalytic mechanism for C–C coupling catalyzed by PVP-capped Pd nanoparticles happened at the metal surface in a heterogeneous fashion.

The studies of the reaction mechanism of Pd nanoparticles provide two different theories regarding the nature of active catalytic species wherein the reaction either proceeds exclusively in a heterogeneous or homogeneous manner. In purely heterogeneous catalysis, the transformation reaction occurred at the surface of metal nanoparticles and, as such, changes in the morphology, crystallinity, and size of the nanocatalyst were minimal. On the other hand, in a homogeneous catalytic mechanism, the active species was abstracted from the nanoparticle surface during the oxidative addition resulting in morphological and crystallinity changes, as well as Pd black formation. These theories presented conflicting explanations regarding the catalytic mechanism of Pd nanoparticles in a C–C coupling reaction. Although both of these explanations are plausible, further studies are required to fully elucidate the catalytic reactivity of Pd nanoparticles for C–C coupling.

Understanding the actual reaction mechanism can provide valuable information for the design of new synthetic routes for the production of more efficient nanocatalysts.

7.4 Pd Nanoparticles in the Stille Coupling Reaction

The Stille coupling process was first reported in 1978 as a reaction between acid chlorides and organostannane reagents in the synthesis of ketones, catalyzed by phosphine-stabilized Pd catalysts.[60] Since then, various modifications to the phosphine ligands have been reported for this reaction,[172–174] which takes advantage of the air- and moisture-stable nucleophilic partner to afford C–C bond formation with substrates containing different functional groups. Typically, this transformation is performed at elevated temperatures (70–120 °C), in organic solvents, and with catalyst loadings of >1 mol% Pd, which can be reduced *in situ*.[175] In the age of nanotechnology, however, milder reaction conditions are employed due to the nature of the ligand employed in material production. For example, using dendritic materials for the synthesis of Pd nanoparticles, Stille coupling was performed in water at room temperature with only 0.10 mol% Pd as the catalyst.[86] In this system, the Crooks group used hydroxyl-terminated fourth generation PAMAM dendrimers to encapsulate Pd nanoparticles using a 40 : 1 molar ratio of the $PdCl_4^-$ metal precursor and the dendrimer (Figure 7.5(a)).[86] Upon reduction, ~40 Pd atoms were contained within the dendrimer, producing a nanoparticle with an average size of 1.7 ± 0.4 nm.[86] These dendrimer-encapsulated Pd nanoparticles (Pd DENs) were employed for Stille coupling between 4-iodobenzoic acid and phenyltin trichloride, generating a new C–C bond in the 4-biphenylcarboxylic acid product, as shown in Figure 7.5(b). Quantitative yields were reported after a 15 h reaction time with constant stirring at room temperature. Aryl halides with different halogen species and functional groups were also employed in the Stille reaction using the Pd DENs with results corresponding to the expected

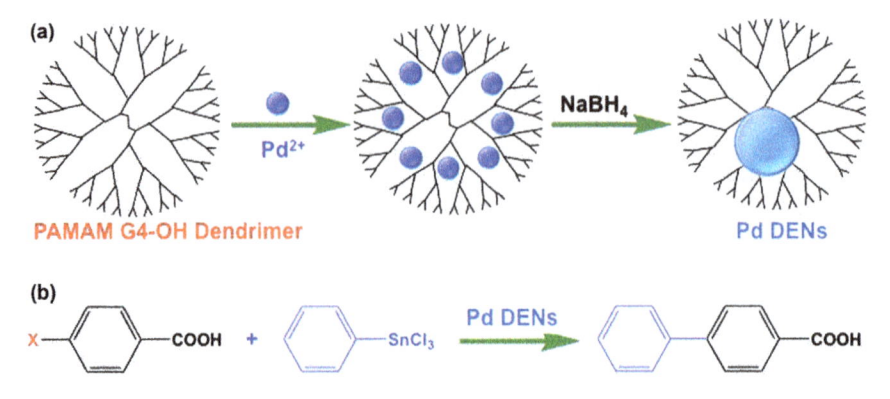

Figure 7.5 (a) Synthesis of dendrimer-encapsulated Pd nanoparticles, (b) Stille coupling reaction using Pd DENs.

reactivity of each substrate. To this end, the aryl halide substrates with electron-withdrawing (–COOH) and iodide groups were more reactive than the corresponding bromide-substituted and phenolic substrates. TOF values of 2000 ± 200 mol product (mol Pd × h)$^{-1}$ for 4-iodobenzoic acid were obtained, suggesting a highly efficient nanocatalyst.[86]

An interesting control reaction was performed using Pd nanoparticles generated without the DENs to evaluate the role of the dendritic cage in catalytic reactivity. As such, K_2PdCl_4 was reduced with $NaBH_4$ in the absence of the dendrimers, thus producing polydisperse particles that generated a TOF value slightly lower than that of Pd DENs (1800 mol product (mol Pd × h)$^{-1}$).[86] This result suggests that the size control provided by the DENs enables the catalyst to have a higher surface area than the particles without DENs to provide better catalytic reactivity.[86] TEM analyses of the Pd DENs before and after the catalytic reaction were performed, which showed a moderate increase in particle size and dispersity from 1.7 ± 0.4 nm to 2.7 ± 1.1 nm.[86] These results indicated partial aggregation of the Pd nanoparticles as a consequence of the catalytic reaction that was attributed to leaching of Pd atoms during the catalytic cycle. To confirm this, the Pd DENs were added to an identical reaction solution but without the substrates and allowed to proceed for 24 h. TEM analyses of the nanoparticles before and after the reaction showed statistically similar particle diameter indicating that the catalytic reaction was responsible for an increase in nanoparticle size.[86]

When the Pd DENs were employed in Stille couplings at elevated temperatures, catalyst aggregation was observed by the appearance of a black precipitate.[86] As such, product yields under these reaction conditions were lower than the reactions at room temperature, which could arise from dendrimer degradation. Finally, the recyclability of the Pd DENs was also evaluated using sequential Stille coupling between two different substrates.[86] For this, the second substrate was added after the reaction of the initial substrate was concluded. Specifically, a reaction solution containing 4-iodophenol and phenyltin trichloride was added to the Pd DENs and allowed to proceed for 15 h. After reaching completion, another reaction solution containing 4-iodobenzoic acid and phenyltin trichloride was added to the first mixture and allowed to proceed for 15 h such that in the end, two possible products can be observed if the Pd DENs remain active. Gas chromatography-mass spectrometry (GC-MS) analysis of the reaction demonstrated the quantitative formation of the two anticipated products (4-phenylphenol and 4-biphenylcarboxylic acid), indicating that the Pd DENs were still catalytically active after the initial cycle.[86] These studies showed that Pd DENs catalyzed Stille coupling under ambient reaction conditions, which could serve as a model for designing catalytic systems.

In a separate study, the effect that the dendrimers possessed over the catalyst reactivity and selectivity was studied by comparing different precatalysts, namely: Pd DENs, Pd(OAc)$_2$, and Pd^{2+}-PAMAM complexes.[90] In Stille coupling reactions employing the same system, which Crooks and coworkers employed at higher temperatures of 40 °C and 80 °C, results indicated that both Pd DENs and Pd(OAc)$_2$ have similar reactivity with quantitative product yields after 24 h; however, the unreduced Pd^{2+}-PAMAM complex only generated 18% product.[90] Furthermore, the DENs showed 100% selectivity for the conversion of 4-iodobenzoic acid to 4-biphenylbenzoic acid, while Pd(OAc)$_2$ generated both the anticipated Stille coupling product and the undesired homocoupling biphenyl. When using less water-soluble substrates, such as activated 4-iodoacetophenone, almost quantitative yields were obtained using the naked precatalyst at 40 °C, while a higher temperature of 80 °C was necessary for the DENs to generate 66% of the product with 100% selectivity. Similar trends were determined using aryl halides with different functional groups, indicating that the simplest catalyst, Pd(OAc)$_2$, is the fastest system, but it was not selective for generating C–C bonds.[90] When the same precatalysts were used in recycling experiments, the materials were inactive after the initial catalytic cycle; however, the DENs were able to complete several reaction cycles with slightly diminished reactivity after multiple reactions.[90] TEM images taken after the fourth catalytic cycle showed an increase in the Pd nanoparticle size from 1.6 ± 0.3 nm to 2.3 ± 0.6 nm as a result of the reaction.[90] These indicated that the dendrimer was able to retain most of the Pd species to prevent bulk Pd black formation, but that migration of Pd materials between the dendrimers occurred.

In contrast with the aqueous-based Pd DENs, organic solvents have also been used in the preparation of dendrimer-stabilized Pd nanoparticles. The structural difference between DENs and dendrimer-stabilized materials is that DENs encapsulate the nanoparticle within the void space of the polymer, while dendrimer-stabilized materials bind multiple dendrimers to the particle surface to impart material stability. This system was prepared by co-dissolving phosphine-oxide dendrimers and a metal precursor, Pd(acac)$_2$, in THF at room temperature as shown in Figure 7.6(a).[93] The resulting complex was then reduced by H$_2$ at 60 °C overnight producing a black solution of G$_n$DenP-Pd (n = 1–3) with sizes of 5.0 ± 0.4 nm, 4.6 ± 0.5 nm and 3.2 ± 0.5 nm for G$_1$DenP-Pd, G$_2$DenP-Pd and G$_3$DenP-Pd nanoparticles, respectively.[93] The dendritic-phosphine-based Pd nanoparticles were then used for Stille coupling using methyl-4-bromobenzoate and tributyl(phenyl)stannane (Figure 7.6(b)) with 1.5 mol% nanocatalyst at 40 °C and DMF–water as the solvent.[93] Excellent yields of 98% after 5.0 h were reported for this model reaction, while 78–98% yields were obtained for an array of aryl bromides containing electron-donating and withdrawing substituents and heteroatom-containing substrates.[93] Reactivity over such a wide variety of reagents signifies the versatility of the G$_n$DenP-Pd system for practical applications in organic synthesis. Furthermore, when the Stille reaction using aryl chlorides was used at 110 °C, formation of products was observed with excellent yields ranging from 78–95%.[93] This was quite remarkable as aryl chlorides are difficult to get to participate in C–C couplings. This suggests that the G$_n$DenP-Pd system presents a robust, highly efficient nanocatalyst with

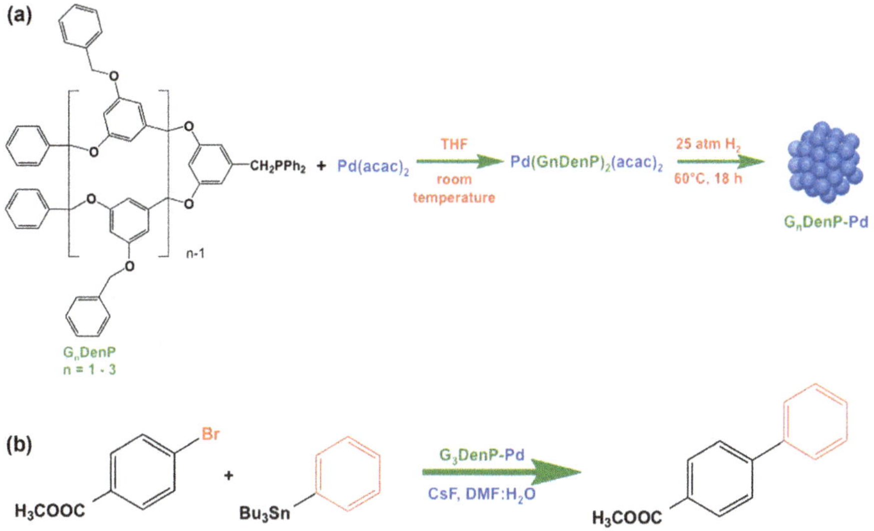

Figure 7.6 (a) Scheme for the synthesis of G$_n$DenP-Pd nanoparticles and (b) their use in the Stille coupling reaction between methyl-4-bromobenzoate and tributyl(phenyl)stannane.[93]

industrial and pharmaceutical applications for organic and materials synthesis.

As a complement to the dendrimer-based systems, an aqueous-based synthesis of Pd nanoparticle catalysts has also been accomplished using polymer ligands such as polyethylene glycol (PEG). In this process, K_2PdCl_4 interacted with the PEG, which was reduced by drop wise addition of a Fisher carbene complex, generating 7–10 nm spherical Pd nanoparticles.[97] For these materials, the Stille coupling between an aryl bromide and phenyltributylstannane was performed in an aqueous solution of K_2CO_3 at 80 °C for 2–3 h. Product yields of ≥88% were obtained using different aryl bromide substrates with varying functional groups, indicating applicability in a wide array of reagents.

Imidazolium-based ionic polymers have also been used to stabilize Pd nanoparticles as a functionalized ionic liquid, producing 5.0 nm particles after $NaBH_4$ reduction.[135] The Pd nanocatalyst (Pd-IP-IL) was used in coupling reactions between different aryl halides and an organostannane reagent at 80 °C, producing quantitative yields for aryl iodides with a variety of functional groups. The recyclability of the Pd-IP-IL nanoparticles was established wherein no loss of activity was reported after five reaction cycles. This indicated that the catalyst was stable at elevated temperatures and is potentially viable for industrial use. Amine-rich ionic liquids, tetrabutylammonium bromide and tetrabutylammonium acetate, were also employed for the preparation of monodisperse, core–shell Pd nanoparticles composed of a 3.3 nm metallic core and stabilized by tetrabutylammonium cations, Br^-, and $[PdBr_4]^{2-}$.[134] These nanoparticles were used for Stille coupling using bromo- and chloro-arenes with tetrabutylphenylstannane. The starting materials were reacted in tetraheptylammonium bromide at elevated temperatures between 90–130 °C.[134] High product yields were generated from aryl bromides and activated aryl chlorides at 90 °C, while higher temperatures of up to 130 °C were needed to generate modest yields for deactivated electron-rich aryl chlorides. Recycling experiments were performed after the initial catalytic cycle by first extracting the Pd nanoparticles with cyclohexane and the resulting viscous mixture was added with fresh starting materials.[134] The nanocatalyst was recycled five times with minimal loss of activity, suggesting a practical, easily recoverable and reusable catalyst, which were the main advantages of ionic liquid based Pd nanoparticles.

Another ligand system employed for nanocatalyst synthesis is Pd-substituted Keggin-type polyoxometallates. The precursor species, $K_5[PdPW_{11}O_{39}] \cdot 12H_2O$, was reacted with acetophenone at 200 °C for 4 h with an applied H_2 pressure of 30 bar, which produced a blue-black solid.[176] Due to the polymer's high anionic charge, the polymer-capped Pd nanoparticles were electrostatically stabilized, forming spherical clusters of ∼15–20 nm in size. The Stille reaction between 4-bromotoluene and tetraphenyltin was performed using these materials in a 50 : 50 DMF–H_2O mixture at 110 °C for 12.0 h.[176] Under these conditions, the reaction generated a nearly quantitative yield of 93% for the 4-methylbiphenyl product. Similarly, coupling between 1-chloro-4-nitrobenzene and tetraphenyltin produced 92% 4-nitrobiphenyl under similar conditions. Together, this suggests that these materials are highly reactive catalysts due to their reactivity with chloro-based reagents.

More recently, biological molecules have been employed for the synthesis of solvent-dispersible nanocatalysts due to their ability to interact and bind with target materials. The biotic–abiotic interaction displayed by the biological molecules and inorganic substrates was also observed in the silaffin proteins that precipitate silica in the cell walls of diatoms. The Kröger group isolated and identified the R5 peptide from the diatom *Cylindrotheca fusiformis* bearing multiple basic residues (sequence = SSKKS GSYSGSKGSKRRIL).[177,178] This peptide self-assembles due to the hydrophobic RRIL motif to form a bioscaffold that can interact and stabilize metal nanomaterials as illustrated in Figure 7.7(a).[87] In this regard, peptide-templated Pd nanostructures were synthesized by co-dissolving Pd^{2+} with the R5, followed by addition of $NaBH_4$ as the reducing agent.[87] Several Pd–peptide ratios were prepared, resulting in the generation of differently shaped materials, as shown in the TEM images of Figure 7.7.[87] Using a 60 : 1 Pd–peptide ratio (Pd60), nearly monodisperse and spherical nanoparticles were formed with an average diameter of 2.9 ± 0.6 nm (Figure 7.7(b)).[87] For a 90 : 1 ratio (Pd90), linear nanoribbons were observed with a width of 3.9 ± 0.8 nm (Figure 7.7(c)), while nanoparticle networks (NPNs) were generated at a 120 : 1 ratio (Pd120) with an average width of 4.1 ± 1.2 nm (Figure 7.7(d)).[87]

Pd Nanoparticles in C–C Coupling Reactions 131

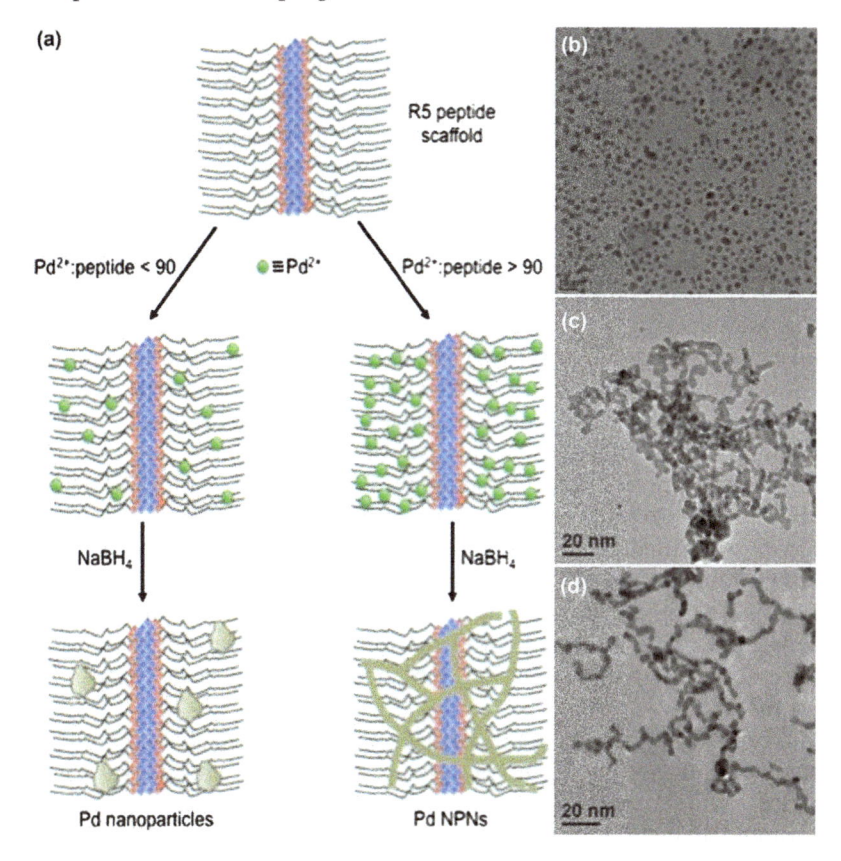

Figure 7.7 (a) Schematic representation of the R5-templated synthesis of Pd NPNs, and TEM images of Pd materials produced using different Pd–peptide ratios: (b) Pd60, (c) Pd90, and (d) Pd120.[87] A. Jakhmola, R. Bhandari, D. B. Pacardo and M. R. Knecht, *J. Mater. Chem.*, 2010, **20**, 1522–1531. Reproduced by permission of The Royal Society of Chemistry.

The catalytic activity of these nanocatalysts was initially tested using the Stille coupling reaction between 4-iodobenzoic acid and phenyltin trichloride. Quantitative yields were observed after 24 h for reactions using low catalyst amounts ($\geq$0.01 mol% Pd for all Pd–peptide ratios), suggesting that these materials were highly efficient under ambient conditions.[87] TOF analysis of the different peptide-templated structures generated modest values of $\sim$450 mol product (mol Pd$\times$h)$^{-1}$ for both Pd60 and Pd120, while a value of $\sim$350 mol product (mol Pd$\times$h)$^{-1}$ was obtained for Pd90.[87] Although these TOF values are lower than those reported by Crooks and colleagues[86] using the same reaction, it is worth noting that for the R5-templated Pd nanoparticles, lower catalyst amounts were required to generate quantitative product yields.

Employing biological molecules as tools for the biomimetic synthesis of metal nanoparticles could generate a model system for green nanocatalysis.

As shown in the R5-templated Pd nanoparticles, highly active nanocatalysts with different morphologies can be generated by exploiting the scaffold-like nature of the biomolecules; however, there are limited numbers of naturally-occurring biomolecules with affinity for inorganic substrates. Furthermore, biological molecules exhibit unspecific interactions with different metal ions as exemplified by the R5 template, which showed templating/stabilizing ability in the synthesis of SiO_2,[179] TiO_2,[180] Pd,[37,87] and Au[181] nanomaterials. To overcome these limitations, bioselection techniques such as phage display[182] have been employed to generate a library of biomolecules having strong and specific interactions to a metal surface. In phage display, a combinatorial library of $\sim 10^9$ peptides are expressed as fusion proteins with the viral minor coat protein.[182] As such, the nearly random peptide sequences were exposed on the outer phage surface enabling their interaction with a target metal surface. After a series of steps to remove non-specific and weakly bound species, peptide sequences with great affinity for the target can be isolated. To this end, a 12-mer peptide termed Pd4 (TSNAVHPTLRHL) has been identified through phage display to bind strongly to Pd metal.[88] Computational studies indicate that the histidine residues at positions 6 and 11 have the strongest interaction with Pd to anchor the biomolecule to the surface, generating a kinked conformation.[183] As a result of this structure, particle stability can be achieved, which provides substrate access to the active metal.

Pd4 peptide-capped Pd nanoparticles were synthesized by chemical reduction of the Pd4-Pd^{2+} complex using $NaBH_4$ to generate 1.9 ± 0.3 nm spherical nanoparticles.[88] The catalytic reactivity of the nanocatalyst was quantitated using the coupling reaction between select aryl halides and phenyltin trichloride in an aqueous solvent at room temperature, using low catalyst loadings. From the reaction employing 4-iodobenzoic acid as the aryl halide substrate, quantitative amounts of the 4-biphenylcarboxylic acid product were generated after 24 h using $\geq$0.005 mol% Pd,[88] similar to the results generated by the R5-Pd nanoparticles.[87] For the catalyst efficiency analysis, however, TOF values of $\sim$2400 mol product (mol Pd$\times$h)$^{-1}$ were observed using the Pd4-capped Pd nanocatalysts, indicating more efficient reactivity.[88]

Further studies of the peptide-capped Pd nanoparticles for Stille coupling indicated its reactivity with aryl iodides containing electron-donating and electron-withdrawing groups. As expected, slower reactivity was observed for aryl bromides and no reaction was obtained using aryl chlorides.[88] Interestingly, when the reaction was performed with a mild temperature increase to 40 °C, conversion of the aryl bromide substrate reached $\sim$75 % after 24 h, suggesting that minor thermal activation may be required to initiate coupling using aryl bromides.[184] As such, the reactivity of the peptide-capped Pd nanoparticles was expanded to elevated temperatures up to 60 °C, after which a further increase in temperature resulted in decreased catalytic efficiency due to the formation of Pd black.[184] In this reaction, the aryl halide substrate was anticipated to interact with the nanoparticles during oxidative addition and abstract the active metal from the particle surface, from which the catalytic cycle was completed, indicating an atom-leaching mechanism. After reductive elimination when the product was released, the free, highly active Pd⁰ generated in solution could proceed along three different paths, as illustrated in Figure 7.8(a): (1) reused catalytically until all starting materials are consumed, (2) quenched by the remaining nanoparticles, or (3) aggregated with other leached Pd atoms to generate Pd black.[184] To probe this catalytic mechanism, TOF analyses using different Pd loadings were performed. Based on the results shown in Figure 7.8(b), the TOF values are constant at $\sim$2400 mol product $(\text{mol Pd} \times \text{h})^{-1}$ when using $\leq$0.05 mol% Pd; however, higher Pd loadings generated a decreasing trend in TOF values.[184] This indicates decreased catalyst efficiency due to the formation of catalytically-inactive Pd black at the higher reaction concentrations.

The fate of the Pd⁰ atoms after the individual Stille catalytic cycle was explored using competition experiments containing combinations of two different aryl halides in a single Stille reaction. For this, 4-iodobenzoic acid and 4-bromobenzoic acid were co-dissolved in water and added with phenyltin trichloride and the peptide-capped Pd nanoparticles at room temperature.[185] The results indicated that only the aryl iodide substrate reacts at room temperature even though the highly reactive Pd⁰ atoms were present in solution after the initial cycle. When the reaction temperature was increased to 40 °C, consumption of the aryl bromide was observed along with an increase in the amount of product generated. Furthermore, when the same bicomponent reaction was heated to 60 °C after the addition of nanocatalyst (*i.e.* no substrate conversion had begun), coupling for both iodo- and bromo-substrates was observed, demonstrating a loss of catalyst selectivity.[185] These results suggest that substrate selectivity can be thermally controlled at rather lower temperatures wherein the iodo-components reacted at room temperature while the bromo-component required thermal activation to

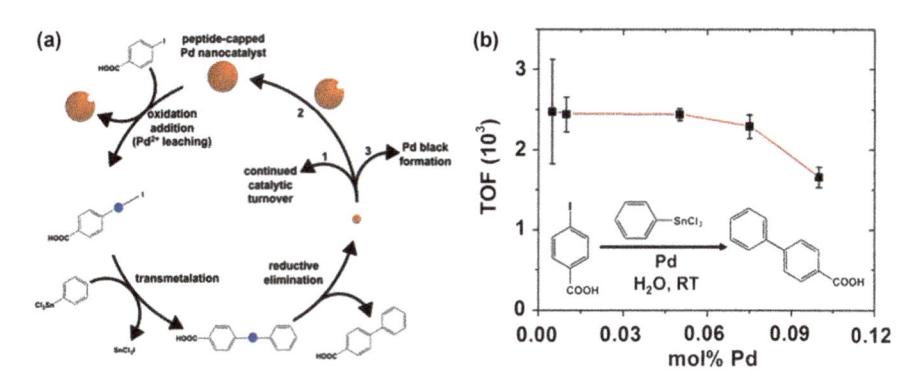

Figure 7.8 (a) Scheme demonstrating the atom-leaching mechanism for peptide-capped Pd nanoparticles and (b) TOF analysis for Stille coupling using different Pd loadings.[184] D. B. Pacardo, J. M. Slocik, K. C. Kirk, R. R. Naik and M. R. Knecht, *Nanoscale*, 2011, **3**, 2194–2201.
Reproduced by permission of The Royal Society of Chemistry.

proceed. Unfortunately, no reactivity was observed for comparable chlorohalides. Similar degrees of selectivity were observed for Stille coupling using aryl dihalides; however, the attachment of electron-donating or withdrawing groups modulated the catalyst selectivity.[185]

7.5 Pd Nanoparticles in the Suzuki Coupling Reaction

The Suzuki reaction was first reported in 1979 where the cross-coupling of alkenylboranes with aryl halides was performed using a phosphine-stabilized Pd catalyst in the presence of a base.[64] This provided a new synthetic strategy to produce stereoselective arylated alkene products from different aryl halides with activated alkenylboranes as the nucleophilic partner. The main advantages of this coupling reaction are the ability to tolerate different functional groups attached to the aryl ring, as well as the relatively low toxicity of the boron reagent.[64] In this reaction, the presence of a base is important as it converts the boronate species into its active form for transmetalation to occur; absence of the base results in no product formation.[64] In a separate paper, Suzuki and coworkers reported the coupling of alkenylboranes with alkenyl halides in the presence of a base and a catalytic amount of tetrakis(triphenylphosphine) Pd to produce decent yields of conjugated dienes and enynes with high regio- and stereoselectivity.[63] From these simple borane compounds, the application of the Suzuki reaction was expanded to different organoboranes such as boronic acids, boronate ester, and arylboranes, providing a wide array of applications in organic synthesis.

The application of Pd nanoparticles for Suzuki coupling was first introduced by the Reetz group.[186] For this, tetrabutylammonium (R_4N^+)- or (PVP)-stabilized Pd nanocatalysts were employed for the reaction between aryl bromides and aryl chlorides with phenylboronic acid producing biaryl products. The R_4N^+ Pd clusters of 2–3 nm were initially employed for the Suzuki coupling reaction between bromobenzene and phenylboronic acid in DMF at 100 °C for 3.5 h.[186] The results of these studies demonstrated 100% conversion of the starting material based on GC and thin layer chromatography (TLC) analyses with a 53% isolated yield for the product. Similar conversion capabilities were obtained from the same reaction using PVP-stabilized Pd materials, although significantly lower isolated yields were achieved.

In separate work, PVP-stabilized Pd nanoparticles were synthesized by El-Sayed and coworkers for applications in Suzuki coupling reactions in aqueous solution.[96] The TEM analysis displayed the formation of ~3.6 nm Pd nanoparticles after heating a solution of H_2PdCl_4 and PVP in 40% EtOH for 3.0 h. Cross-coupling reactions between aryl iodides and arylboronic acids were performed in 40% EtOH and in the presence of Na_3PO_4 as the base under reflux conditions for 12 h using a lower catalyst loading of 0.3 mol% Pd.[96] In the reaction involving iodobenzene and phenylboronic acid, 95% yield of the biphenyl product was generated. On the other hand, using sulfur-containing starting materials such as 2-iodothiophene or 2-thiopheneboronic acid, resulted in decreased product yields ranging from 26% to 92%. The decrease in product yield can be attributed to poisoning of the nanocatalyst by the thiol groups in the reactants. Although these reactions showed decent product yields for most of the substrates used, precipitation of Pd black was noted during the reaction, which was attributed to the high reaction temperatures. In another experiment, the initial rate of the coupling reaction between phenylboronic acid and iodobenzene in acetonitrile–water solvent was determined.[96] The results indicated that the reaction rate was directly proportional to the concentration of PVP-Pd nanocatalyst, from which the authors suggested that C–C bond formation occurred at the catalyst surface. While the PVP-Pd nanoparticles were effective Suzuki coupling catalysts, the formation of Pd black decreased the overall efficiency of the catalyst.

Another polymer ligand used for the synthesis of Pd nanoparticles is the block copolymer polystyrene-*b*-poly(sodium acrylate) (PS-*b*-PANa).[187] The polymer structure was composed of a hydrophobic polystyrene component and a hydrophilic poly(sodium acrylate) component. This amphiphilic polymer generates a micellar structure in solution that provides steric stabilization of nanoparticles. The block copolymer-stabilized Pd nanoparticles were synthesized by refluxing PS-*b*-PANa with H_2PdCl_4 in EtOH, producing a clear dark brown solution containing 3.0 ± 0.7 nm particles based upon TEM analysis.[187] Suzuki coupling reactions between 2-thiopheneboronic acid and iodobenzene were performed using a 0.6 mol% Pd catalyst loading with NEt_3 or NaOAc as the base, generating yields of 85% or 88%, respectively after 12.0 h under reflux. Incidentally, the base influenced the stability of the Pd nanoparticles; NEt_3 provided additional nanoparticle stabilization, while NaOAc resulted in particle precipitation. As such, NEt_3 was used as a base in subsequent Suzuki couplings between phenylboronic acid and bromothiophene using 3 : 1 acetonitrile–water as the solvent. In this reaction, competitive formation of the product (2-phenylthiophene) and homocoupling product (bithiophene) was observed, generating similar yields of 16% with no precipitation of Pd black.[187] In contrast, when PVP-Pd nanoparticles were employed for a similar reaction between bromothiophene and phenylboronic acid, only homocoupling products were generated; however, comparing the catalyst stability as a function of ligands, the PS-*b*-PANa-based nanoparticles were more stable than the PVP at these conditions as Pd black was not produced in the reaction.[187]

PEG-stabilized Pd nanoparticles have also been employed for Suzuki coupling reactions in an aqueous medium between aryl halides and phenylboronic acid using K_2CO_3 as the base.[97] Initially, the reaction between *p*-bromoacetophenone and phenylboronic acid in a 1 : 5 dimethoxyethane (DME)–water solvent at room temperature generated the expected product with 95% isolated yield after 3.0 h. Subsequent reactions were performed in pure water using aryl halides with different electron-donating and electron-withdrawing groups, producing excellent yields within 2.0 h of reaction.[97] For instance, near quantitative yields were obtained for aryl iodides within 0.5 h of reaction at room temperature, indicating a superior catalytic reactivity for the nanocatalyst. On the other hand, higher temperatures were required for conversion of sterically hindered aryl bromides, while aryl chlorides were unreactive. Recycling of the catalyst was studied using the aqueous extract retained from product purification. Unfortunately, significant loss of catalytic activity was recorded from these systems with decreased yields from 90% to 40%.[97] This was attributed to metal leaching or incomplete extraction of the nanocatalyst.

Imidazolium-based ionic polymer-stabilized Pd nanoparticles in the functionalized Pd-IP-IL described earlier were also used in Suzuki coupling at 100 °C.[135] Aryl iodides, containing electron-donating or electron-withdrawing groups, were reacted with phenylboronic acid generating coupled products from 82% up to >99% using a 1.0 mol% Pd loading. As expected, lower yields were observed for substituted aryl bromides using similar reaction conditions, ranging from 6% to 99%. On the other hand, when iodobenzene and bromobenzene were used, 90% and 55% yields were produced after 6 h of reaction. Consequently, when the reaction was performed in water at 35 °C using aryl benzoic acids, the amount of Pd nanocatalyst required was reduced from 1.0 mol% to 0.5 mol%. Interestingly, the yields of the expected products for the aqueous-based reactions were similar to the yields of reactions in ionic liquids such that >99% yield was generated for iodobenzoic acid and 57% yield for bromobenzoic acid. This suggests that the solvent played only a minor role in controlling the reactivity.

Seed-mediated nanoparticle synthesis methods in the ionic liquids were also used to prepare larger diameter ($\sim$10–20 nm) Pd nanoparticles (termed Pd-IP-IL-2).[135] These materials were then employed for Suzuki coupling between aryl iodides and phenylboronic acid wherein almost quantitative yields were obtained, similar to the yields noted for the initial Pd-IP-IL nanocatalysts. As expected, significantly lower yields were obtained when aryl bromides were employed as the substrate instead of the iodo-derivatives. TEM analysis of the nanocatalyst after the coupling reaction revealed that the particle size decreased to $\sim$5.0 nm, suggesting that the reaction was driven *via* the catalytic leaching mechanism.

A unique polymer-stabilized Pd nanocatalyst was prepared using polyaniline nanofibers (PANI), shown in Figure 7.9(a), that function as both the reducing agent and particle support.[188] The synthesis was accomplished by incubating the metal precursor ($Pd(NO_3)_2$) with an aqueous dispersion of PANI nanofibers for one day. This reaction results in the formation of polycrystalline particles (Figure 7.9(b)) with a bimodal size distribution with two populations centered at $\sim$2 nm and $\sim$75 nm. Once fully characterized, the Pd-PANI nanocatalyst system was then used for the coupling of aryl chlorides and phenylboronic acid with NaOH as the base in water at 80–100 °C for 2–6 h. Using a highly activated substrate, 4-chlorobenzaldehyde, almost quantitative yields were obtained in the Suzuki reaction with phenylboronic acid. On the

Pd Nanoparticles in C–C Coupling Reactions 137

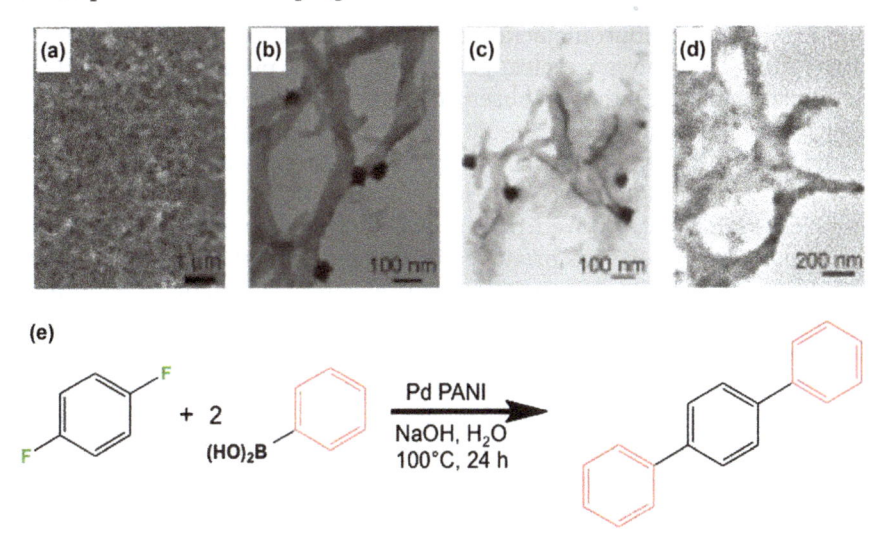

Figure 7.9 (a) SEM image of PANI nanofibers, TEM images of Pd-PANI nanoparticles (b) before the reaction, (c) halfway through the reaction, and (d) after two reaction cycles. (e) Suzuki coupling reaction between 1,4-difluorobenzene and phenylboronic acid using Pd-PANI nanoparticles.[188]
Reproduced with permission from B. J. Gallon, R. W. Kojima, R. B. Kaner and P. L. Diaconescu, *Angew. Chem., Int. Ed.*, 2007, **46**, 7251–7254. Copyright © 2007 WILEY-VCH Verlag GmbH & Co. KGaA, Weinheim.

other hand, when a deactivated substrate, such as 4-chlorophenol, was used for the reaction, a significant yield of 88% was generated. Hetero-substituted substrates, such as pyridyl chloride, also generated yields of 89% and 96% of the expected coupling products from 4-chloropyridine and 2,6-dichloropyridine substrates, respectively. The effect of changing the boronic acid substituent was also explored using the Pd-PANI nanoparticles such that when activated 3,5-di(trifluoromethyl)phenylboronic acid was used, a 92% yield was generated, while using 2,4-dimethoxyphenylboronic acid only produced a 70% product yield.[188] Taken together, these results indicate that this approach generated a highly reactive nanocatalyst, as aryl chlorides do not usually undergo coupling reactions under such mild conditions.

To further test the reactivity of the Pd-PANI nanoparticles, 1,4-difluorobenzene was used as a substrate for coupling with phenylboronic acid as illustrated in Figure 7.9(e). As expected, the Suzuki reactions using the fluoro-substituted substrates were much slower compared with the aryl chlorides due to the stronger C–F bond. Remarkably, after 24 h at 100 °C, 60% product yield was obtained, whereas when fluorobenzene and phenylboronic acid were used for the coupling reaction, 80% yield of the biphenyl product was generated.[188] Finally, recycling of the nanoparticles was also accomplished, as studied using the coupling of 4-acetylphenyl chloride and phenylboronic acid. For this, minimal loss of activity was observed after ten reaction cycles, generating ≥89% product yields with no noted catalyst aggregation as observed in the TEM analysis of the particles after two reaction cycles.[188] The decreased product yield was attributed to the degradation of the nanofiber support after several catalytic cycles.

While synthetic PANI was an excellent support for the catalytic nanoparticles, biologically derived nanofibers, such as bacterial cellulose (BC), can also be used.[189] BC was produced by fermentation of *Acetobacter xylinum* with glucose as the carbon source. The three-dimensional nanofibers interacted with metal ions to serve as a support for nanoparticle production. The one-pot synthesis involved mixing of metal precursors with BC nanofibers under an N_2 atmosphere at 140 °C.[189] Once complete, $NaBH_4$ was added, producing polydispersed, ~20 nm Pd nanoparticles deposited on the nanofibers (termed Pd-BC). Initial catalytic analysis of the materials was performed using the coupling of iodobenzene and phenylboronic acid, generating almost quantitative yields of 98% after 3.5 h at 85 °C. Recycling experiments using the same reactants showed no loss of activity after two reaction cycles with no visible Pd aggregation as observed by TEM; however, similar to the Pd-PANI nanoparticles, a decline in the quality of the nanofiber support was observed after extended recycling runs that resulted in lower product yields.[189]

Suzuki reactions using aryl iodide and substituted phenylboronic acids were also studied with the Pd-BC nanocatalyst, which generated excellent yields between 88–99% of the expected coupling products depending on the substituent attached on the nucleophile.[189] To this end, when 3-methoxyphenylboronic acid was used in the reaction with iodobenzene, a 99% yield was obtained, whereas using 4-cyanophenylbenzoic acid generated a yield of only 88%. Recycling experiments were again performed for the Pd-BC nanoparticles using the substituted nucleophiles wherein product yields >86% were obtained after five catalytic cycles. To further explore the reactivity of Pd-BC nanoparticles, the least reactive substrate, aryl chloride, was employed for Suzuki coupling with different substituted phenylboronic acids. Surprisingly, product yields between 75–92% were recorded, indicating that an extremely efficient nanocatalyst was developed.[189] Finally, leaching of the active catalyst was also studied using inductively coupled plasma-atomic emission spectroscopy (ICP-AES) analysis wherein the amount of Pd species before and after the five reaction cycles was quantitated. The results showed that the Pd concentration before the reaction was 5.29% and 5.26% after the reaction, indicating a negligible amount of Pd was leached during the reaction.[189] As such, the results of elemental analysis suggested that the BC nanofibers are good protecting ligands for the nanoparticles.

The Pd-substituted Keggin-type polyoxometallate-based nanoparticles, described earlier, were also used for the Suzuki reaction.[176] For this, aryl bromides and phenylboronic acid were reacted in an EtOH–water solvent with diisopropylamine as the base. Nearly quantitative yields of the expected biaryl coupling products were obtained for a variety of aryl bromide substrates in the reactions, which were heated to temperatures between 80–85 °C. For example, the coupling reaction between 1-bromo-4-nitrobenzene and phenylboronic acid generated >99% yield of 4-phenylnitrobenzene product, while using 4-bromotoluene gave an 89% yield of 4-phenyltoluene, revealing the effects of electron-withdrawing and electron-donating groups on substrate reactivity.[176] On the other hand, when 1-chloro-4-nitrobenzene was employed as the starting material under similar reaction conditions, a 98% product yield was generated, demonstrating the reactivity of the materials toward difficult to use substrates.

The effects of different stabilizers on the reactivity of Pd nanoparticles for the Suzuki reaction were also investigated using hydroxyl-terminated PAMAM dendrimers.[187] The preparation of dendrimer-encapsulated Pd nanoparticles was similar to the process described by Crooks and coworkers, albeit with some modifications.[187] To this end, the hydroxyl-terminated PAMAM-dendrimer solution was added with an appropriate amount of K_2PdCl_4; prior to reduction, the pH of the solution was adjusted to 4.0 using HCl to ensure the coordination of Pd^{2+} ions with the interior tertiary amines of the polymer. Upon reduction using $NaBH_4$, 1.4 nm particles were formed using third (G3-OH(Pd)) and fourth generation (G4-OH(Pd)) hydroxyl-terminated dendrimers, while bigger nanoparticles (3.6 nm) were formed from second generation dendrimers (G2-OH(Pd)). The Pd nanoparticles were then employed in the Suzuki coupling of phenylboronic acid (or 2-thiopheneboronic acid) and iodobenzene in 40% EtOH under reflux for 24.0 h. The results demonstrated that the G4-OH(Pd) nanoparticles were less efficient, generating ≤35% of the expected products, compared with the G3-OH(Pd) and G2-OH(Pd) nanoparticles, which produced ≥69% yields. Interestingly, precipitation of inactive Pd black was generated during the reaction when using G3-OH(Pd) and G2-OH(Pd) nanoparticles; however, no such catalyst aggregation was observed for the G4-OH(Pd) nanoparticles' catalytic reaction. These results suggested that a trade off between catalyst stability and efficiency existed, wherein the fourth generation nanoparticles were more stable with limited substrate access to the metal surface; however, for the lower generation structures, higher catalytic efficiency and diminished stability was noted.[187] Among these materials, the third generation dendrimer provided the optimal mix of material stability and catalytic efficiency, although Pd black formation was observed when using 2-thiopheneboronic acid as the substrate. This effect is possibly due to Pd leaching during the reaction or through the thiol moiety of the substrate coordinating to the metallic material.

In another study of the hydroxyl-terminated dendrimer materials, Christensen and colleagues prepared the same dendrimer-based Pd nanoparticles using fourth generation dendrimers, where particles of ~3 nm were generated.[190] These structures were then employed for the coupling of iodobenzene and p-tolylboronic acid at 78 °C using EtOH as the solvent. From this study, product yields of 98% were determined, thus a more efficient catalytic system was obtained using the nearly identical dendrimer-based materials, which may arise from the different reagents employed. Interestingly, no precipitation of Pd black was noted during the reaction. Aryl bromides were also studied using this system; however, they were not as reactive as the iodo-based substrates, requiring higher temperatures and longer reaction times. Aryl chlorides were also shown to be completely unreactive.[190] Finally, the recyclability of the nanocatalyst was studied using the coupling of ethyl-4-iodobenzoate and p-tolylboronic acid wherein formation of Pd black during the subsequent reactions was observed resulting in decreased product yield, from 99% to 80%, after the third reaction cycle.[190]

The Astruc group has developed a unique dendrimer-stabilized Pd nanoparticle system using 1,2,3-triazolylsulfonate dendrimers, synthesized through click chemistry.[191] In this system, Pd^{2+} was introduced to the triazole ligands producing two absorption bands in the UV-vis spectrum at 208 and 235 nm. Upon addition of $NaBH_4$, Pd nanoparticles of $\sim$2–3 nm in diameter were generated using three different dendrimer generations (Pd-DSN-G_0, Pd-DSN-G_1, Pd-DSN-G_2).[191] TEM analysis showed that the nanoparticles' sizes were larger than the size of the ligands, thus leading to a large degree of polydispersity from the materials as the generation of the dendrimers increases. Room temperature Suzuki coupling reactions between iodobenzene and phenylboronic acid in a water–ethanol solvent were used to examine the reactivity of the dendrimer-stabilized materials. Using a 0.1 mol% Pd loading, product yields of 96%, 95%, and 70% were obtained using Pd-DSN-G_0, Pd-DSN-G_1 and Pd-DSN-G_2, respectively, suggestive of a decrease in catalytic efficiency with respect to increasing dendrimer generation.[191] Similar trends were observed when lower Pd loadings (0.01 mol%) were used in the reaction. TOF analysis employing the Pd-DSNs indicated that there was no significant difference in the values based upon dendrimer generation; however, significantly higher TOFs of 1167–1533 mol PhI (mol Pd×h)$^{-1}$ were obtained for the reaction employing 0.01 mol% Pd, while TOF values of 117–167 mol PhI (mol Pd×h)$^{-1}$ were generated when using 0.1 mol% Pd, demonstrating an effect based upon the Pd loading.[191] From these results, the catalytic mechanism of Pd-DSNs was suggested to follow an atom-leaching process wherein the TOF values of the reaction decrease as the amount of Pd loading increases. Quantitative yields were also obtained in the coupling reaction using bromobenzene, but the reaction needed to be heated to 100 °C.[191]

The phosphine-based dendrimers, G_nDenP-Pd, whose synthesis was discussed earlier, have also been used for Suzuki coupling.[93] The reaction between bromobenzene and phenylboronic acid was initially tested with different bases such as NEt_3, NaOAc, K_2CO_3, and K_3PO_4 to optimize the reaction conditions. As such, K_3PO_4 was determined to be the best base for the Suzuki reaction using G_nDenP-Pd in dioxane as the solvent under refluxing conditions. Aryl iodides and aryl bromides with electron-withdrawing or electron-donating groups were reacted with different phenylboronic acids, which generated nearly quantitative yields (≥94%) of the expected coupling products using G$_2$DenP-Pd nanoparticles; however, when the reaction was performed using electron-deficient arylboronic acids, slightly lower yields of 88% to 92% were obtained.[93] Interestingly, sterically hindered *ortho*-substituted aryl halides generated excellent product yields between 86–99%. Expanding the scope of G$_n$DenP-Pd catalytic activity to coupling reactions between different aryl chlorides and phenylboronic acid using G$_3$DenP-Pd nanoparticles resulted in good to excellent yields (81–94%) even with deactivated or *ortho*-substituted substrates.[93] Using heteroaryl bromides and chlorides as substrates in Suzuki coupling also generated excellent yields using G$_3$DenP-Pd nanoparticles, such as those shown in the coupling between pyridine-derived aryl bromide and phenylboronic acid to generate a phenylpyridine product. Interestingly, even pyridine-derived aryl chloride produces the expected product with an 82% yield.[93]

The recyclability of the G$_3$DenP-Pd nanoparticles was also explored using 0.2 mol% of the catalyst in a model reaction between 4-hydroxyphenyl iodide and phenylboronic acid.[93] Using these materials, no loss of activity was observed during the first four cycles. Lower reaction yields were recorded after the fifth run (79% yield), but when the subsequent cycles were allowed to proceed at longer reaction times (from 20 h to 48 h), excellent product yields (94%) were generated for three cycles before decreased activity was observed, generating only a 75% yield after the ninth cycle. Furthermore, only a very small amount of Pd was leached during the recycling activity, as measured by inductively coupled plasma X-ray fluorescence (ICP-XRF); however, TEM analysis of the nanoparticles after nine runs showed an increased particle size with irregular shape, which was suggestive of particle aggregation.

The role of G$_n$DenP-Pd nanoparticles in the coupling reaction was further probed by using a bulky dendritic aryl bromide substrate and 4-acetophenylboronic acid (Figure 7.10) to determine the catalytic mechanism.[93] The bulky substrate was anticipated to ascertain if the catalytic reaction occurred on the metal surface inside the dendritic cage or through potentially leached Pd species. To this end, the formation of the coupling product will be minimized if the reaction mainly occurs on the nanoparticles' surface. On the other hand, Suzuki coupling between the dendritic aryl halide and phenylboronic acid will proceed smoothly if the reaction follows the atom-leaching mechanism. Interestingly, the results generated 96% yield of the coupling product, suggesting that the Suzuki reaction could not occur completely inside the dendritic box due to the bulky aryl substituent.[93] As such, the catalytically active Pd atom species must be leached from the G$_3$DenP-Pd nanoparticles during the reaction, resulting in near quantitative yields for the dendritic aryl halide substrate. To validate the results, a homogeneous catalytic complex, Pd(PPh$_3$)$_4$, was employed to drive a control analysis using the same bulky aryl halide and phenylboronic acid under similar reaction conditions. The results from this study provided a 79% yield of the expected product, which was lower than the yield for

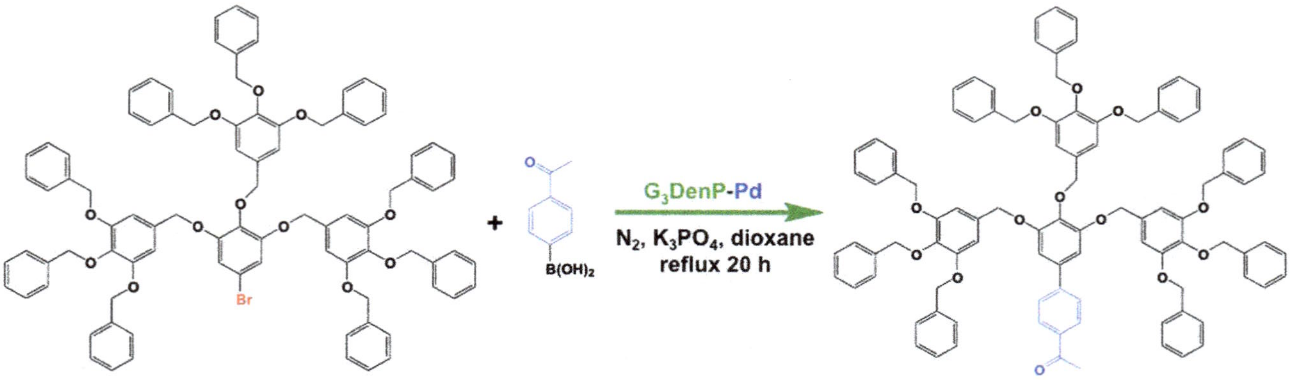

142 Chapter 7

Figure 7.10 Suzuki coupling reaction using dendritic bromobenzene and 4-acetylphenylboronic acid.[93]

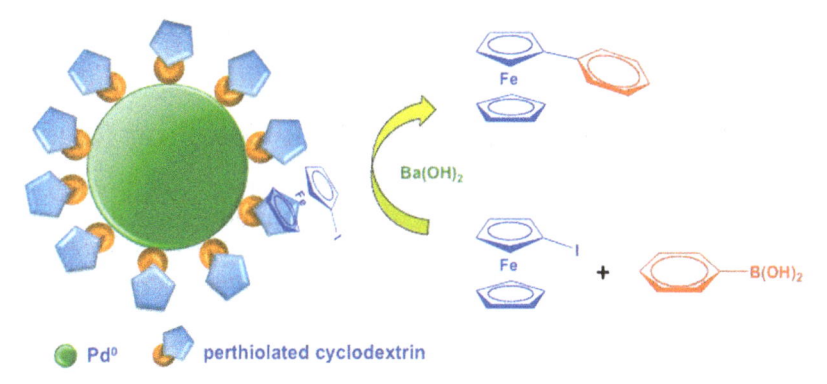

Figure 7.11 Perthiolated cyclodextrin-capped Pd nanoparticles used in the Suzuki coupling reaction between iodoferrocene and phenylboronic acid.[98]

G$_3$DenP-Pd nanoparticles.[93] These results suggested that the catalytic mechanism for G$_3$DenP-Pd nanoparticles was similar to the homogeneous catalyst although the former was much more efficient than the latter.

In a different synthetic approach, Pd nanoparticles were generated using perthiolated cyclodextrin (β-SH-CD) as surface passivants, where the materials were reactive for Suzuki coupling (Figure 7.11).[98] In this process, the metal precursor was added to a solution containing NaBH$_4$ and β-SH-CD in DMF, which resulted in the formation of 3.5 nm CD-capped Pd nanoparticles. The Suzuki coupling between aryl halides and arylboronic acid in a 1 : 1 (v/v) acetonitrile–water solvent were performed using the Pd nanoparticles under reflux for 2 h. Nearly quantitative yields were obtained for aryl iodides with different functional groups and, as expected, lower yields were obtained from aryl bromides even after longer reaction times.[98] TOF analysis demonstrated values ranging from 7.8 to 48 mol product (mol Pd×h)$^{-1}$. These decreased values were likely due to the covalently bound CD ligands, which covered ~50% of the active surface, thus making it inaccessible for substrate interaction.[98]

Another unique Suzuki reaction using these materials was performed using iodoferrocene, which can interact to form stable inclusion complexes with the surface-bound CD, as illustrated in Figure 7.11.[98] In the presence of Ba(OH)$_2$ as the base under an N$_2$ atmosphere, the coupling reaction between iodoferrocene and phenylboronic acid was performed using 1 mol% CD-capped Pd nanoparticles, generating a 70% isolated product yield of the product, phenylferrocene.[98] The attachment of the iodoferrocene to the CD receptor on the nanoparticle surface trapped the substrate at the active metal site, thus generating high product yields. In this regard, a control reaction was performed wherein ferrocene was added to the system to compete with iodoferrocene for available CD receptors, resulting in a much lower product yield. This nanoparticle system employed CD receptors as both a nanoparticle stabilizer and as a reagent gatekeeper wherein the specific binding between the receptor and substrates controlled the reactivity, leading to high degrees of specificity.

7.6 Pd Nanoparticles in the Heck Coupling Reaction

Development of Pd nanoparticles for Heck coupling was pioneered by Reetz and coworkers wherein they used tetraalkylammonium salts such as $(C_8H_{17})_4NBr$ instead of organic phosphine, as stabilizers for soluble Pd clusters.[186] Synthesis of the ligand-stabilized Pd nanoparticles was described wherein 2–3 nm particles were produced.[186] Heck coupling between iodobenzene and the acrylic acid n-butyl ester was performed using these materials in DMF at 30 °C to produce cinnamic acid ester. Nearly quantitative product yields were generated after 14 h, providing initial catalytic results for the phosphine-free Pd nanocatalysts.

In expanding the scope of reactivity in Heck coupling using Pd nanocatalysts, another stabilizer was employed for the synthesis of nanoparticles using an electrochemical method and thermolysis.[192] For this, propylene carbonate (PC) was used for the production of Pd nanoclusters through electrochemical methods involving anodic dissolution of a Pd sheet at 60 °C, which formed monodisperse, nearly spherical, highly crystalline particles of 8–10 nm in diameter.[192] In a separate process, $Pd(O_2CMe)_2$ was mixed with PC at 100 °C while sonicating to form 8–10 nm particles with similar shapes through thermal decomposition. The catalytic activities for the PC-Pd nanoparticles were tested using the Heck reaction between aryl halides and styrene in the presence of a base at temperatures ranging from 130–160 °C.[192] The results showed excellent product yields of 96% and 97% for the activated 4-bromonitrobenzene substrate in the presence of PC-Pd nanoparticles using NEt_3 and NaO_2CMe as the base, respectively. On the other hand, chlorobenzene was also able to generate significant product yields between 40% to 58%, but longer reaction times were required.[192]

Dendrimer-based Pd nanoparticles have also been used for the Heck reaction; however, the higher temperatures that were required to accomplish the coupling necessitated the use of highly stable dendrimers.[193] As such, poly(propylene imine) (PPI) dendrimers were employed to passivate the Pd nanoparticles for reactions that require refluxing. The perfluorinated polyether-derivatized PPI DENs (third and fifth generation) were exposed to Pd^{2+} ions to coordinate to the dendrimer interior. Addition of a reducing agent resulted in the production of nearly monodisperse Pd nanoparticles of ~2–3 nm in diameter according to TEM analysis.[193] The PPI-Pd nanoparticles were then used for Heck coupling between different aryl halides and n-butylacrylate to generate n-butyl-formylcinnamate using a fluorocarbon–hydrocarbon solvent and Et_3N as the base at 90 °C. The results indicated that the PPI-Pd nanocatalyst was able to drive the coupling reaction generating modest yields; however, regioselectivity of the product was established to be 100% selective toward trans-n-butyl-formylcinnamate.[193] Interestingly, the fifth generation PPI-Pd nanoparticles catalyzed higher product yields compared with the third generation PPI-Pd nanoparticles even though the diameters of the particles were similar. These results suggested that the fifth generation dendrimers provided larger encapsulated space wherein the coupling reaction could occur leading to the observed regioselectivity and higher product yields. In contrast, the third generation dendrimers contained less interior room such that slower reactions occurred, lowering the product yields. Nevertheless, product regioselectivity was maintained. As anticipated, lower yields were obtained when using aryl bromides compared with aryl iodides and no activation of aryl chlorides was noted. Finally, *para*-nitro-substituted aryl bromides generated lower product yields compared with unsubstituted aryl reagents, which was attributed to the repulsive forces between the electron-donating nitro-group and the perfluorinated polyether dendrimer groups, making substrate incorporation difficult.[193] Taken together, these results suggested that Heck coupling using the PPI-Pd nanoparticles occurred at the dendrimer interior.

Since the reaction using the dendrimer nanocatalysts was performed in a fluorocarbon–hydrocarbon solvent, separation and recovery of the PPI-Pd nanoparticles was easily achieved.[193] To this end, the nanocatalysts remained dissolved in the fluorous phase of the solvent, as shown by the retention of the dark color of the catalyst, while the organic phase that contained the product was colorless. Recyclability of the nanocatalyst was performed using the separated particles in the fluorous phase wherein significant loss of activity was observed with each reaction cycle. This loss in reactivity was attributed to changes in the morphology of the Pd nanoparticles due to the redox cycling of Pd atoms in the Heck reaction. Control reactions were also performed to determine the effect of base identity on the reaction. These results showed that even in the absence of Et$_3$N in the reaction, Heck coupling still occurred due to the interior tertiary amines in the dendrimers that acted as the base. Overall, the results of these initial studies of Pd DENs for Heck coupling showed highly regioselective intradendrimer catalytic reactivity under relatively ambient reaction conditions.

Further DEN effects for Pd nanoparticles were performed by Kaneda *et al.*, wherein the Pd complex was immobilized inside the dendrimer through ionic bonds (Figure 7.12).[91] Initially, the peripheral amino groups of PPI dendrimers were functionalized with decanoyl chloride and 3,4,5-triethoxybenzoyl chloride to produce alkylated and arylated dendritic materials, respectively. On the other hand, Pd nanoparticles were exposed to 4-diphenylphosphinobenzoic acid as ligands, which can then form ionic bonds between the carboxylic acid moiety and the internal amino groups of the dendrimers. To this end, the phosphine-stabilized Pd nanoparticles were encapsulated at the interior of the functionalized PPI dendrimers and employed as nanocatalysts in the Heck reaction.[91] Initially, the coupling between iodobenzene and *n*-butyl acrylate in toluene with KOAc was performed with the PPI modified Pd DENs. The results of the reaction showed that the dendrimer generation directly affected the reactivity where the reaction rate increased with higher generations; however, when the molar ratio of

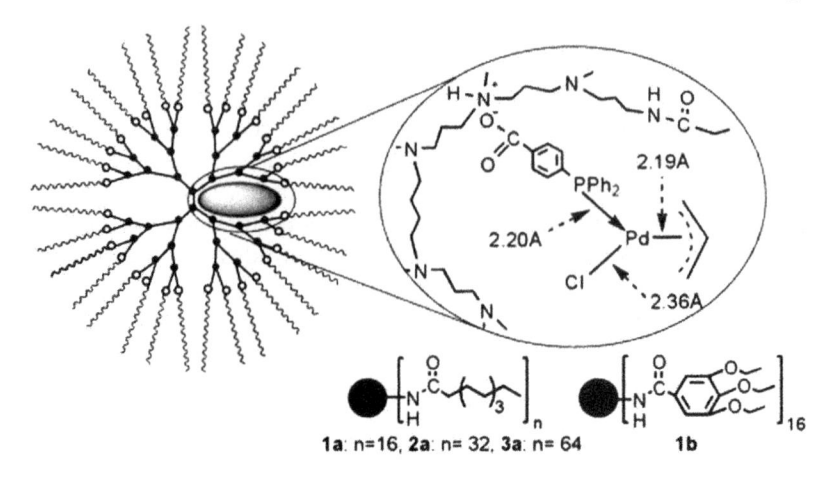

Figure 7.12 Schematic illustration of the Pd complex immobilized inside the dendrimer.[91]
Reprinted (adapted) with permission from M. Ooe, M. Murata, T. Mizugaki, K. Ebitani and K. Kaneda, *J. Am. Chem. Soc.*, 2004, **126**, 1604–1605. Copyright 2004 American Chemical Society.

phosphine ligand–Pd in the 5th-generation PPI Pd nanoparticles was increased from 1 to 4, product yields decreased from 92% to 19%.[91] These results suggested that the large amounts of amino groups in higher generation dendrimers provided complexation with the active species and as such led to high stability and catalytic reactivity.

To further probe the reactivity, as well as the selectivity of the nanocatalyst, a second Heck coupling reaction was employed using *p*-diiodobenzene and *n*-butylacrylate with two possible products, mono- and di-substituted benzene.[91] The results showed that in the presence of the nanocatalyst, only *n*-butyl-*p*-iodocinnamate, the mono-substituted product, was formed, indicating a highly reactive (86% yield) and selective (92%) nanocatalyst. Furthermore, the allylic amination of cinnamyl methyl carbonate was also catalyzed by the Pd DENs, demonstrating that the catalytic materials were quite versatile. Finally, the recyclability of the catalyst was established wherein minimal loss of activity was observed after four reaction cycles.

Expanding the scope of reactivity of the PPI-Pd nanoparticles for Heck coupling under environmentally friendly conditions was achieved using supercritical CO_2 (scCO_2) as the solvent.[194] The PPI dendrimer end groups were functionalized with a perfluoro-2,5,8,11-tetramethyl-3,6,9,12-tetraoxapentadecanoyl perfluoropolyether chain to enhance solubility with scCO_2, generating 1–2 nm Pd particles after reduction.[194] The modified PPI-Pd nanoparticles were then used as the coupling catalyst between iodobenzene and methylacrylate with Et$_3$N as the base at 75 °C and 5000 psi.[194] Formation of the possible products, methyl-2-phenylacrylate and *cis/trans*-cinnamate, was monitored using GC wherein 57% of the aryl iodide was consumed after 24 h.[194] Characterization of the reaction solution by ^{1}H NMR revealed that only the methyl-2-phenylacrylate product was formed and no cinnamate products or impurities were generated. These results indicated high selectivity for the nanocatalyst, which was attributed to the steric environment of the dendrimer system.[194]

Calculation of the turnover number (TON) generated a value of ~22 mol substrate/mol Pd only, which was suggestive of a slow reaction using the modified-PPI Pd nanoparticles.[194] A notable change in color of the scCO$_2$ solution containing the nanoparticles from dark-brown to dull-orange was observed after the reaction, which likely arises from Pd^0 oxidized to Pd^{2+}.[194] Upon treatment of this solution with NaBH$_4$, the formation of a dark-brown solution was observed, consistent with metal reduction and nanoparticle formation. The oxidation of the catalytically active Pd species explained the low catalytic conversion driven by the modified PPI-Pd nanoparticles. When the reaction was extended to 52 h, no visible Pd precipitation was observed.[194]

In contrast to the low TON values obtained in the Heck coupling reaction using the modified PPI-Pd nanoparticles, drastically higher TONs were generated using Pd-DENs with hydroxyl-terminated PAMAM dendrimers; however, the reaction was performed using DMA as the solvent instead of scCO$_2$.[92] In this system, the reaction between aryl halides and acrylic acid was employed to determine the reactivity of Pd@[PAMAM G4-OH] in the presence of NaOAc as the base at 140 °C. Two different Pd–dendrimer ratios, Pd$_{60}$ and Pd$_{40}$, were used for the reaction, where similar product yields were obtained for both systems.[92] The results of the coupling reaction showed significantly higher TON values depending on the halogen group in aryl halides such that iodobenzene gave a value of 30 000 mol product/mol Pd, while bromobenzene generated a TON of 14 000 mol product/mol Pd.[92] No product formation was observed for chlorobenzene. As expected, electron-withdrawing substituents on the aryl halide generated higher product yields and TONs, while decreasing the catalyst loading resulted in the corresponding decrease in yields and TON values.

Aside from Pd DENs, other ligand stabilized Pd catalysts have also been employed for Heck coupling. For instance, PEG-Pd nanoparticles were used for the coupling of *t*-butylstyrene with aryl iodides and aryl bromides in the presence of K$_2$CO$_3$ in water under refluxing conditions (80–100 °C).[97] The results demonstrated excellent yields of 90–92% for the reaction using activated iodo- and bromo-arenes. Interestingly, even sterically hindered *ortho*-substituted aryl bromides generated relatively good yields of 72–80%, while the *meta*-substituted substrates produced poor (5%) product yields when coupled with *t*-butylstyrene.[97] Overall, the PEG-Pd nanoparticles showed high reactivity for Heck coupling under relatively mild reaction conditions with water as the solvent, indicating a positive development towards environmentally friendly, but highly efficient nanoparticle catalysts.

Another polymer-based catalytic system used for Heck coupling was the ionic polymer-ionic liquid stabilized Pd nanoparticles, Pd-IP-IL.[135] The 5 nm Pd-IP-IL nanoparticles were used in the Heck reaction between aryl halides and ethyl acrylate using tributylamine as the base in the ionic liquid at 80 °C.

The results displayed that quantitative yields (>99%) of the expected products were obtained from activated aryl halides after only 2 h. Iodobenzene, on the other hand, only managed to generate 49% of the product after 1.5 h, while increasing the reaction time to 4 h showed a corresponding yield increase to 81%.[135] These results indicated that the ionic liquid was a great solvent for the Heck reaction and the use of Pd-IP-IL nanoparticles shows excellent reactivity under ambient reaction conditions.

Finally, the Keggin-type polyoxometallate-stabilized Pd nanoparticles were also employed for the Heck coupling reaction.[176] Initially, the reaction between bromotoluene and styrene was performed employing diisopropylamine as the base in a water–ethanol solvent. Interestingly, excellent yields of 95% were obtained for the Heck coupling product, stilbene, at 80–85 °C after 16 h. When activated aryl chloride, 1-chloro-4-nitrobenzene, was used under similar reaction conditions, a 91% product yield was obtained.[176] Similarly, for the Heck reaction between aryl halides and methyl acrylate, an almost quantitative yield (97%) of methyl cinnamate was generated from aryl bromide, while a 94% product yield was obtained using aryl chloride.[176] The mild reaction conditions employed in the coupling reaction suggested a highly efficient nanocatalyst that was operational under environmentally-friendly conditions.

7.7 Summary and Conclusions

In this chapter, we presented the recent advances in Pd nanoparticle research with regards to their applications in select C–C coupling reactions. Significant numbers of studies have been devoted to the development of new nanomaterials over the past few decades, leading to the production of remarkably efficient nanocatalysts. The advantages of nanomaterials in catalytic transformations include the use of very low metal loading, their ability to catalyze reactions under ambient conditions, their use in energy-efficient reactions, and their recyclability with minimal activity loss; however, the catalytic mechanism driven by these Pd nanoparticles remains a subject of debate with two different theories suggested for C–C couplings. The surface-based mechanism invokes a truly heterogeneous catalyst in which the reaction occurs directly at the surface of the nanoparticles. As such, minimal morphological changes should be observed for these nanoparticles. In contrast, the atom-leaching mechanism indicates that the active Pd^0 atoms are abstracted from the nanoparticle surface during the reaction. In this process, changes in size, shape and crystal structure should be observed during the catalytic cycle. Although each of these theories can be supported by the available evidence, additional studies are necessary in order to verify the reaction mechanism. Compounding this issue, different reaction mechanisms may be possible for different materials, which may give rise to the disparate set of results present in the literature. Nevertheless, Pd nanoparticles have demonstrated excellent reactivities for C–C coupling catalytic reactions, which may serve as model systems for the development of sustainable materials for future chemical transformations.

References

1. M. R. Kember and C. K. Williams, *J. Am. Chem. Soc.*, 2012, **134**, 15676–15679.
2. N. Hanada, T. Ichikawa and H. Fujii, *J. Phys. Chem. B*, 2005, **109**, 7188–7194.
3. A. Ganguly, P. Trinh, K. V. Ramanujachary, T. Ahmad, A. Mugweru and A. K. Ganguli, *J. Colloid Interface Sci.*, 2011, **353**, 137–142.
4. A. B. Patil and B. M. Bhanage, *Catal. Commun.*, 2013, **36**, 79–83.
5. C. López and A. Corma, *ChemCatChem*, 2012, **4**, 751–752.
6. J. F. Sonnenberg, N. Coombs, P. A. Dube and R. H. Morris, *J. Am. Chem. Soc.*, 2012, **134**, 5893–5899.
7. H. M. Torres Galvis, J. H. Bitter, C. B. Khare, M. Ruitenbeek, A. I. Dugulan and K. P. de Jong, *Science*, 2012, **335**, 835–838.
8. S. Lastella, Y. J. Jung, H. Yang, R. Vajtai, P. M. Ajayan, C. Y. Ryu, D. A. Rider and I. Manners, *J. Mater. Chem.*, 2004, **14**, 1791–1794.
9. Y. Moglie, C. Vitale and G. Radivoy, *Tetrahedron Lett.*, 2008, **49**, 1828–1831.
10. K. Shimizu, K. Kon, W. Onodera, H. Yamazaki and J. N. Kondo, *ACS Catal.*, 2012, **3**, 112–117.
11. H. Y. Wang and A. C. Lua, *J. Phys. Chem. C*, 2012, **116**, 26765–26775.
12. P.-Z. Li, A. Aijaz and Q. Xu, *Angew. Chem., Int. Ed.*, 2012, **51**, 6753–6756.
13. F. Alonso, I. Osante and M. Yus, *Adv. Synth. Catal.*, 2006, **348**, 305–308.
14. A. Wang, H. Yin, H. Lu, J. Xue, M. Ren and T. Jiang, *Catal. Commun.*, 2009, **10**, 2060–2064.
15. W. Zhang, H. Qi, L. Li, X. Wang, J. Chen, K. Peng and Z. Wang, *Green Chem.*, 2009, **11**, 1194–1200.
16. I. Geukens, J. Fransaer and D. E. De Vos, *ChemCatChem*, 2011, **3**, 1431–1434.
17. Y. Chen, P. Crawford and P. Hu, *Catal. Lett.*, 2007, **119**, 21–28.
18. L. Guczi, D. Horváth, Z. Pászti, L. Tóth, Z. E. Horváth, A. Karacs and G. Petö, *J. Phys. Chem. B*, 2000, **104**, 3183–3193.
19. M. Kidwai, V. Bansal, A. Kumar and S. Mozumdar, *Green Chem.*, 2007, **9**, 742–745.
20. M. Kidwai and S. Bhardwaj, *Appl. Catal., A*, 2010, **387**, 1–4.
21. T. Mitsudome, A. Noujima, T. Mizugaki, K. Jitsukawa and K. Kaneda, *Chem. Commun.*, 2009, 5302–5304.
22. L. Pasquato, F. Rancan, P. Scrimin, F. Mancin and C. Frigeri, *Chem. Commun.*, 2000, 2253–2254.
23. L. Wang, X. Meng, B. Wang, W. Chi and F.-S. Xiao, *Chem. Commun.*, 2010, **46**, 5003–5005.
24. X. Jiang, Y. Xie, J. Lu, L. Zhu, W. He and Y. Qian, *Langmuir*, 2001, **17**, 3795–3799.

25. Y. Mikami, A. Noujima, T. Mitsudome, T. Mizugaki, K. Jitsukawa and K. Kaneda, *Tetrahedron Lett.*, 2010, **51**, 5466–5468.
26. Y. Mikami, A. Noujima, T. Mitsudome, T. Mizugaki, K. Jitsukawa and K. Kaneda, *Chem. Lett.*, 2010, **39**, 223–225.
27. S. Saha, A. Pal, S. Kundu, S. Basu and T. Pal, *Langmuir*, 2010, **26**, 2885–2893.
28. A. Sutton, G. Franc and A. Kakkar, *J. Polym. Sci., Part A: Polym. Chem*, 2009, **47**, 4482–4493.
29. M. Steffan, A. Jakob, P. Claus and H. Lang, *Catal. Commun.*, 2009, **10**, 437–441.
30. A. Y. Kim, H. Bae, S. Park, S. Park and K. Park, *Catal. Lett.*, 2011, **141**, 685–690.
31. T. Ueno, M. Suzuki, T. Goto, T. Matsumoto, K. Nagayama and Y. Watanabe, *Angew. Chem., Int. Ed.*, 2004, **43**, 2527–2530.
32. O. M. Wilson, M. R. Knecht, J. C. Garcia-Martinez and R. M. Crooks, *J. Am. Chem. Soc.*, 2006, **128**, 4510–4511.
33. Y. Niu, L. K. Yeung and R. M. Crooks, *J. Am. Chem. Soc.*, 2001, **123**, 6840–6846.
34. T. H. Galow, U. Drechsler, J. A. Hanson and V. M. Rotello, *Chem. Commun.*, 2002, 1076–1077.
35. N. Semagina and L. Kiwi-Minsker, *Catal. Lett.*, 2009, **127**, 334–338.
36. Y. Mei, Y. Lu, F. Polzer, M. Ballauff and M. Drechsler, *Chem. Mater.*, 2007, **19**, 1062–1069.
37. R. Bhandari and M. R. Knecht, *ACS Catal.*, 2011, **1**, 89–98.
38. S. Harish, J. Mathiyarasu, K. L. N. Phani and V. Yegnaraman, *Catal. Lett.*, 2009, **128**, 197–202.
39. M. Faticanti, N. Cioffi, S. De Rossi, N. Ditaranto, P. Porta, L. Sabbatini and T. Bleve-Zacheo, *Appl. Catal., B*, 2005, **60**, 73–82.
40. G. Glaspell, L. Fuoco and M. S. El-Shall, *J. Phys. Chem. B*, 2005, **109**, 17350–17355.
41. J.-N. Park, A. J. Forman, W. Tang, J. Cheng, Y.-S. Hu, H. Lin and E. W. McFarland, *Small*, 2008, **4**, 1694–1697.
42. T. Schalow, B. Brandt, M. Laurin, S. Schauermann, J. Libuda and H.-J. Freund, *J. Catal.*, 2006, **242**, 58–70.
43. H. Wang, Z. Jusys and R. J. Behm, *J. Phys. Chem. B*, 2004, **108**, 19413–19424.
44. B. Xu, J. Guo, H. Jia, X. Yang and X. Liu, *Catal. Today*, 2007, **125**, 169–172.
45. P. Inkaew, W. Zhou and C. Korzeniewski, *J. Electroanal. Chem.*, 2008, **614**, 93–100.
46. S. Mostafa, J. R. Croy, H. Heinrich and B. R. Cuenya, *Appl. Catal., A*, 2009, **366**, 353–362.
47. J. Li, X. Liang, D. M. King, Y.-B. Jiang and A. W. Weimer, *Appl. Catal., B*, 2010, **97**, 220–226.
48. N. V. Long, M. Ohtaki, M. Nogami and T. D. Hien, *Colloid Polym. Sci.*, 2011, **289**, 1373–1386.

49. H. Wu, R. Tang, Q. He, X. Liao and B. Shi, *J. Chem. Technol. Biotechnol.*, 2009, **84**, 1702–1711.
50. J. N. Armor, *Appl. Catal., A*, 2000, **194–195**, 3–11.
51. G. Centi, P. Ciambelli, S. Perathoner and P. Russo, *Catal. Today*, 2002, **75**, 3–15.
52. D. Astruc, *Inorg. Chem.*, 2007, **46**, 1884–1894.
53. C. Barnard, *Platinum Met. Rev.*, 2008, **52**, 38–45.
54. T. J. Colacot, *Platinum Met. Rev.*, 2009, **53**, 183–188.
55. R. Jin, *Nanotechnol. Rev.*, 2012, **1**, 31–56.
56. J. Yin, *Applications of Transition Metal Catalysis in Drug Discovery and Development*, ed. M. L. Crawley and B. M. Trost, John Wiley & Sons, Inc., Hoboken, NJ, 1st edn, 2012, pp. 97–163.
57. R. F. Heck and J. P. Nolley, *J. Org. Chem.*, 1972, **37**, 2320–2322.
58. A. Balanta, C. Godard and C. Claver, *Chem. Soc. Rev.*, 2011, **40**, 4973–4985.
59. K. Sonogashira, Y. Tohda and N. Hagihara, *Tetrahedron Lett.*, 1975, **16**, 4467–4470.
60. D. Milstein and J. K. Stille, *J. Am. Chem. Soc.*, 1978, **100**, 3636–3638.
61. E. Negishi, A. O. King and N. Okukado, *J. Org. Chem.*, 1977, **42**, 1821–1823.
62. T. Hayashi, M. Konishi and M. Kumada, *Tetrahedron Lett.*, 1979, **20**, 1871–1874.
63. N. Miyaura, K. Yamada and A. Suzuki, *Tetrahedron Lett.*, 1979, **20**, 3437–3440.
64. N. Miyaura and A. Suzuki, *J. Chem. Soc., Chem. Commun.*, 1979, 866–867.
65. T. J. Colacot, *Platinum Met. Rev.*, 2011, **55**, 84–90.
66. *Nat. Mater.*, 2011, **10**, 333.
67. F. Qing, Y. Sun, X. Wang, N. Li, Y. Li, X. Li and H. Wang, *Polym. Chem.*, 2011, **2**, 2102–2106.
68. N. C. Thanh, M. Ikai, T. Kajioka, H. Fujikawa, Y. Taga, Y. Zhang, S. Ogawa, H. Shimada, Y. Miyahara, S. Kuroda and M. Oda, *Tetrahedron*, 2006, **62**, 11227–11239.
69. H. A. M. van Mullekom, J. A. J. M. Vekemans and E. W. Meijer, *Chem.-Eur. J.*, 1998, **4**, 1235–1243.
70. P. D. Thornton, N. Brown, D. Hill, B. Neuenswander, G. H. Lushington, C. Santini and K. R. Buszek, *ACS Comb. Sci.*, 2011, **13**, 443–448.
71. P. S. Deore and N. P. Argade, *J. Org. Chem.*, 2011, 77, 739–746.
72. W. Bentoumi, J. Helhaik, G. Plé and Y. Ramondenc, *Tetrahedron*, 2009, **65**, 1967–1970.
73. F. Zeng and E. Negishi, *Org. Lett.*, 2001, **3**, 719–722.
74. S. J. Danishefsky, J. J. Masters, W. B. Young, J. T. Link, L. B. Snyder, T. V. Magee, D. K. Jung, R. C. A. Isaacs, W. G. Bornmann, C. A. Alaimo, C. A. Coburn and M. J. Di Grandi, *J. Am. Chem. Soc.*, 1996, **118**, 2843–2859.
75. C. Y. Hong, N. Kado and L. E. Overman, *J. Am. Chem. Soc.*, 1993, **115**, 11028–11029.

76. J. Barluenga, M. Tomás-Gamasa, P. Moriel, F. Aznar and C. Valdés, *Chem.–Eur. J.*, 2008, **14**, 4792–4795.
77. K. Pomeisl, A. Holy, R. Pohl and K. Horska, *Tetrahedron*, 2009, **65**, 8486–8492.
78. C. Burda, X. Chen, R. Narayanan and M. A. El-Sayed, *Chem. Rev.*, 2005, **105**, 1025–1102.
79. J. Peréz-Juste, I. Pastoriza-Santos, L. M. Liz-Marzán and P. Mulvaney, *Coord. Chem. Rev.*, 2005, **249**, 1870–1901.
80. Y. Xia, Y. Xiong, B. Lim and S. E. Skrabalak, *Angew. Chem., Int. Ed.*, 2009, **48**, 60–103.
81. M. Rycenga, C. M. Cobley, J. Zeng, W. Li, C. H. Moran, Q. Zhang, D. Qin and Y. Xia, *Chem. Rev.*, 2011, **111**, 3669–3712.
82. H. You, S. Yang, B. Ding and H. Yang, *Chem. Soc. Rev.*, 2013, **42**, 2880–2904.
83. R. Coppage, J. M. Slocik, B. D. Briggs, A. I. Frenkel, R. R. Naik and M. R. Knecht, *ACS Nano*, 2012, **6**, 1625–1636.
84. M.-C. Daniel and D. Astruc, *Chem. Rev.*, 2003, **104**, 293–346.
85. X. Huang, S. Tang, X. Mu, Y. Dai, G. Chen, Z. Zhou, F. Ruan, Z. Yang and N. Zheng, *Nat. Nanotechnol.*, 2011, **6**, 28–32.
86. J. C. Garcia-Martinez, R. Lezutekong and R. M. Crooks, *J. Am. Chem. Soc.*, 2005, **127**, 5097–5103.
87. A. Jakhmola, R. Bhandari, D. B. Pacardo and M. R. Knecht, *J. Mater. Chem.*, 2010, **20**, 1522–1531.
88. D. B. Pacardo, M. Sethi, S. E. Jones, R. R. Naik and M. R. Knecht, *ACS Nano*, 2009, **3**, 1288–1296.
89. D. Astruc, *Tetrahedron: Asymmetry*, 2010, **21**, 1041–1054.
90. M. Bernechea, E. de Jesús, C. López-Mardomingo and P. Terreros, *Inorg. Chem.*, 2009, **48**, 4491–4496.
91. M. Ooe, M. Murata, T. Mizugaki, K. Ebitani and K. Kaneda, *J. Am. Chem. Soc.*, 2004, **126**, 1604–1605.
92. E. H. Rahim, F. S. Kamounah, J. Frederiksen and J. B. Christensen, *Nano Lett.*, 2001, **1**, 499–501.
93. L. Wu, Z.-W. Li, F. Zhang, Y.-M. He and Q.-H. Fan, *Adv. Synth. Catal.*, 2008, **350**, 846–862.
94. R. Coppage, J. M. Slocik, M. Sethi, D. B. Pacardo, R. R. Naik and M. R. Knecht, *Angew. Chem., Int. Ed.*, 2010, **49**, 3767–3770.
95. P. Dutta and A. Sarkar, *Adv. Synth. Catal.*, 2011, **353**, 2814–2822.
96. Y. Li, X. M. Hong, D. M. Collard and M. A. El-Sayed, *Org. Lett.*, 2000, **2**, 2385–2388.
97. S. Sawoo, D. Srimani, P. Dutta, R. Lahiri and A. Sarkar, *Tetrahedron*, 2009, **65**, 4367–4374.
98. L. Strimbu, J. Liu and A. E. Kaifer, *Langmuir*, 2003, **19**, 483–485.
99. C. Liang, W. Xia, M. van den Berg, Y. Wang, H. Soltani-Ahmadi, O. Schlüter, R. A. Fischer and M. Muhler, *Chem. Mater.*, 2009, **21**, 2360–2366.
100. C. Liang, W. Xia, H. Soltani-Ahmadi, O. Schlüter, R. A. Fischer and M. Muhler, *Chem. Commun.*, 2005, 282–284.

101. A. Binder, M. Seipenbusch, M. Muhler and G. Kasper, *J. Catal.*, 2009, **268**, 150–155.
102. G. Cristoforetti, E. Pitzalis, R. Spiniello, R. Ishak, F. Giammanco, M. Muniz-Miranda and S. Caporali, *Appl. Surf. Sci.*, 2012, **258**, 3289–3297.
103. S. Z. Mortazavi, P. Parvin, A. Reyhani, A. N. Golikand and S. Mirershadi, *J. Phys. Chem. C*, 2011, **115**, 5049–5057.
104. C.-B. Hwang, Y.-S. Fu, Y.-L. Lu, S.-W. Jang, P.-T. Chou, C. R. C. Wang and S. J. Yu, *J. Catal.*, 2000, **195**, 336–341.
105. C. E. Allmond, A. T. Sellinger, K. Gogick and J. M. Fitz-Gerald, *Appl. Phys. A: Mater. Sci. Process.*, 2007, **86**, 477–480.
106. T. Fujimoto, S. Terauchi, H. Umehara, I. Kojima and W. Henderson, *Chem. Mater.*, 2001, **13**, 1057–1060.
107. K. Okitsu, A. Yue, S. Tanabe and H. Matsumoto, *Chem. Mater.*, 2000, **12**, 3006–3011.
108. N. Kijima, Y. Takahashi, J. Akimoto, T. Tsunoda, K. Uchida and Y. Yoshimura, *Chem. Lett.*, 2005, **34**, 1658–1659.
109. N. Arul Dhas and A. Gedanken, *J. Mater. Chem.*, 1998, **8**, 445–450.
110. K. Okitsu, Y. Mizukoshi, H. Bandow, T. A. Yamamoto, Y. Nagata and Y. Maeda, *J. Phys. Chem. B*, 1997, **101**, 5470–5472.
111. W. Pan, X. Zhang, H. Ma and J. Zhang, *J. Phys. Chem. C*, 2008, **112**, 2456–2461.
112. C.-C. Wu, D.-S. Chan, C.-C. Yang, H.-P. Chiou, H.-R. Chen and C.-L. Lee, *J. Electrochem. Soc.*, 2011, **158**, D616–D620.
113. P. Zhang and T. K. Sham, *Appl. Phys. Lett.*, 2003, **82**, 1778–1780.
114. J. Liu, F. He, T. M. Gunn, D. Zhao and C. B. Roberts, *Langmuir*, 2009, **25**, 7116–7128.
115. Y. Zhang, G. Chang, S. Liu, J. Tian, L. Wang, W. Lu, X. Qin and X. Sun, *Catal. Sci. Technol.*, 2011, **1**, 1636–1640.
116. J. Kou and R. S. Varma, *RSC Adv.*, 2012, **2**, 10283–10290.
117. J. Chen, Z. Li, D. Chao, W. Zhang and C. Wang, *Mater. Lett.*, 2008, **62**, 692–694.
118. J. Han, Z. Zhou, Y. Yin, X. Luo, J. Li, H. Zhang and B. Yang, *CrystEngComm*, 2012, **14**, 7036–7042.
119. R. W. J. Scott, H. Ye, R. R. Henriquez and R. M. Crooks, *Chem. Mater.*, 2003, **15**, 3873–3878.
120. M. Zhao and R. M. Crooks, *Angew. Chem., Int. Ed.*, 1999, **38**, 364–366.
121. Z. Li, J. Gao, X. Xing, S. Wu, S. Shuang, C. Dong, M. C. Paau and M. M. F. Choi, *J. Phys. Chem. C*, 2010, **114**, 723–733.
122. D. H. Turkenburg, A. A. Antipov, M. B. Thathagar, G. Rothenberg, G. B. Sukhorukov and E. Eiser, *Phys. Chem. Chem. Phys.*, 2005, 7, 2237–2240.
123. V. Mazumder and S. Sun, *J. Am. Chem. Soc.*, 2009, **131**, 4588–4589.
124. F. Lu, J. Ruiz and D. Astruc, *Tetrahedron Lett.*, 2004, **45**, 9443–9445.
125. S. Chen, K. Huang and J. A. Stearns, *Chem. Mater.*, 2000, **12**, 540–547.

126. D. E. Cliffel, F. P. Zamborini, S. M. Gross and R. W. Murray, *Langmuir*, 2000, **16**, 9699–9702.
127. F. P. Zamborini, S. M. Gross and R. W. Murray, *Langmuir*, 2001, **17**, 481–488.
128. S.-W. Kim, J. Park, Y. Jang, Y. Chung, S. Hwang, T. Hyeon and Y. W. Kim, *Nano Lett.*, 2003, **3**, 1289–1291.
129. M. Tamura and H. Fujihara, *J. Am. Chem. Soc.*, 2003, **125**, 15742–15743.
130. S. U. Son, Y. Jang, K. Y. Yoon, E. Kang and T. Hyeon, *Nano Lett.*, 2004, **4**, 1147–1151.
131. E. Drinkel, F. D. Souza, H. D. Fiedler and F. Nome, *Curr. Opin. Colloid Interface Sci.*, 2013, **18**, 26–34.
132. E. Zapp, F. D. Souza, B. S. Souza, F. Nome, A. Neves and I. C. Vieira, *Analyst*, 2013, **138**, 509–517.
133. B. S. Souza, E. C. Leopoldino, D. W. Tondo, J. Dupont and F. Nome, *Langmuir*, 2012, **28**, 833–840.
134. V. Calò, A. Nacci, A. Monopoli and F. Montingelli, *J. Org. Chem.*, 2005, **70**, 6040–6044.
135. X. Yang, Z. Fei, D. Zhao, W. H. Ang, Y. Li and P. J. Dyson, *Inorg. Chem.*, 2008, **47**, 3292–3297.
136. Y. Zeng, Y. Wang, Y. Xu, Y. Song, J. Jiang and Z. Jin, *Catal. Lett.*, 2013, **143**, 200–205.
137. J. Huang, T. Jiang, H. Gao, B. Han, Z. Liu, W. Wu, Y. Chang and G. Zhao, *Angew. Chem., Int. Ed.*, 2004, **43**, 1397–1399.
138. S.-K. Oh, Y. Niu and R. M. Crooks, *Langmuir*, 2005, **21**, 10209–10213.
139. X. Huang, Y. Wang, X. Liao and B. Shi, *Chem. Commun.*, 2009, 4687–4689.
140. A. P. Umpierre, G. Machado, G. H. Fecher, J. Morais and J. Dupont, *Adv. Synth. Catal.*, 2005, **347**, 1404–1412.
141. N. Karousis, G.-E. Tsotsou, F. Evangelista, P. Rudolf, N. Ragoussis and N. Tagmatarchis, *J. Phys. Chem. C*, 2008, **112**, 13463–13469.
142. J.-P. Tessonnier, L. Pesant, G. Ehret, M. J. Ledoux and C. Pham-Huu, *Appl. Catal., A*, 2005, **288**, 203–210.
143. B. Yoon and C. M. Wai, *J. Am. Chem. Soc.*, 2005, **127**, 17174–17175.
144. J. Y. Kim, Y. Jo, S.-K. Kook, S. Lee and H. C. Choi, *J. Mol. Catal. A: Chem.*, 2010, **323**, 28–32.
145. R. W. J. Scott, O. M. Wilson and R. M. Crooks, *Chem. Mater.*, 2004, **16**, 5682–5688.
146. N. G. Willis and J. Guzman, *Appl. Catal., A*, 2008, **339**, 68–75.
147. S. Moussa, V. Abdelsayed and M. S. El-Shall, *Chem. Phys. Lett.*, 2011, **510**, 179–184.
148. H.-F. Wang, W. E. Kaden, R. Dowler, M. Sterrer and H.-J. Freund, *Phys. Chem. Chem. Phys.*, 2012, **14**, 11525–11533.
149. S.-S. Lee, B.-K. Park, S.-H. Byeon, F. Chang and H. Kim, *Chem. Mater.*, 2006, **18**, 5631–5633.
150. K. Mori, K. Furubayashi, S. Okada and H. Yamashita, *RSC Adv.*, 2012, **2**, 1047–1054.

151. J. D. Webb, S. MacQuarrie, K. McEleney and C. M. Crudden, *J. Catal.*, 2007, **252**, 97–109.
152. S. Komhom, O. Mekasuwandumrong, P. Praserthdam and J. Panpranot, *Catal. Commun.*, 2008, **10**, 86–91.
153. S. Kidambi and M. L. Bruening, *Chem. Mater.*, 2005, **17**, 301–307.
154. H. Feng, J. A. Libera, P. C. Stair, J. T. Miller and J. W. Elam, *ACS Catal.*, 2011, **1**, 665–673.
155. K. Mori, Y. Miura, S. Shironita and H. Yamashita, *Langmuir*, 2009, **25**, 11180–11187.
156. M. E. Domine, M. C. Hernández-Soto, M. T. Navarro and Y. Pérez, *Catal. Today*, 2011, **172**, 13–20.
157. F. Li, Q. Zhang and Y. Wang, *Appl. Catal., A*, 2008, **334**, 217–226.
158. M. Ghiaci, F. Ansari, Z. Sadeghi and A. Gil, *Catal. Commun.*, 2012, **21**, 82–85.
159. A. Drelinkiewicza, A. Waksmundzka, W. Makowski, J. W. Sobczak, A. Król and A. Zieba, *Catal. Lett.*, 2004, **94**, 143–156.
160. J. H. Ding and D. L. Gin, *Chem. Mater.*, 2000, **12**, 22–24.
161. R. Xing, Y. Liu, H. Wu, X. Li, M. He and P. Wu, *Chem. Commun.*, 2008, 6297–6299.
162. H. R. Choi, H. Woo, S. Jang, J. Y. Cheon, C. Kim, J. Park, K. H. Park and S. H. Joo, *ChemCatChem*, 2012, **4**, 1587–1594.
163. K. Anderson, S. Cortiñas Fernandez, C. Hardacre and P. C. Marr, *Inorg. Chem. Commun.*, 2004, **7**, 73–76.
164. M. Chtchigrovsky, Y. Lin, K. Ouchaou, M. Chaumontet, M. Robitzer, F. Quignard and F. Taran, *Chem. Mater.*, 2012, **24**, 1505–1510.
165. Z. Zhang and Z. Wang, *J. Org. Chem.*, 2006, **71**, 7485–7487.
166. N. T. S. Phan, M. Van Der Sluys and C. W. Jones, *Adv. Synth. Catal.*, 2006, **348**, 609–679.
167. F. Zhao, B. M. Bhanage, M. Shirai and M. Arai, *Chem.-Eur. J.*, 2000, **6**, 843–848.
168. M. B. Thathagar, J. E. ten Elshof and G. Rothenberg, *Angew. Chem., Int. Ed.*, 2006, **45**, 2886–2890.
169. A. V. Gaikwad, A. Holuigue, M. B. Thathagar, J. E. ten Elshof and G. Rothenberg, *Chem.-Eur. J.*, 2007, **13**, 6908–6913.
170. Z. Niu, Q. Peng, Z. Zhuang, Q. He and Y. Li, *Chem.-Eur. J.*, 2012, **18**, 9813–9817.
171. P. J. Ellis, I. J. S. Fairlamb, S. F. J. Hackett, K. Wilson and A. F. Lee, *Angew. Chem., Int. Ed.*, 2010, **49**, 1820–1824.
172. K. Menzel and G. C. Fu, *J. Am. Chem. Soc.*, 2003, **125**, 3718–3719.
173. H. Zhao, Y. Wang, J. Sha, S. Sheng and M. Cai, *Tetrahedron*, 2008, **64**, 7517–7523.
174. A. Pal, R. Ghosh, N. N. Adarsh and A. Sarkar, *Tetrahedron*, 2010, **66**, 5451–5458.
175. J. K. Stille, *Angew. Chem., Int. Ed.*, 1986, **25**, 508–524.
176. V. Kogan, Z. Aizenshtat, R. Popovitz-Biro and R. Neumann, *Org. Lett.*, 2002, **4**, 3529–3532.

177. N. Kröger, R. Deutzmann and M. Sumper, *Science*, 1999, **286**, 1129–1132.
178. N. Kröger, S. Lorenz, E. Brunner and M. Sumper, *Science*, 2002, **298**, 584–586.
179. M. R. Knecht and D. W. Wright, *Chem. Commun.*, 2003, 3038–3039.
180. S. L. Sewell and D. W. Wright, *Chem. Mater.*, 2006, **18**, 3108–3113.
181. R. Bhandari and M. R. Knecht, *Catal. Sci. Technol.*, 2012, **2**, 1360–1366.
182. R. R. Naik, S. E. Jones, C. J. Murray, J. C. McAuliffe, R. A. Vaia and M. O. Stone, *Adv. Funct. Mater.*, 2004, **14**, 25–30.
183. R. B. Pandey, H. Heinz, J. Feng, B. L. Farmer, J. M. Slocik, L. F. Drummy and R. R. Naik, *Phys. Chem. Chem. Phys.*, 2009, **11**, 1989–2001.
184. D. B. Pacardo, J. M. Slocik, K. C. Kirk, R. R. Naik and M. R. Knecht, *Nanoscale*, 2011, **3**, 2194–2201.
185. D. B. Pacardo and M. R. Knecht, *Catal. Sci. Technol.*, 2013, **3**, 745–753.
186. M. T. Reetz, R. Breinbauer and K. Wanninger, *Tetrahedron Lett.*, 1996, **37**, 4499–4502.
187. Y. Li and M. A. El-Sayed, *J. Phys. Chem. B*, 2001, **105**, 8938–8943.
188. B. J. Gallon, R. W. Kojima, R. B. Kaner and P. L. Diaconescu, *Angew. Chem., Int. Ed.*, 2007, **46**, 7251–7254.
189. P. Zhou, H. Wang, J. Yang, J. Tang, D. Sun and W. Tang, *RSC Adv.*, 2012, **2**, 1759–1761.
190. M. Pittelkow, K. Moth-Poulsen, U. Boas and J. B. Christensen, *Langmuir*, 2003, **19**, 7682–7684.
191. C. Ornelas, J. Ruiz, L. Salmon and D. Astruc, *Adv. Synth. Catal.*, 2008, **350**, 837–845.
192. M. T. Reetz and G. Lohmer, *Chem. Commun.*, 1996, 1921–1922.
193. L. K. Yeung and R. M. Crooks, *Nano Lett.*, 2000, **1**, 14–17.
194. L. K. Yeung, C. T. Lee Jr, K. P. Johnston and R. M. Crooks, *Chem. Commun.*, 2001, 2290–2291.

CHAPTER 8

Metal Salt-based Gold Nanocatalysts

ZHEN MA*[a] AND FRANKLIN (FENG) TAO*[b]

[a] Department of Environmental Science and Engineering, Fudan University, Shanghai, 200433, P.R. China; [b] Department of Chemistry and Biochemistry, University of Notre Dame, Notre Dame, IN 46556, USA
*Email: zhenma@fudan.edu.cn; Franklin.Tao.7@nd.edu

8.1 Introduction

Gold was traditionally regarded as non-reactive in chemistry and useless in catalysis. Indeed, the chemical industry relies heavily on other metal catalysts, such as supported platinum, palladium, and silver catalysts. The fact that bulk gold is stable in air does not mean that gold is not reactive in other reaction media. Likewise, the fact that gold catalysts prepared inappropriately (with large gold particles dispersed on certain supports) are inactive in catalysis does not mean that gold catalysts prepared properly (with small and finely divided gold nanoparticles) are useless. In the 1980s, Haruta and co-workers found that gold catalysts prepared by co-precipitation or deposition–precipitation can be highly active for CO oxidation below room temperature.[1,2] Many gold nanocatalysts have been developed ever since,[3–7] and some can catalyze CO oxidation below −80 °C.[8,9]

Catalytic CO oxidation is an important reaction in saving lives and for purifying indoor air. The fact that a catalyst can catalyze CO oxidation below −80 °C does not mean we have to use a cooler in practical applications. Rather, a high activity of a catalyst at a very low temperature usually dilates an even higher activity of the catalyst at or above room temperature, *i.e.* it can convert a larger volume of contaminated air at room temperature or it can achieve complete conversion at room temperature with a smaller amount of catalyst. Room-temperature catalysis is energy-efficient. In addition, a high activity of a catalyst at low temperature hints that the catalyst has extraordinary active sites. These active sites have been demonstrated to be active in CO oxidation, and they may (or may not) be active in other reactions. Much can be done to explore the potential applications of these gold nanocatalysts.[10–21]

Supported gold catalysts are portrayed as gold nanoparticles dispersed on solid supports. They are usually prepared by impregnation, co-precipitation, deposition–precipitation, or colloidal deposition.[3] Solid supports frequently used for loading gold are usually oxides (*e.g.* TiO_2, ZrO_2, Al_2O_3, CeO_2, SiO_2) and carbon.[3] It was initially proposed that supports with redox properties (*e.g.* TiO_2, ZrO_2, CeO_2) can activate oxygen and are therefore good for making gold catalysts active for CO oxidation, whereas non-reducible supports (*e.g.* Al_2O_3, SiO_2) are not good for making gold catalysts active for CO oxidation.[22] However, later evidence showed that Au/Al_2O_3[23] and Au/SiO_2[24,25] prepared appropriately can also be active for CO oxidation below room temperature.

Metal salts are seldom used as supports for loading gold due to some pitfalls. Firstly, many metal salts (*e.g.* metal chlorides, metal nitrates, metal sulfates) are soluble in water, whereas the preparation, storage, and application of supported catalysts often entails water or moisture. Secondly, some metal salts (*e.g.* metal nitrates, metal sulfates) can decompose at elevated temperatures. However, it is still interesting to develop metal salt-based gold nanocatalysts because some metal salts (*e.g.* metal phosphates) are solid acids useful in heterogeneous acid catalysis.[26] The combination of acidic supports and gold nanoparticles may lead to the development of bifunctional catalysts that may achieve high activity and selectivity in organic catalysis. Such an advantage is seldom seen with metal oxide-supported gold catalysts. Some metal salt supports used to prepare supported gold catalysts include metal carbonates, metal phosphates, metal sulfates, and heteropolyacid salts. Here, we summarize the recent progress in this area.

8.2 Metal Salt-based Gold Nanocatalysts

8.2.1 Metal Carbonate-based Gold Catalysts

Metal carbonates have different solubilities in water and different decomposition temperatures. For instance, $MgCO_3$ decomposes at 300 °C to form MgO and CO_2, whereas $CaCO_3$ decomposes at 900 °C. It is possible to prepare metal carbonate-supported gold catalysts, if the support is insoluble in water and the preparation and operation temperatures of the gold catalysts are lower than the decomposition temperatures of selected metal carbonate supports.

Zhang and co-workers prepared metal carbonate-supported gold catalysts by a co-precipitation method, and studied their catalytic activities in CO oxidation under dry and wet conditions.[27] The type of metal carbonate support and the calcination temperature affect the activity significantly. The activity followed the sequence of Au/BaCO$_3$ >Au/SrCO$_3$ >Au/CaCO$_3$, and the presence of moisture further promoted the activity of these catalysts. However, catalyst deactivation as a function of time on stream was still observed, even though the testing period was short (6 h, Figure 8.1).[27] Although potential deactivation mechanisms were not elucidated, this work nicely showed that metal carbonates can also be used as supports for making active gold catalysts.

Karimi and co-workers studied the aerobic oxidation of alcohols using NaAuCl$_4$ in the presence of Cs$_2$CO$_3$ in toluene at room temperature.[28] Cs$_2$CO$_3$ was found to be a better additive than K$_2$CO$_3$, and toluene was a better solvent than TFT, TFT–water, CH$_3$CN, and CH$_2$Cl$_2$. The catalyst system was not only effective for the aerobic oxidation of benzyl alcohol, but also good for the oxidation of a wide range of alcohols. Although NaAuCl$_4$ is not a heterogeneous catalyst, the authors found that it converted to metallic gold nanoparticles (15–80 nm) on Cs$_2$CO$_3$ in the reaction mixture after a prolonged reaction time. The recovered Au/Cs$_2$CO$_3$ matrix exhibited high activity for the reaction, but it deactivated after prolonged use due to extensive agglomeration.

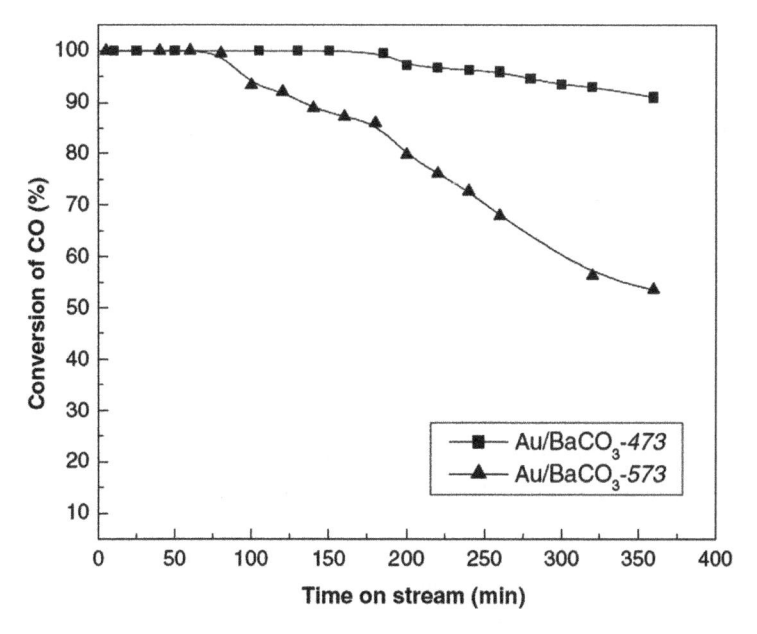

Figure 8.1 CO conversions on Au/BaCO$_3$-T catalysts under wet conditions at 25 °C.[27] Reproduced with permission of Elsevier.

We can see from these examples that certain metal carbonate-based gold catalysts can catalyze CO oxidation and aerobic oxidation of alcohols. However, the scope of application is very narrow, and the application in a wide range of catalytic applications has not been demonstrated. Besides, in the second example,[28] the Au/Cs$_2$CO$_3$ was formed *in situ*, and the sizes of the formed gold nanoparticles were quite large. It would be interesting to prepare a series of metal carbonate-supported gold nanocatalysts and test these materials in the aerobic oxidation of alcohols in future research. Finally, there has been some interest in the synthesis of nanosized metal carbonates such as CaCO$_3$, but the application of these materials in making gold catalysts has not been demonstrated.

8.2.2 Metal Phosphate-based Gold Catalysts

Metal phosphates are generally insoluble in water and have good thermal stability. Some of them are considered to be solid acids but they also have varied basic and redox properties.[26] Therefore, metal phosphates themselves can be used as solid catalysts. Alternatively, they can be used as supports for loading metal nanoparticles. That way, the prepared catalysts may have both the functionality originated from the support and the functionality bestowed by the supported gold. The combination of dual functionalities is amazing, considering that metal particles can catalyze oxidation and hydrogenation reactions.

Dai and co-workers prepared Au/LaPO$_4$ catalysts.[29] LaPO$_4$ nanoparticles (6.8 nm, 111 m^2 g^{-1}) were prepared by a sonication method, and gold was loaded onto the support *via* deposition–precipitation. For comparison, a commercial LaPO$_4$ (10 nm, 55 m^2 g^{-1}) was also used for loading gold. Both catalysts, in the as-synthesized forms, showed high CO conversions below 0 °C, whereas the Au/LaPO$_4$-nanoparticle catalyst exhibited better thermal stability and activity after calcination at 500 °C. This work is interesting because it showed that LaPO$_4$, a metal phosphate, can also be used as a support for making active gold catalysts.

Ma *et al.* developed an array of Au/metal phosphate (denoted as Au/M-P-O) catalysts.[30] Zirconium phosphate (denoted as Zr-P-O) was prepared by precipitation; other metal phosphates were purchased from Aldrich. Gold was then loaded onto these supports *via* deposition–precipitation. It was found that 200 °C-pretreated Au/M-P-O (M = Fe, Co, Y, La, Pr, Nd, Sm, Eu, Ho, Er) showed high CO conversions below 50 °C (Figure 8.2),[30] and 500 °C-pretreated Au/M-P-O (M = Ca, Y, La, Pr, Nd, Sm, Eu, Ho, Er) showed high CO conversions below 100 °C. On the other hand, other Au-M-P-O (M = Mg, Al, Zn, Zr) catalysts were not particularly active in CO oxidation (Figure 8.2).[30] This paper showed that metal phosphates other than LaPO$_4$ can be used to load gold.[29] Nevertheless, the gold loadings on different supports were different, and further optimization of catalysts is still needed.

Ma *et al.* studied the influence of preparation methods on the performance of metal phosphate-supported gold catalysts in CO oxidation.[31] Several

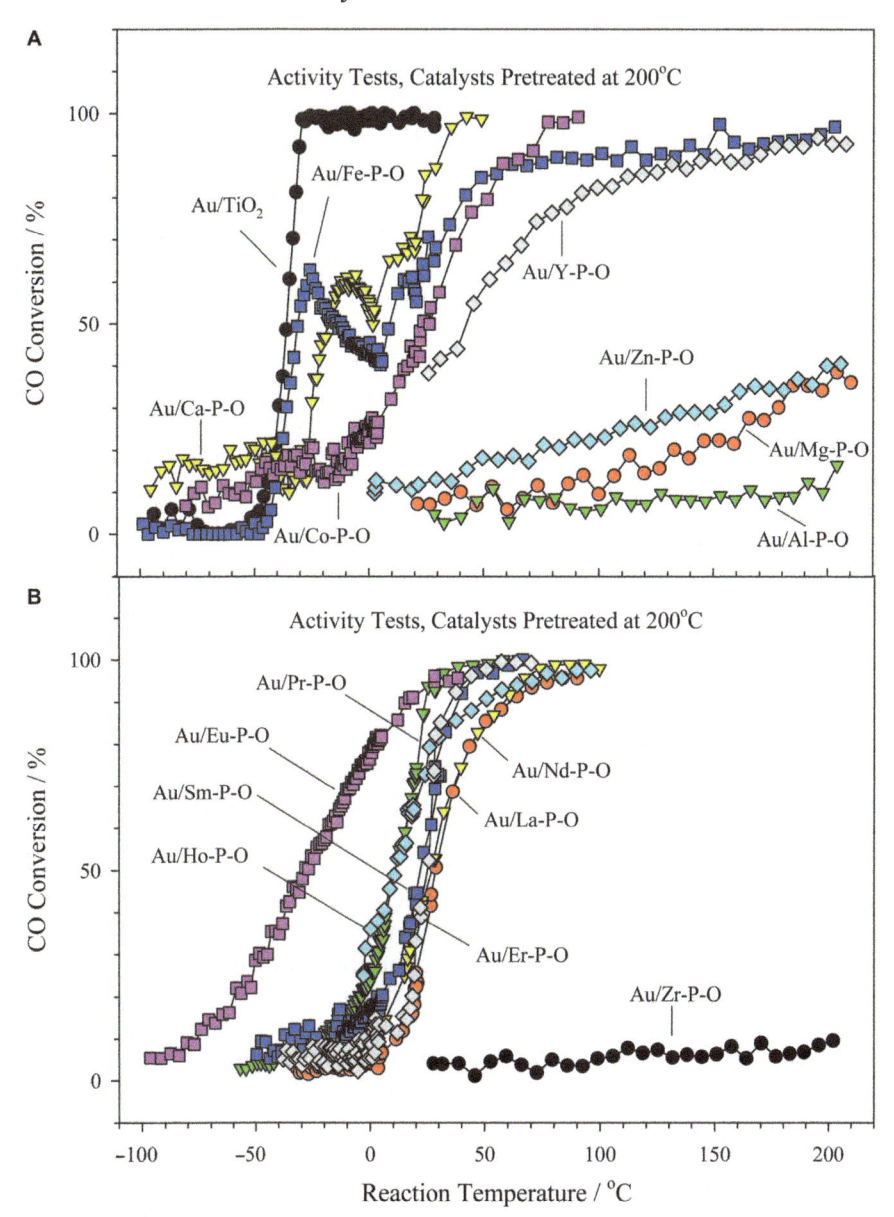

Figure 8.2 CO conversion curves of Au/TiO$_2$ and Au/M-P-O catalysts pretreated at 200 °C.[30]
Reproduced with permission of Springer.

metal phosphates, either regarded as active supports (M-P-O, M = Fe, La, Eu, Ho) or inactive supports (M-P-O, M = Al, Zn) were picked as typical metal phosphate supports. Two methods were selected to load gold. In one, dodecanethiol-capped gold nanoparticles were loaded onto these supports via colloidal deposition. In another, Au(en)$_2$Cl$_3$ (en = ethylenediamine) was used as the precursor. By comparing with the data reported previously, several empirical conclusions were made.[31] (a) Au/Al-P-O and Au/Zn-P-O always showed low activities in CO oxidation, no matter whether these catalysts were prepared by deposition–precipitation, colloidal deposition, or using Au(en)$_2$Cl$_3$ as the precursor. (b) Au/La-P-O was always active for CO oxidation, regardless of the preparation methods. (c) Au/Fe-P-O, Au/Eu-P-O, and Au/Ho-P-O showed some activities in CO oxidation when the catalysts were prepared by deposition–precipitation, whereas the catalysts were not quite active when prepared using the other two methods. It seems that deposition–precipitation is the best method to make metal phosphate-supported gold catalysts active for CO oxidation.

More recently, Dai and co-workers developed Au/LnPO$_4$-MCFs catalysts for CO oxidation.[32] Here, LnPO$_4$ represents LaPO$_4$, CePO$_4$, and EuPO$_4$ nanorods dispersed on the walls of mesostructured cellular foams (MCFs). The LnPO$_4$/MCFs supports were prepared by the controlled heterogeneous reaction of highly dispersed lanthanide oxides embedded in MCFs with phosphate ions in solution. Gold was loaded onto these supports via deposition–precipitation. The resulting catalysts showed high activities in CO oxidation below room temperature (Figure 8.3).[32]

Overbury and co-workers carried out thorough work on the fundamental aspects of CO oxidation on Au/FePO$_4$.[33] Their results indicated that cationic gold species dominated the catalyst surface after pretreatment in O$_2$ at 200 °C, whereas H$_2$-pretreated Au/FePO$_4$ had metallic gold but not cationic gold. The cationic gold caused by O$_2$-pretreatment can be partially reduced upon room temperature CO adsorption. The authors identified two parallel reaction pathways: (1) a redox pathway in which FePO$_4$ furnishes active oxygen, and (2) a direct pathway in which CO and O$_2$ react via either a Langmuir–Hinshelwood or Eley–Rideal mechanism.[33] However, it is not clear which pathway dominates the reaction mechanism.

Overbury and co-workers did further work on the reaction mechanisms of CO oxidation on Au/LaPO$_4$.[34] No cationic gold was observed under both oxidative and reductive conditions. Instead, metallic gold always dominated, indicating that metallic gold was the active site. The non-reducible LaPO$_4$-supported gold catalyst catalyzed CO oxidation via two pathways: (1) a direct pathway, and (2) a hydroxyl-mediated pathway in which surface hydroxyls react with CO to form CO$_2$ and H$_2$. Evidence for the latter pathway came from the evolution of CO$_2$ during room temperature CO adsorption on the catalyst and from the formation of CO$_2$ and H$_2$ during CO-TPR and isotope labeling adsorption studies. By comparing the pathways and activities of Au/FePO$_4$ and Au/LaPO$_4$ catalysts, the authors concluded that the most important reaction pathway is the direct reaction pathway catalyzed by metallic gold.[34]

We can learn from these examples that some metal phosphate-based gold catalysts have recently been found to be active for CO oxidation, and the

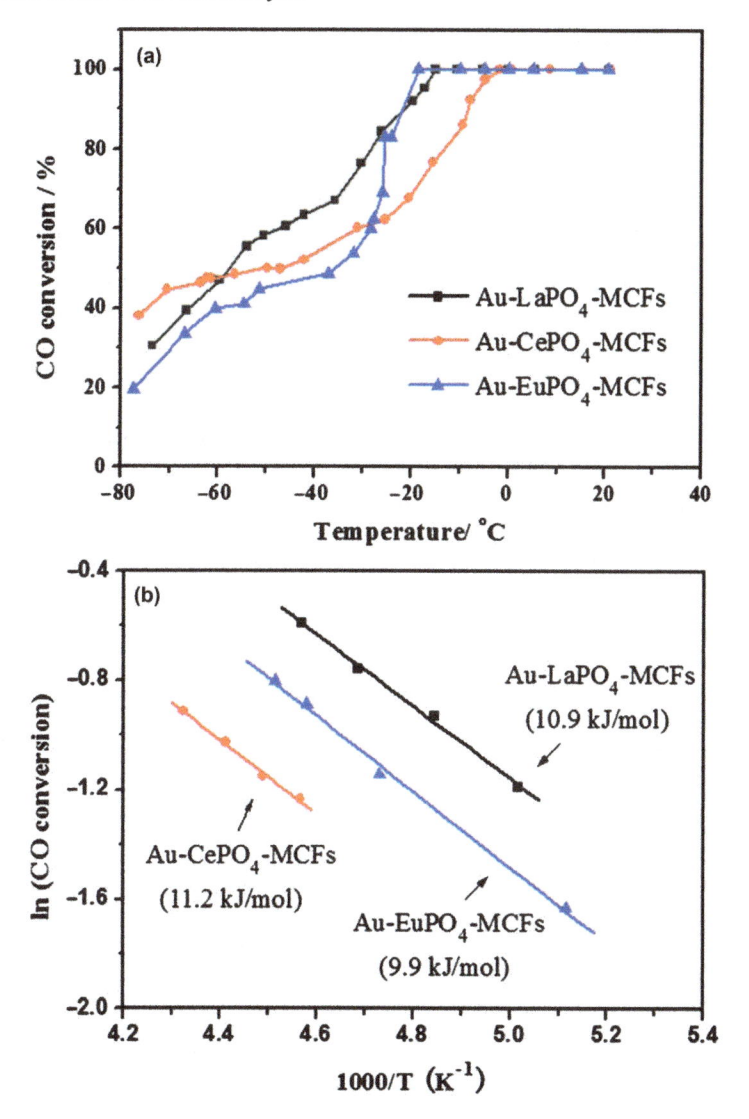

Figure 8.3 Light-off curves (a) and Arrhenius plots (b) for CO oxidation over Au/LnPO$_4$-MCFs pretreated at 300 °C.[32]
Reproduced with permission of the Royal Society of Chemistry.

reaction mechanism was studied in certain cases. However, the scope of research is still quite narrow. It is worthwhile to use these new catalysts for other more useful reactions, such as organic catalysis. The research along this direction is still very preliminary. The influence of various preparation details was not studied sufficiently, and the reaction mechanism on a number of metal phosphate-based gold catalysts was not studied in detail.

8.2.3 Hydroxyapatite-based Gold Catalysts

Hydroxyapatite (HAP), with the formula $Ca_{10}(PO_4)_6(OH)_2$, is non-reducible, insoluble in water, and has good thermal stability. The OH- ion can be replaced by F- or Cl- to produce fluorapatite or chlorapatite. HAP can be used as a catalyst.[35] Alternatively, it can also be used to support metal nanoparticles.

Scurrell and co-workers compared the performance of Au/HAP and Ru/HAP catalysts in the water–gas shift reaction $(CO + H_2O = CO_2 + H_2)$.[36] As shown in Figure 8.4,[36] the activity of Au/HAP was higher than that of Ru/HAP, given that the gold loadings of these catalysts were the same (3%). In addition, the WGS activity of Au/HAP was stable as a function of time on stream.

Au/HAP catalysts were found to be active for CO oxidation at or above room temperature.[30,37–40] Cao and co-workers loaded gold on hydroxyapatite *via* deposition–precipitation with urea, and investigated the influence of calcination atmosphere on the performance of the catalyst in CO oxidation.[39] The catalyst pretreated in He (denoted as Au/HAP-He) showed the highest initial activity, due to the presence of very small gold nanoparticles, but it deactivated obviously on stream. On the other hand, the catalyst pretreated in O_2 (denoted as Au/HAP-O_2) showed the best stability on stream, and afforded the highest steady-state conversion, due to its limited surface basicity.

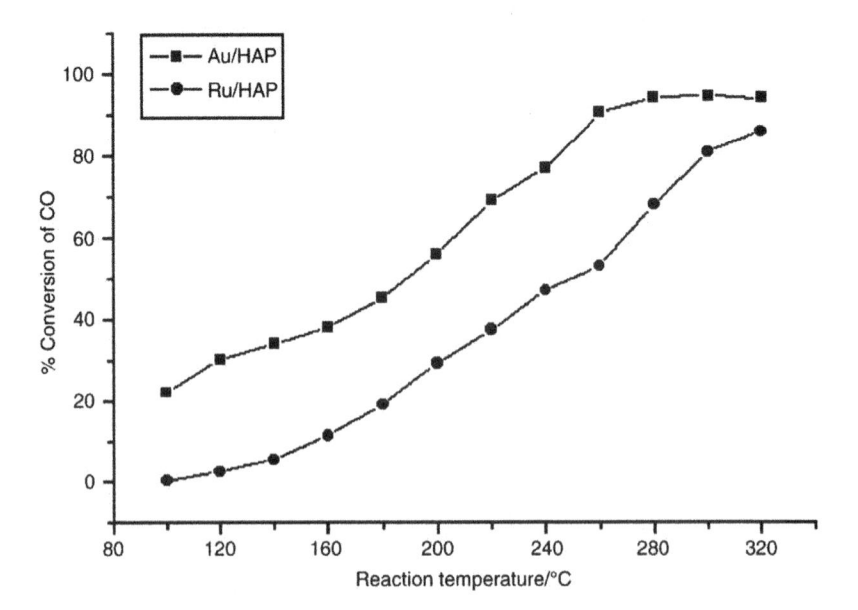

Figure 8.4 CO conversion in the water–gas shift reaction on Au/HAP and Ru/HAP catalysts, as a function of reaction temperature.[36]
Reproduced with permission of Elsevier.

Kaneda and co-workers prepared an Au/HAP catalyst.[41] This catalyst was not only active for the deoxygenation of various tertiary and secondary amides to the corresponding amines using silanes as reductants, but was also able to catalyze the deoxygenation of sulfoxides and pyridine N-oxides to the corresponding sulfides and pyridines, respectively. By conducting FTIR experiments, the authors found evidence for the interaction between Au/HAP and dimethylphenylsilane, and proposed that the deoxygenation may proceed through a pathway initiated by the Au–silane interaction.

Tsukuda and co-workers used a HAP support to load gold clusters.[42–44] $Au_{25}(SC_2H_4Ph)_{18}$ clusters were deposited onto HAP, and the resulting gold catalyst, after proper pretreatment to remove the glutathionate ligands, was active for the selective epoxidation of styrene using TBHP as an oxidant.[42] The authors also prepared gold catalysts by loading $Au_{10}(SC_2H_4Ph)_{10}$, $Au_{18}(SC_2H_4Ph)_{14}$, $Au_{25}(SC_2H_4Ph)_{18}$, and $Au_{39}(SC_2H_4Ph)_{24}$ on HAP.[43] The average gold cluster sizes were estimated by HAADF-STEM to be 1.0–1.1 nm. For comparison, large gold clusters (1.4 nm, corresponding to Au_{85}) were loaded onto HAP by a conventional adsorption method. The conversion on these catalysts in the aerobic oxidation of cyclohexane increased as the n went from 10 to 39, and was the lowest for the Au_{85} catalyst (Figure 8.5).[43]

Cao and co-workers explored the application of HAP-supported gold catalysts in the direct tandem synthesis of imines and oximes.[45] The HAP support alone has an abundance of basic sites of medium strength and a limited number of Lewis acid sites, whereas the loading of gold significantly increases the overall number of both acidic and basic sites. A delicate co-operation between metallic gold and the acid/base sites of the support was

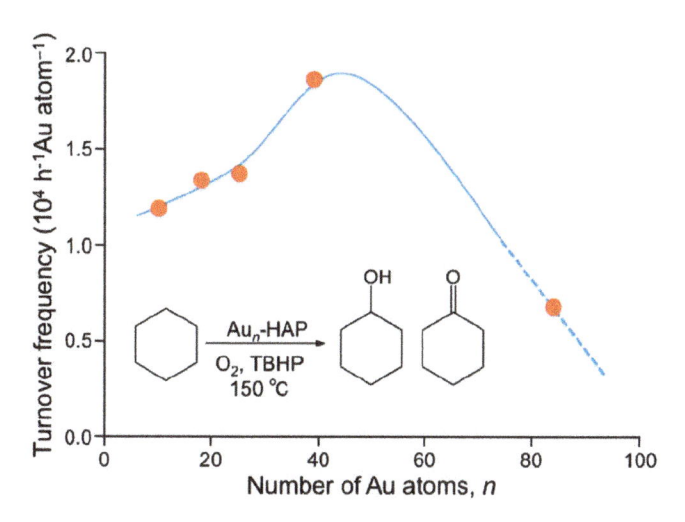

Figure 8.5 TOF values of Au_n/HAP catalysts in the aerobic oxidation of cyclohexane as a function of the cluster size n.[43]
Reproduced with permission of the American Chemical Society.

proposed by the authors to account for the tandem reaction that involves alcohol oxidation and a subsequent condensation step (Figure 8.6).[45]

We can learn from this summary that the research on hydroxyapatite-supported gold catalysts is quite interesting. This kind of catalyst not only catalyzes CO oxidation, it can also catalyze the conversion of various organic substrates. The influences of pretreatment conditions, gold cluster sizes, and organic substrates were investigated in detail, and some reaction mechanisms were proposed. It is worthwhile to expand the scope of the research and demonstrate the application of this kind of catalyst in other reactions.

8.2.4 Hydroxylated Fluoride-based Gold Catalysts

Metal fluorides are barely used as supports for making gold catalysts. Tomska-Foralewska and co-workers found that Au/MgF$_2$ was not active for CO oxidation between 30 and 300 °C.[46] Coman and co-workers developed Au/MgF$_{2-x}$(OH)$_x$ catalysts for the one-pot synthesis of menthol.[47] The synthesis of menthol from citronellal usually involves two steps: (1) the cyclization of citronellal to isopulegol, and (2) the hydrogenation of isopulegol to menthol (Figure 8.7).[47] Here, MgF$_{2-x}$(OH)$_x$ was prepared by the reaction of

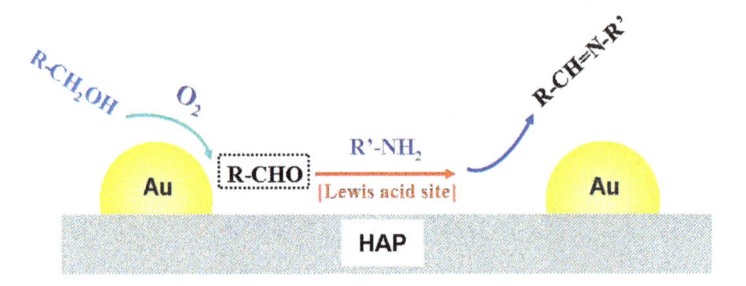

Figure 8.6 Scheme proposed for the tandem process on Au/HAP.[45]
Reproduced with permission of Wiley-VCH.

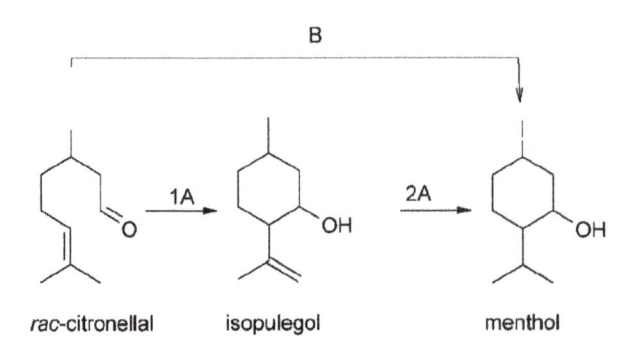

Figure 8.7 Scheme for the synthesis of menthol from citronellal in two steps (route 1A + 2A) and in one pot (route B).[47]
Reproduced with permission of Wiley-VCH.

magnesium methoxide with a methanolic hydrogen fluoride solution. The OH content can be controlled by the water content in the HF solution. When the OH content is very low ($x<0.1$), the catalyst contains both Lewis (Mg^{2+}) and Brønsted (OH) sites responsible for the diastereoselective isomerization of citronellal to isopulegol. On the other hand, the catalyst that calcined at 100 °C contained cationic gold species responsible for the hydrogenation of isopulegol to menthol. Therefore, the bifunctional catalyst was able to catalyze the diastereoselective one-pot synthesis of menthol.

8.2.5 Metal Sulfate-based Gold Catalysts

Some metal sulfates can be used as solid acids, but they are seldom used as supports to prepare gold catalysts. Dai and co-workers prepared $BaSO_4$-MCFs supports by dispersing BaO onto MCFs by impregnation, followed by sulfation of BaO using sodium dodecyl-benzenesulfonate (SDBS) as the sulfate source.[48] Highly dispersed $BaSO_4$ nanocrystals were formed during calcination in air at 500 °C. For comparison, $BaSO_4$-MCFs were also prepared using $KHSO_4$ as the sulfate source. The $BaSO_4$ formed that way is in the form of isolated rod-like particles with bigger sizes. Gold nanoparticles with small sizes were highly dispersed on $BaSO_4$-MCFs (SDBS), whereas the sintering of gold nanoparticles on $BaSO_4$-MCFs ($KHSO_4$) was more obvious. The former gold catalyst showed higher activity in CO oxidation than the latter (Figure 8.8).[48]

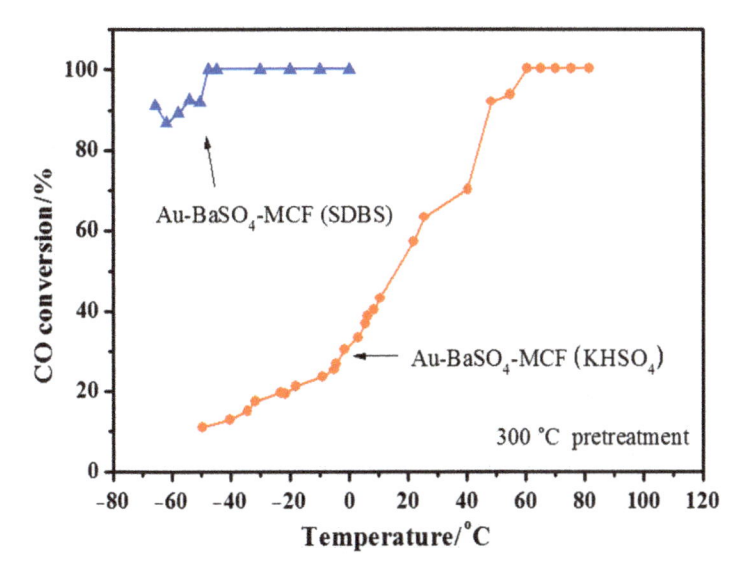

Figure 8.8 CO conversions on Au/$BaSO_4$-MCFs (SDBS) and Au/$BaSO_4$-MCFs ($KHSO_4$) pretreated at 300 °C.[48]
Reproduced with permission of the Royal Society of Chemistry.

168   Chapter 8

8.2.6 Heteropolyacid Salt-based Gold Catalysts

Heteropolyacids are a class of acids made up of a particular combination of hydrogen and oxygen with certain metals and non-metals. Typical examples include $H_4SiW_{12}O_{40}$, $H_3PW_{12}O_{40}$, and $H_6P_2W_{18}O_{62}$. They can be used as homogeneous catalysts and heterogeneous catalysts.[49,50] In particular, those with the Keggin structure have good thermal stability, high acidity, and high oxidizing ability. However, heteropolyacids are generally soluble in water, which makes them unsuitable for preparing supported gold catalysts. Only certain heteropolyacid salts that are insoluble in water can be used as heterogeneous catalysts or supports.

Han and co-workers prepared $Au/Cs_2HPW_{12}O_{40}$ for the conversion of cellobiose to gluconic acid in water.[51] The reaction involves two steps: (1) the hydrolysis of cellobiose to glucose, and (2) the oxidation of glucose to gluconic acid. The $Cs_2HPW_{12}O_{40}$ support, insoluble in water, has strong acidity, and therefore can catalyze the hydrolysis of cellobiose. On the other hand, the supported gold nanoparticles can catalyze the oxidation of glucose. Figure 8.9 shows the steps for the conversion on the $Au/Cs_2HPW_{12}O_{40}$ catalyst.[51] The $Cs_2HPW_{12}O_{40}$ support also modulates the oxidative activity

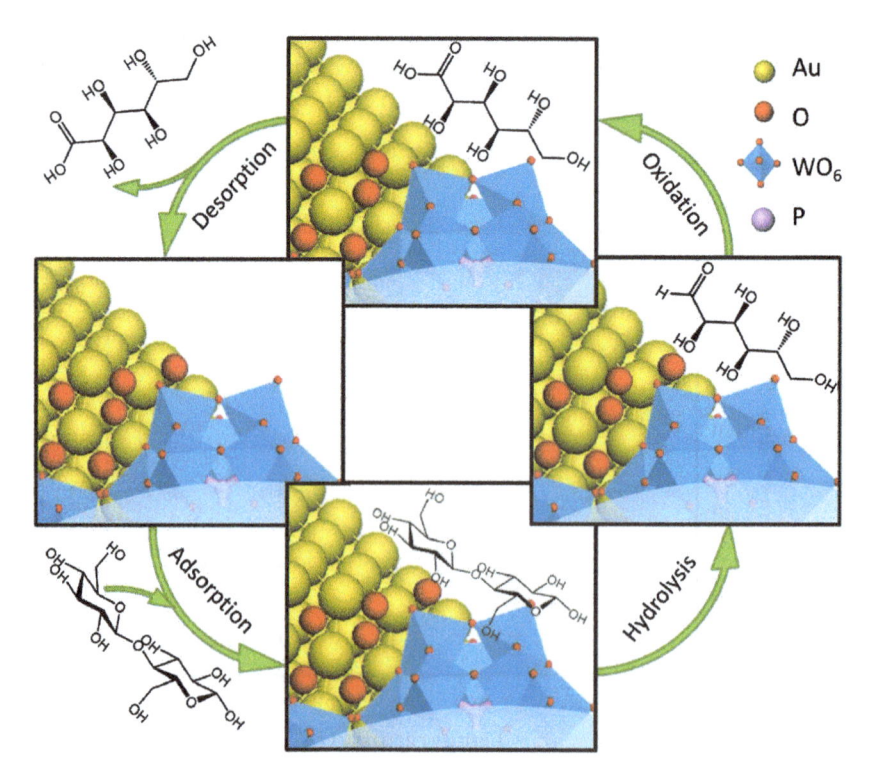

Figure 8.9 Selective oxidation of cellobiose to gluconic acid with molecular oxygen over $Au/Cs_2HPW_{12}O_{40}$.[51]
Reproduced with permission of Wiley-VCH.

of gold nanoparticles, thus avoiding the further oxidation of the formed gluconic acid.

8.3 Summary

Catalysis by supported gold nanoparticle has been a hot topic recently. Most supported gold catalysts have been prepared by loading gold on oxides or carbon supports, but metal salts have been seldom used as supports to load gold. Here, we have highlighted the development and applications of gold nanoparticles supported on metal salts, including metal carbonates, metal phosphates, hydroxyapatite, hydroxylated fluorides, metal sulfates, and heteropolyacid salts. Compared with metal oxides, these metal salts are used less frequently to make supported gold catalysts. Nevertheless, the results summarized in this chapter have demonstrated that some metal salts can be good supports to make active gold catalysts. Not only do they catalyze CO oxidation, they can also catalyze some organic reactions. The latter aspect is particularly interesting, considering that metal salt-supported gold catalysts may be bifunctional, *i.e.* they not only possess the functionality furnished by gold, they also possess the functionality furnished by supports. Thorough experiments are still needed in the future to fine tune the functionalities *via* changes in the composition of catalysts and preparation details, and the potential of these catalysts in catalyzing a variety of organic reactions should be studied in more detail. Reaction mechanisms can also been understood based on thorough experiments. This will certainly entail a lot of work in the future, and good advances are sure to come.

Acknowledgments

Z. Ma thanks the National Natural Science Foundation of China (Grant Nos. 21007011 and 21177028) and the overseas returnees start-up research fund of the Ministry of Education in China for financial support. F. Tao acknowledges the University of Notre Dame, the U. S. Department of Energy Basic Energy Sciences Catalysis Science Program and the ACS Petroleum Research Fund for funding support.

References

1. M. Haruta, T. Kobayashi, H. Sano and N. Yamada, *Chem. Lett.*, 1987, 405.
2. M. Haruta, N. Yamada, T. Kobayashi and S. Iijima, *J. Catal.*, 1989, **115**, 301.
3. G. C. Bond, C. Louis and D. T. Thompson, *Catalysis by Gold*, Imperial College Press, London, 2006.
4. Z. Ma and S. Dai, *Nano Res.*, 2011, **4**, 3.
5. Z. Ma and S. Dai, *ACS Catal.*, 2011, **1**, 805.
6. L.-F. Gutiérrez, S. Hamoudi and K. Belkacemi, *Catalysts*, 2011, **1**, 97.
7. L. Parti and A. Villa, *Catalysts*, 2012, **2**, 24.

8. W. F. Yan, S. M. Mahurin, B. Chen, S. H. Overbury and S. Dai, *J. Phys. Chem. B*, 2005, **109**, 15489.
9. Z. Ma, S. H. Overbury and S. Dai, *J. Mol. Catal. A*, 2007, **273**, 186.
10. G. C. Bond and D. T. Thompson, *Catal. Rev.: Sci. Eng.*, 1999, **41**, 319.
11. M. Haruta and M. Daté, *Appl. Catal. A*, 2001, **222**, 427.
12. T. V. Choudhary and D. W. Goodman, *Top. Catal.*, 2002, **21**, 25.
13. A. S. K. Hashmi and G. J. Hutchings, *Angew. Chem., Int. Ed.*, 2006, **45**, 7896.
14. M. C. Kung, R. J. Davis and H. H. Kung, *J. Phys. Chem. C*, 2007, **111**, 11767.
15. B. K. Min and C. M. Friend, *Chem. Rev.*, 2007, **107**, 2709.
16. C. Della Pina, E. Falletta, L. Prati and M. Rossi, *Chem. Soc. Rev.*, 2008, **37**, 2077.
17. A. Corma and H. Garcia, *Chem. Soc. Rev.*, 2008, **37**, 2096.
18. J. C. Fierro-Gonzalez and B. C. Gates, *Chem. Soc. Rev.*, 2008, **37**, 2127.
19. Y. Zhang, X. J. Cui, F. Shi and Y. Q. Deng, *Chem. Rev.*, 2012, **112**, 2467.
20. J. L. Gong, *Chem. Rev.*, 2012, **112**, 2987.
21. M. Stratakis and H. Garcia, *Chem. Rev.*, 2012, **112**, 4469.
22. M. M. Schubert, S. Hackenberg, A. C. van Veen, M. Muhler, V. Plzak and R. J. Behm, *J. Catal.*, 2001, **197**, 113.
23. W. F. Yan, Z. Ma, S. M. Mahurin, J. Jiao, E. W. Hagaman, S. H. Overbury and S. Dai, *Catal. Lett.*, 2008, **121**, 209.
24. H. G. Zhu, C. D. Liang, W. F. Yan, S. H. Overbury and S. Dai, *J. Phys. Chem. B*, 2006, **110**, 10842.
25. H. G. Zhu, Z. Ma, J. C. Clark, Z. W. Pan, S. H. Overbury and S. Dai, *Appl. Catal. A*, 2007, **326**, 89.
26. K. Tanabe, M. Misono, Y. Ono and H. Hattori, *New Solid Acids and Bases*, Elsevier, Amsterdam, 1989.
27. H. L. Lian, M. J. Jia, W. C. Pan, Y. Li, W. X. Zhang and D. Z. Jiang, *Catal. Commun.*, 2005, **6**, 47.
28. B. Karimi and F. K. Esfahani, *Chem. Commun.*, 2009, 5555.
29. W. F. Yan, S. Brown, Z. W. Pan, S. M. Mahurin, S. H. Overbury and S. Dai, *Angew. Chem., Int. Ed.*, 2006, **45**, 3614.
30. Z. Ma, H. F. Yin, S. H. Overbury and S. Dai, *Catal. Lett.*, 2008, **126**, 20.
31. Z. Ma, H. F. Yin and S. Dai, *Catal. Lett.*, 2010, **138**, 40.
32. C. C. Tian, S.-H. Chai, X. Zhu, Z. L. Wu, A. Binder, J. A. Bauer, S. Brown, M. F. Chi, G. M. Veith, Y. L. Guo and S. Dai, *J. Mater. Chem.*, 2012, **22**, 25227.
33. M. J. Li, Z. L. Wu, Z. Ma, V. Schwartz, D. R. Mullins, S. Dai and S. H. Overbury, *J. Catal.*, 2009, **266**, 98.
34. M. J. Li, Z. L. Wu and S. H. Overbury, *J. Catal.*, 2011, **278**, 133.
35. J. Xu, T. White, P. Li, C. H. He and Y.-F. Han, *J. Am. Chem. Soc.*, 2010, **132**, 13172.
36. A. Venugopal and M. S. Scurrell, *Appl. Catal. A*, 2003, **245**, 137.
37. N. Phonthammachai, Z. Y. Zhong, J. Guo, Y. F. Han and T. J. White, *Gold Bull.*, 2008, **41**, 42.

38. M. I. Domínguez, F. Romero-Sarria, M. A. Centeno and J. A. Odriozola, *Appl. Catal. B*, 2009, **87**, 245.
39. J. Huang, L.-C. Wang, Y.-M. Liu, Y. Cao, H.-Y. He and K.-N. Fan, *Appl. Catal. B*, 2011, **101**, 560.
40. K. F. Zhao, B. T. Qiao, J. H. Wang, Y. J. Zhang and T. Zhang, *Chem. Commun.*, 2011, **47**, 1779.
41. Y. Mikami, A. Noujima, T. Mitsudome, T. Mizugaki, K. Jitsukawa and K. Kaneda, *Chem.–Eur. J.*, 2011, **17**, 1768.
42. Y. M. Liu, H. Tsunoyama, T. Akita and T. Tsukuda, *Chem. Commun.*, 2010, **46**, 550.
43. Y. M. Liu, H. Tsunoyama, T. Akita, S. H. Xie and T. Tsukuda, *ACS Catal.*, 2011, **1**, 2.
44. H. Tsunoyama, Y. M. Liu, T. Akita, N. Ichikuni, H. Sakurai, S. H. Xie and T. Tsukuda, *Catal. Surv. Asia*, 2011, **15**, 230.
45. H. Sun, F.-Z. Su, J. Ni, Y. Cao, H. Y. He and K. N. Fan, *Angew. Chem., Int. Ed.*, 2009, **48**, 4390.
46. I. Tomska-Foralewska, W. Przystajko, M. Pietrowski, M. Zieliński and M. Wojciechowska, *React. Kinet., Mech. Catal.*, 2010, **100**, 111.
47. A. Negoi, S. Wuttke, E. Kemnitz, D. Macovei, V. I. Parvulescu, C. M. Teodorescu and S. M. Coman, *Angew. Chem., Int. Ed.*, 2010, **49**, 8134.
48. C. C. Tian, S.-H. Chai, D. R. Mullins, X. Zhu, A. Binder, Y. L. Guo and S. Dai, *Chem. Commun.*, 2013, **49**, 3464.
49. M. Misono, *Catal. Rev.: Sci. Eng.*, 1987, **29**, 269.
50. T. Okuhara, N. Mizuno and M. Mizuno, *Adv. Catal.*, 1996, **41**, 113.
51. J. Z. Zhang, X. Liu, M. N. Hedhili, Y. H. Zhu and Y. Han, *ChemCatChem*, 2011, **3**, 1294.

CHAPTER 9

Catalysis with Colloidal Metallic Hollow Nanostructures: Cage Effect

MAHMOUD A. MAHMOUD

Laser Dynamics Laboratory, Department of Chemistry and Biochemistry, Georgia Institute of Technology, 901 Atlantic Dr, Atlanta, Georgia 30332, USA
Email: mmahmoud@gatech.edu

9.1 Introduction

Noble metallic nanoparticles have been used in many applications such as catalysis,[1] sensing,[2–4] drug delivery,[5,6] optical switching,[7,8] magnetization switching in magneto-plasmonics,[9,10] and the potential diagnosis and *in vivo* treatment of cancer.[11–13] The great versatility of these nanoparticles comes from the ability to fine-tune their optical,[14–16] catalytic,[17] photo-thermal,[15,18–22] photoelectromagnetic,[23] and photoacoustic[24] properties with changing their shape and size.

Nanoparticles of various shapes, sizes, and structures have been prepared to be used in many applications. These shapes are isotropic such as spheres,[17] cubes,[25] triangles,[26] shells,[27] hollow nanospheres,[28] nanocages,[25] and frames,[29] while the anisotropic shapes are rods,[30] wires,[31,32] and stars.[33] Unlike regular synthetic techniques for the synthesis of solid, shaped, nanoparticles, which are based on the direct reduction of ions into metallic nanoparticles, hollow nanostructures have been prepared by special techniques such as template mediated synthesis,[34] Kirkendall effect[35] and galvanic replacement[25] reactions.

Catalysis with nanoparticles is a rapidly growing field driven by progress in materials research.[36-39] Important applications of nanocatalysis include chemical production,[40] sustainable energy (hydrogen production and fuel cell catalysts),[41] pollution reduction,[42] and clean photocatalysis.[43] The nanoscale size of the catalyst causes two effects: changing the electronic structure of the nanocatalyst (*i.e.* increasing the Fermi energy of the nanocatalyst leading to a lower reduction potential on the surface of the nanocatalyst)[1,44-46] and increasing the number of chemically unsaturated and thermodynamically active surface atoms.[38,39,47] It has also been suggested that the type of exposed crystal planes, which can be controlled by the nanocatalyst shape, can influence the activity and selectivity of the nanocatalyst.[48] Although catalysts of nanoscale dimensions possess advantages, problems that limit the future applications of using nanoparticles in catalysis result from the following: reshaping of the nanocatalyst during or after the catalysis reaction,[49,50] products depositing on the surface of the nanocatalyst decreasing their stability and limiting their recycling,[51] capping materials on the surface of colloidal nanocatalysts reduce their activity and can affect their Fermi energy level,[52] and aggregation of the nanocatalyst especially in organic solvents.[53] Many efforts have been put forth to overcome these problems and improve the efficiency and selectivity of the nanocatalysts. These effects include: loading the nanocatalysts on a support (micro-scale materials,[54] micro-gels,[55] or zeolite supports[56,57]) for improved stability and recyclability, and partially coating the surface with porous silica or a brush type polymer to improve the stability of the nanocatalysts in different media.[58,59] Assembling the nanocatalyst on the surface of a substrate by the Langmuir–Blodgett technique[29,60] or fabrication of the nanocatalyst on the surface of a substrate by lithographic techniques[61,62] are also improving the applicability of the nanocatalysts. The surface of the nanocatalysts prepared by these methods is then cleaned by oxygen plasma or UV irradiation to improve the interface with reacting materials.

Usage of the nanoparticles in catalysis can reduce the production cost as small quantities of material are needed to obtain the same surface area and activity as conventional bulk catalysts. Different techniques, such as DC sputtering,[63] laser deposition, the colloidal chemical method,[64,65] and lithography,[61,62] have been used for the preparation of nanocatalysts with different sizes starting from small clusters to a tenth of a nanometre. The colloidal chemical method has been used successfully to prepare nanocatalysts with different shapes, sizes, and crystal structures.[37,64,65]

An important question that arises in the field of nanocatalysis is whether catalysis with nanoparticles is heterogeneous[66-68] or homogeneous.[38,39,69] The common and acceptable definition of colloidal nanocatalysis is heterogeneous when the reaction occurs on the surface of nanoparticles and forms a complex on the surface atoms themselves. If the complex dissolves away from the surface of the nanocatalyst, the reaction is homogeneous.

The efficiency of any nanocatalyst is based on the large surface to volume ratio and the geometric shape. Nanoparticle shapes with sharp corners and edges showed high activity due to the high thermodynamic activity of these atoms on the corners and edges.[49] When the nanocatalysts are fixed inside an inert support with voids, such as porous metal–organic frameworks (MOFs),[70] empty polymer nanofibers,[71] or yolk–shell structures,[72] their catalytic activity is improved due to the confinement of the reactant inside the hollow structure. In fact, similar activity was observed when the nanocatalyst itself was designed with hollow structures with porous walls made of a single shell[67,73,74] or double shells.[67,75–80]

9.2 Synthetic Approaches to Hollow Metallic Nanocatalysts

Solid nanoparticles with different shapes, sizes, and compositions have been prepared by seedless[81] and seed-mediated[82] techniques. These two synthetic methods involve the formation of small seeds, either *in situ*[30] in the case of seedless,[17] or prepared separately, which are then allowed to grow in a new solution. These techniques allow the seed of the nanoparticles to grow into different shapes and sizes depending on the preferential binding of the capping materials to the crystal faces and the thermodynamic stability of the plane facets. These two variables guide the nanoparticle seed growth to proceed in different ways, forming different shapes. The size of the nanoparticle is controlled by the rate of growth of the seed and the rate of capping by the capping agent. The temperature plays an important role in the synthesis of nanoparticles, as higher temperatures induce thermodynamic control.

Hollow nanoparticles have pores on their surfaces and an empty cavity.[25] Therefore, it is not possible to use traditional synthetic techniques for solid nanoparticles to prepare the hollow structure nanoparticles. Three different techniques are applied to prepare hollow nanoparticles. The first method is the template-mediated method, which was introduced by Mohwald's group.[34] This technique involves depositing metals on the surface of a template made up of polymer or silica beads. The core is later etched away using H_2O_2 and HCl to produce a hollow structure. Post-synthetic treatment of the nanoparticle to remove the template limits the applicability of this technique. The nanoparticles prepared by this method are not smooth and their sizes depend on the sizes of the underlying bead templates, which are typically no smaller than 100 nm. The second method is based on the common metallurgical Kirkendall effect, which involves non-equilibrium mutual diffusion of counter atoms through a reaction interface.[35,83] Since the diffusion flow rates and sizes of the elements are not similar, an internal vacancy is formed inside the nanoparticle. This technique is limited to the synthesis of composite hollow nanoparticles. Finally, the most applicable technique for the synthesis of hollow nanoparticles is the galvanic

Catalysis with Colloidal Metallic Hollow Nanostructures: Cage Effect 175

replacement approach, which was introduced by Sun and Xia.[25] This technique involves the galvanic oxidation of two or more atoms of template nanoparticles by the metal ions of the cage materials, which will be reduced to metal on the surface of the template. A typical galvanic replacement example is the synthesis of gold nanocages (AuNCs) using silver nanocubes (AgNCs) as a template. AgNCs are mixed with gold salt in a boiling solution. The gold ions oxidize the metallic silver and deposit on the nanoparticles'

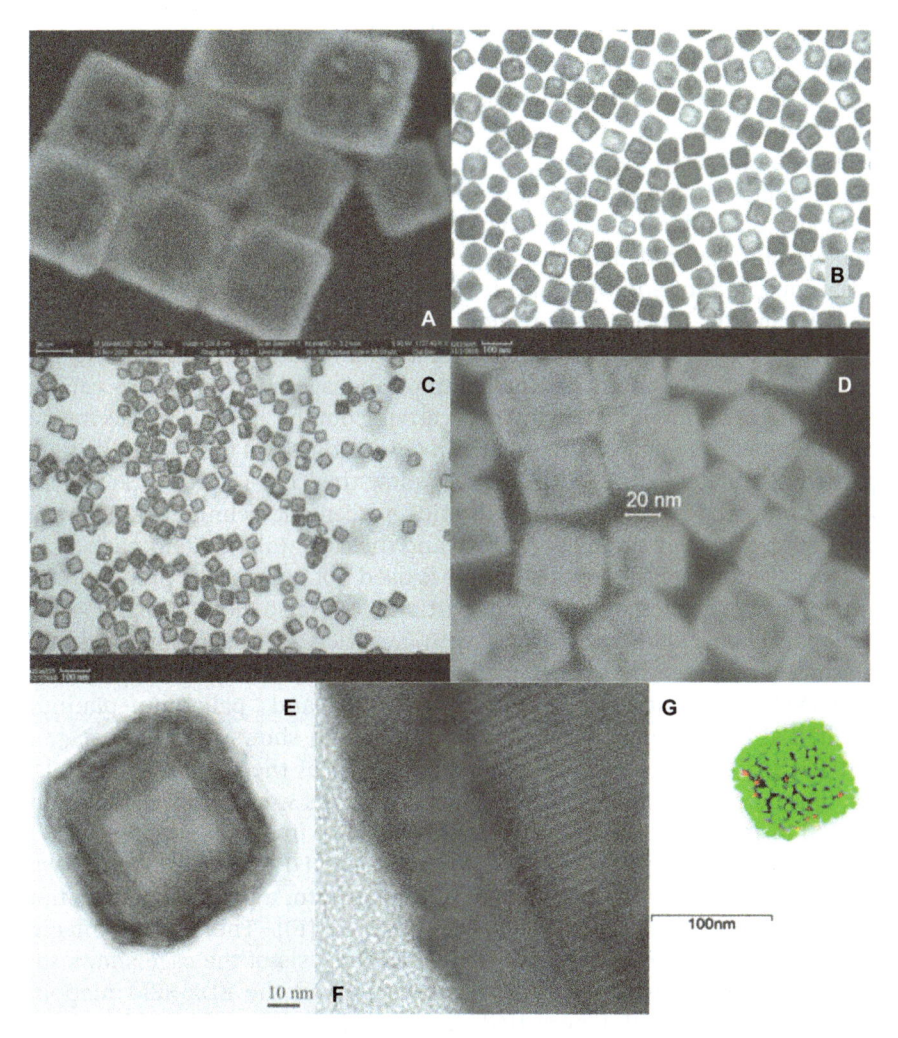

Figure 9.1 (A) SEM image of gold nanocages. (B) TEM image of platinum nanocages. (C) TEM image of palladium nanocages. (D) SEM image of gold–palladium double shell nanocages. (E) Magnified TEM image of platinum–palladium double shell nanocages. (F) High-resolution TEM mage of platinum–palladium nanocages. (G) EDX-SEM mapping of gold–platinum double shell hollow nanoparticles, gold (green) and platinum (red).

surface. Figure 9.1(A) shows the SEM image of 53 nm gold nanocages (AuNCs) prepared by galvanic replacement. The AuNCs possess a hollow interior and have pores on their surfaces.

In this technique, the following variables have to be controlled during the synthesis of hollow nanostructures. (1) The shape of the template controls the final hollow nanoparticles' shape. (2) The capping material of the template should be compatible with the nanocage material in order to avoid the aggregation of the hollow nanoparticles. (3) The lattice parameters of the template and the deposited material should be similar to obtain a smooth surface. (4) The oxidation potential of the atoms of the template has to be higher than that of the hollow nanomaterial to make the galvanic replacement thermodynamically allowed. (5) To prevent a change in shape of the hollow nanomaterials during the galvanic replacement, the surface of the template should be cleared of excess capping materials. (6) The rate of the galvanic replacement controls the pore size and the roughness of the hollow nanomaterial. This rate depends on the activity of hollow nanomaterial ions, rate of addition, and reaction temperature.

The galvanic replacement technique was modified to prepare different transition metal nanocages. Some of these hollow nanoparticles, such as platinum nanocages (PtNCs) and palladium nanocages (PdNCs), have catalytic potential.[67] This modification involves carrying out the galvanic replacement at room temperature and controlling the rate of addition of the salt to the nanoparticle template. Figure 9.1(B) and (C) show the TEM images of PtNCs and PdNCs prepared by the modified galvanic replacement technique. As shown in the figure, the outer surface is smooth and the cage walls contain pores.

The galvanic replacement technique is used not only for the preparation of single metal shelled hollow nanoparticles but also double shell hollow nanoparticles. Hollow nanocages with a variety of plasmonic and non-plasmonic metals including gold–platinum (AuPtNCs), gold–palladium (AuPdNCs), platinum–palladium (PtPdNCs), and palladium–platinum (PdPtNCs) have been prepared.[67,74] Figure 9.1(D) shows the SEM image of AuPdNCs. In order to analyze the composition of the double shell hollow nanoparticles accurately, two imaging techniques were used (HR-TEM and EDX-SEM mapping). Figure 1(E) shows the TEM image of single PtPdNCs. The nanoparticles appear to have two shells. The HR-TEM image of the wall shows three layers: an outer layer of pure platinum, a pure palladium inner layer, and an alloy layer in between (Figure 1(F)). The second imaging technique is EDX-SEM mapping. Elemental analysis of the cage shows that two metals are present in different amounts and the EDX-SEM mapping shows the outer metal. Figure 1(G) shows the EDX-SEM elemental mapping of AuPtNCs revealing that the outer surface is pure gold.

9.3 Assembling the Nanocatalysts on Substrates

The main challenge that limits the usefulness of nanoparticles in catalysis is their recycling, which is difficult due to their small size. Different solutions were proposed to overcome the handling problems. One of the suggested solutions is to fabricate the nanoparticles on the surface of the substrate by lithography such as electron beam lithography (EBL)[84] or nanosphere lithography (NSL).[85] The lithography technique involves fabrication of a polymer template of the nanoparticle on the surface of the substrate. The nanomaterials are deposited inside these polymer templates, which are then dissolved, leaving the nanoparticles on the surface of the substrate. In EBL, the electron beam is used to make these templates on a polymer thin film that coats the surface of the substrate. Although use of this technique led to the fabrication of nanoparticles on the surface of a substrate, the high cost, technical difficulty, limited morphological scope, and the polycrystallinity of the product limits the scale-up of the technique. NSL involves the monolayer self-assembly of nano and microsphere polymer beads on the surface of a substrate. This produces a prismatic template in between the polymer beads. This technique can only produce prisms but does allow for size control by varying the size of the polymer beads.

Traditional colloidal chemical methods are the most efficient approach for synthesizing nanoparticles. Many shapes, sizes, compositions, and crystal structures have been prepared by these techniques. However, this method presents problems when dealing with the stability and handling. The following two methods are used to make colloidal nanoparticles more applicable.

(1) Loading the nanoparticles on the surface of micro-size particle supports such as solid state particles or polymer beads. This method can be carried out by two different techniques: *in situ* reduction of metal ions onto the surface of the support or loading of already synthesized nanoparticles onto the support. Reduction of the metal ions on the support produces different shapes and sizes of nanoparticles. Recently, the solvent-controlled swelling and heterocoagulation technique was introduced[86–88] to load different kinds of nanoparticles onto the surface of a polystyrene polymer bead (PS) support. This method involves swelling the PS by an organic solvent such as tetrahydrofuran or chloroform. Pores and channels are formed on the surface of the PS beads due to the increase in size (size expansion is up to 200%). The capping materials of the nanoparticles anchor inside the pores of the PS. After the PS is washed in a non-swelling solvent, it shrinks and the nanoparticles are fixed on the surface of the PS through the capping materials. This technique was used to coat the PS with different nanoparticles such as 30 nm and 80 nm polyvinylpyrrolidone (PVP) capped gold nanospheres. Figures 9.2(A) and (B) show the SEM images of 10 µm PS coated with 30 and 80 nm gold nanoparticles, respectively. Figures 9.2(E) and (F) are the dark field scattering images for the PS coated with 30 and 80 nm gold nanospheres, respectively. This technique is applicable to other shapes, such as silver nanocubes capped with PVP as seen in Figure 9.2(C) (SEM) and Figure 9.2(G) (dark field). This technique is not only valid for nanoparticles coated with polymers, but also for other cationic capping materials such as trimethyltetradecylammonium bromide (TTAB). Figure 9.2(D) shows the

178 Chapter 9

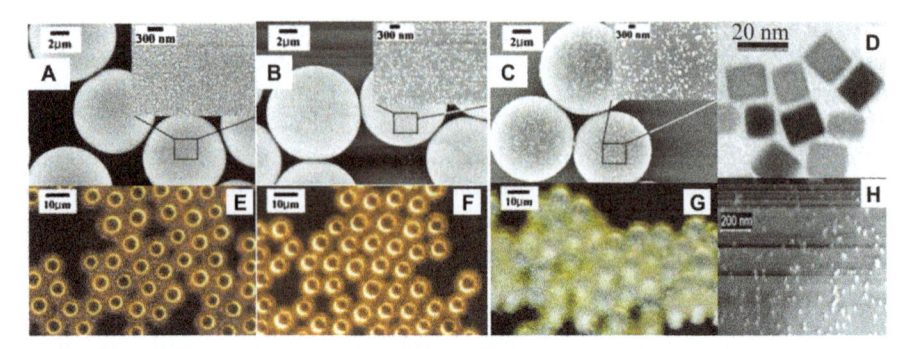

Figure 9.2 SEM images of nanoparticles coating 10 μm polystyrene beads: (A) 30 nm gold nanospheres, (B) 80 nm gold nanospheres, (C) 60 nm silver nanocubes, (H) platinum nanocubes. (D) TEM image of platinum nanocubes. Dark field images of 10 μm PS beads covered by metal NPs: (E) 30 nm AuNPs, (F) 80 nm AuNPs, and (G) 60 nm AgNCs.

TEM image of 20 nm platinum nanocubes and the SEM image of the PS coated with platinum nanocubes capped with TTAB.

(2) The Langmuir–Blodgett (LB) technique is another method for assembling colloidal nanoparticles onto substrates. This technique produces a particle monolayer but requires that the nanoparticles be dispersed in a volatile solvent immiscible with the sub-layer liquid filling the LB trough (such as chloroform solvent and a water sub-layer). The nanoparticles in a volatile solvent are sprayed over the liquid sub-layer by micro-syringe. Due to the surface tension of the sub-layer liquid, the nanoparticles will arrange into a monolayer on the surface of the sub-layer as the solvent evaporates (as postulated by Langmuir). The inter-particle separation distance between the nanoparticles can be controlled by varying the available area that the nanoparticles are dispersed on. The LB film can be transferred to a substrate by the dipping method. Figure 9.3(A) shows AuNCs (dispersed in chloroform) sprayed over the surface of the water sub-layer of the LB trough. The mechanical barrier separates the nanoparticles, which are blue in color (left), from the cleaned water surface (right). The AuNCs are transferred to the surface of a glass slide substrate by dipping the slide into the end of the LB trough and slowly pulling it out. Figure 9.3(B) shows the glass substrate after coating with the AuNCs from the SEM image shown in Figure 9.3(C). The concentration of nanoparticles on the surface affects the surface tension of the water sub-layer. Thus the coverage density can be measured by a Wilhelmy plate attached to a pressure sensor. However, the relationship between the surface pressure and the value of the sub-layer area that the nanoparticles are distributed (isotherm) over determines how the nanoparticles interact with the sub-layer surface and with one another.[89]

Although recycling of the nanocatalyst at the end of the reaction is economically useful, keeping the activity of the catalyst after assembling it on the surface of the substrate is also an important issue. There is no change in the

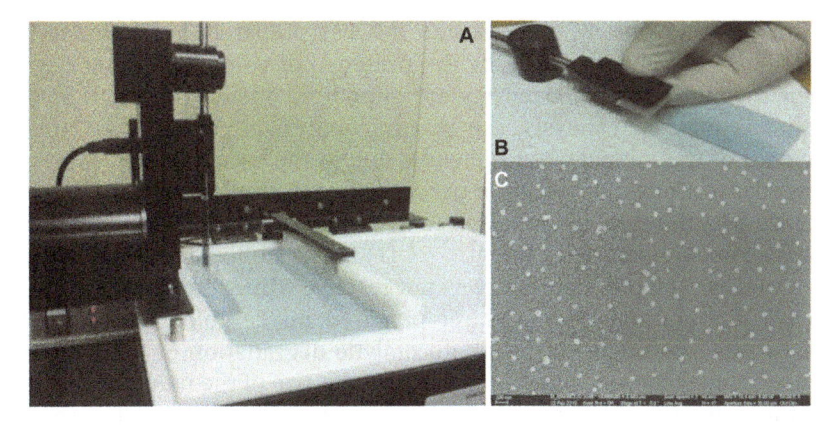

Figure 9.3 (A) Picture of a Langmuir–Blodgett trough after depositing a monolayer of AuNCs on top of a water sub-layer. (B) AuNC monolayer deposited on the surface of a glass slide. (C) SEM image of an AuNC monolayer assembled on the surface of a silicon wafer at a surface pressure of 0 mN m^{-1}, corresponding to 4% coverage of the substrate with nanoparticles.

activation energy of the catalysis reaction catalyzed by platinum nanocubes in colloids and supported on the surface of a PS bead substrate.[88] However, the rate of the reaction is decreased since one-sixth of the surface of the cube is covered by the PS polymer. Unlike the silica and alumina support, this could affect the activity of the catalyst supported on their surfaces.[54]

9.4 Hollow Nanostructures are Different in Catalysis

The efficiency of any catalyst depends on two factors: the surface to volume ratio (number of surface atoms) and the potential activity of the surface atoms (number of saturated surface atoms surrounded by the fewest neighbouring atoms). Therefore, nanoparticles with sharp corners and edges show good catalytic efficiency.[39,49,90]

Hollow metallic nanocatalysts have attracted the attention of many researchers due to their excellent physical and chemical properties.[25,67] The high catalytic efficiency of the hollow nanocatalysts is due to three different possibilities: (1) the hollow nanocatalysts include two surfaces (internal and external), and pores on their walls, which increase the area to volume ratio. (2) The inner surface could be rough and not covered by the capping agent that is usually used during the colloidal nanoparticle synthesis as much as the outer surface; this increases the available surface area. (3) The cage effect of the hollow nanocatalyst, however, the reactant is confined inside the cavity of the nanocatalyst, which increases the rate of collision with the inner surface of the catalyst. Moreover, if the reaction is homogeneous, the cavity of the hollow catalyst keeps the intermediate species inside it so it does not diffuse into the bulk of the solution, so increasing the efficiency of the nanocatalyst.

Porous metal–organic frameworks (MOFs) prepared by Yaghi and his coworkers[70] are characterized by the presence of voids in their structures. When the metallic nanocatalysts are supported inside the MOF and the catalyst is used in catalysis, the reacting materials get confined and the efficiency of the nanocatalyst increases due to the cage effect.

In order to study the reason for the activity of the hollow nanocatalysts and to prove that the catalysis takes place inside the nanocatalyst, the following experiments were carried out:[67,74,77,91] (1) Using hollow nanocatalysts with two surfaces. The outer surface was not reactive while the inner surface was reactive, for example, a Ag_2O layer prepared inside gold nanocages. This catalyst was examined for the photocatalytic degradation of methyl orange dye and its activity was found to depend on the size of the nanocatalyst cavity and the pore size. (2) Comparing the efficiency of the spherical shape. The solid and the hollow nanoparticle show a difference in the reduction reaction pathway of Eosin Y from a two-electron reduction in the absence of the nanoparticles to a one-electron reduction. (3) Synthesis of two hollow double shell nanoparticles made with an inner shell of platinum and an outer shell of palladium and *vice versa* in addition to pure PtNCs and PdNCs.[67,91] These nanocatalysts were used to catalyze the reduction reaction of 4-nitrophenol by borohydride. The kinetic parameters of the double shell catalysts were comparable to those of the pure nanocage made of the same metal as that of the inner shell.[67] (4) Hollow nanoparticles having an inner plasmonic gold surface and an outer non-plasmonic catalytic layer have been reported.[74,91] The shift in the surface plasmon spectrum during the catalytic reaction was used to prove the cage effect.[74]

9.4.1 Hollow Nanostructures with a Catalytically Active Inner Surface and an Inactive Outer Surface

The catalyst is supported on the inner wall of an inert support of hollow structures and does not affect the catalysis reaction. This support could be silica, a polymer, or even a metal. In this case, the reacting materials are confined with the catalyst inside the hollow support. The confinement of the reactants inside polymer nanofibers with platinum and palladium nanocatalysts bound to their inner surface increases their catalytic activity in the hydrogenation of organic compounds.[92] An improvement in the catalytic properties of nickel, cobalt, iron and their oxides was observed when they were prepared inside a SiO_2 shell in a yolk–shell structure. In addition to the excellent catalytic activity, the catalysts showed high thermal stability.[93] Several semiconductor nanoparticles were used to photocatalyze the photodegradation of dyes *via* a free radical mechanism.[94] The photocatalysis reaction proceeds in an aqueous medium through the initial formation of hydroxyl radicals, which attack the double bond of the organic dye.[95] AuNCs were shown to efficiently catalyze the photodegradation of methyl orange dye (MO), after exposure to oxygen gas. Thus the AuNCs would have remaining

Ag on their interior walls, which are oxidized to silver oxide (Ag$_2$O) upon exposure to dissolved O$_2$ in water. Figure 9.4(A) shows the schematic diagram of AuNCs after oxidation of the residual silver into a Ag$_2$O layer.[76] Upon excitation of Ag$_2$O, electrons are excited from the valence band to the conduction band, generating a hole in the valence band. Water molecules in the solution are oxidized into hydroxyl radicals and protons by the holes in the valence band, while the electrons in the conduction band reduce water molecules into hydroxyl ions and peroxide radicals (Figure 9.4(B)). The generated radicals attack MO molecules, resulting in MO radicals that undergo a series of intramolecular fragmentations. The hydroxyl radical generation process is depicted in Figure 9.4(B) and (C).[76]

There are a number of observations that strongly suggest the nanoreactor cage effect. The Ag$_2$O molecules that photocatalyze the reaction are present only on the inner walls of the nanoreactor. The rate of the photocatalytic reactions was found to be greatly dependent on the pore size in the wall of the nanocage, due to diffusion of reactant in and product out through the pores of the cage, and the fact that solid Ag cubes have low significant catalytic activity after exposure to O$_2$ gas. Therefore, the optimum reaction rate requires: a high surface area of Ag (which is oxidized to Ag$_2$O) on the inner wall of the AuNC; pores on the surface of the AuNC that are large enough to allow the reactants and products to diffuse in and out of the cavity but small enough to keep the radical steady state concentration high; and the cavity inside the cage being an appropriate size to allow an optimum collision rate between the reactants. When the pore sizes and surface area of silver oxide are in balance, the cavity can displaying a "cage effect".[76] This means that the concentration of the dye molecules, or the reaction rate determining species formed from them, has built up its concentration to drive the observed kinetics of the reaction (Figure 9.4(C)). The hollow structure nanocatalysts and the hybrid semiconductor metallic structures showed high catalytic activity *i.e.* ruthenium–cuprous sulfide hybrid inorganic nanocages proved to have efficient electrocatalytic properties due to their high surface area and exciting electronic properties.[75] In fact, the electronic and catalytic properties of the hybrid material nanocatalysts are

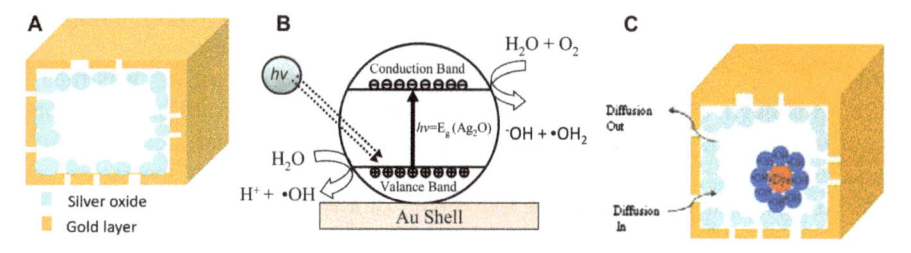

Figure 9.4 Schematic diagram of: (A) AuNC with a remaining layer of silver oxide at the inner wall, (B) the mechanism of photo-formation of the radicals by silver oxide inside the AuNC nanoreactor, (C) the reaction and the cage effect inside the AuNCs-Ag$_2$O nanoreactor.

improved by the ruthenium and cuprous sulfide combination. Therefore, one of the reasons for the efficient catalytic properties of Ag$_2$O in the AuNCs–Ag$_2$O double shell nanocatalyst is the gold outer shell. The gold in this case increases the electron–hole charge separations, scavenges the electrons and increases the amount of generated hydroxyl radicals.

9.4.2 Comparing the Activity of Hollow and Solid Nanocatalysts of Similar Shapes

The reactions catalyzed by the hollow nanocatalysts display efficient activity for two possible reasons. Firstly, the confinement effect of the cage increases the steady state concentration of the species in the rate determining step of the reaction, and secondly, in some cases, the inner surface might not be as well capped as the outer surface and thus is more catalytically active. Since the wall thickness of the nanoreactor is small (a few nanometres), the electron can transfer across the wall during the catalysis of electron transfer reactions.[73] The surface to volume ratio of the hollow nanocatalyst is higher than any solid nanocatalyst, because the surface area of the cavity adds to the outer surface area of the nanocatalyst. This is another factor enhancing the catalytic property of the hollow nanoparticles. Nanoporous platinum nanoparticles, which are prepared by de-alloying of a Ni-Pt nanoparticle alloy, showed excellent cathodic catalytic activity for the oxygen reduction reaction, due to the cage effect.[96]

Eosin Y dye (EY^{2-}) has multiple reduced forms; however, it is reduced to EY^{4-} by borohydride (BH) accepting two electrons. In the presence of metallic nanocatalysts, such as AuNSs or AuHSs, the reduction reaction pathway changes from accepting two electrons to accepting one electron to form EY^{3-}.[97,98] The two-electron reduction is possible also in the presence of the nanoparticles since some of the EY^{2-} molecules get reduced in the bulk of the solution far from the surface of the nanocatalyst. The two reduction pathways of EY^{2-} by BH in the presence and in the absence of gold nanocatalysts are summarized in Figure 9.5(A) and (B). In the presence of the nanocatalyst, the BH- reduction leads to the formation of a mixture of EY^{3-} and EY^{4-}. The efficiency of the gold nanocatalysts increases by increasing the amount of the one-electron reduction species EY^{3-} compared with the two-electron reduction species EY^{4-}. In order to compare the efficiency of AuHSs and AuNSs, the percentage conversion of EY^{2-} into EY^{3-} was measured for each catalyst at different nanocatalyst concentrations. The concentrations of EY^{2-} and EY^{3-} were determined from optical measurements and extinction coefficients. The relationship between the concentrations of the AuNSs and AuHSs and the EY^{3-}–EY^{2-} conversion ratios are shown in Figure 9.5(C). This conversion relationship is linear for both catalysts with slopes at 0.044 ± 0.006 and 0.016 ± 0.006 for AuHSs and AuNSs, respectively. The slope for the relationship between the EY^{3-}–EY^{2-} conversion ratio and nanocatalyst concentration for AuHSs is three times that for AuNSs, which supports the idea that nanocages have an effect.

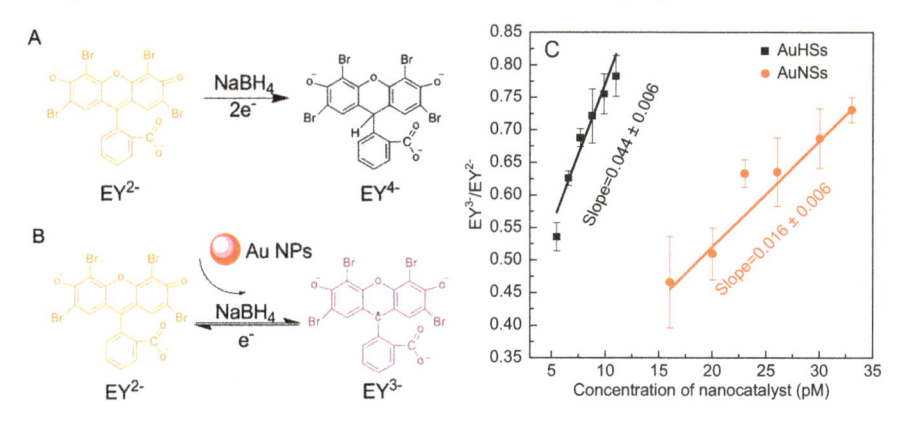

Figure 9.5 The reduction pathways of EY^{2-} by BH: (A) two electrons without a nanocatalyst, (B) one-electron reduction in the presence of a gold nanocatalyst. (C) The relationship between ratios of the one-electron reduction species and the two-electron reduction species and concentrations of AuNSs and AuHSs.

9.4.3 Comparing the Activity of a Single Shell Hollow Nanocatalyst with a Double Shell Consisting of a Similar Inner Shell Metal

The catalytic efficiency of a bimetallic nanocatalyst is better than that of a catalyst containing a single metal. If the nanoalloy catalyst has a hollow structure, the efficiency will be enhanced further. Catalytic oxygen reduction by a Pd–Pt alloy hollow nanocatalyst proved to have enhanced activity.[99] The effect of the geometry and structure on the catalytic properties was confirmed by density functional theoretical calculations, where the results proved that the electronic confinement greatly affects catalytic efficiency.[100] When a palladium nanocatalyst made up of nanotube structures was examined using the Suzuki reaction, an exciting improvement in catalytic efficiency was seen.[101]

The values of kinetic parameters (rate constants, activation energies, frequency factors and entropies of activation) for any catalysis reaction are characteristic for both the catalysis reaction and the surface of the catalyst. Hollow nanocatalysts, such as PtNCs, PdNCs, PtPdNCs, and PdPtNCs, of similar concentration were examined in a model catalysis reaction, the reduction of 4-nitrophenol (4NP) with BH.[67] The kinetic parameter values for the single shell metallic nanocatalysts (PtNCs and PdNCs) were compared to those obtained with the double shell nanocatalysts (PtPdNCs and PdPtNCs) with similar metallic inner shells. The results show that the kinetic parameter values of PdNCs and PtPdNCs (both have an inner shell made of Pd) are comparable as shown in Table 9.1. Close kinetic parameter values were obtained when using PtNC and PdPtNC nanocatalysts. Moreover, the reduction reaction initially proceeds at a slow rate, which then increases. During the first 5–10 minutes, the reaction takes place on the outer surface

Table 9.1 The kinetic parameters of the reduction of 4NP with BH catalyzed by PdNCs, PtPdNCs, PdPtNCs, and PtNCs of similar concentrations.

Nanocatalyst	Rate constant at 25 °C (min^{-1})	Activation energy (kcal mole^{-1})	Entropy of activation (cal/mol K^{-1})	Frequency factor (min^{-1})
PdNCs	$0.0190 \pm 8.7 \times 10^{-4}$	22.6 ± 1.5	67.8 ± 5.0	5.10×10^{14}
PtPdNCs	$0.0035 \pm 1.3 \times 10^{-4}$	20.7 ± 1.8	61.4 ± 6.0	2.13×10^{13}
PdPtNCs	$0.0190 \pm 2.0 \times 10^{-4}$	18.5 ± 1.3	50.4 ± 4.2	8.80×10^{9}
PtNCs	$0.0036 \pm 2.0 \times 10^{-4}$	16.2 ± 1.1	43.2 ± 3.6	2.31×10^{9}

and the reactants diffuse inside the catalyst. These results prove that catalysis using a hollow nanocatalyst takes place inside the nanocatalyst and the enhanced activity is due to the inner surface.[67]

9.4.4 Following the Optical Properties of Plasmonic Nanocatalysts During Catalysis

Plasmonic nanoparticles interact with the electromagnetic radiation of a resonance frequency; this leads to the generation of a localized surface plasmon resonance (LSPR) spectrum.[2–4] The resonance frequency of the plasmonic nanoparticles is based on the dielectric function of the surrounding medium. When the dielectric constant of the medium is changed, the LSPR spectrum peak red- or blue-shifts according to the value of the dielectric constant.[2–4] Hollow nanoparticles have two plasmonic surfaces, the inner and outer.[29] When the dielectric constant of the surrounding medium located on the outer surface of the hollow plasmonic nanoparticles changes, the LSPR peak will shift to a certain position. On changing the dielectric constant of the medium on the inner surface, a further shift in the LSPR is expected. This optical phenomenon of plasmonic nanocatalysts can thus be used to confirm that catalysis with hollow nanoparticles takes place on the inner surface of the nanocages, which is responsible for the activity of the cage nanocatalyst. For this study, AuNCs (with inner and outer plasmonic surfaces) and AuPtNCs (with a gold outer plasmonic surface and a non-plasmonic inner surface) were used to catalyze the reduction of 4NP with BH, as a model reaction. This reaction proceeds through the formation of hydrogen gas, therefore, during the catalysis reaction, the water molecules (refractive index ~1.33) on the surface of the nanoparticles will be replaced with H$_2$ gas (refractive index ~1) and if the surface is plasmonic, the LSPR will be blue-shifted. Figures 9.6(A) and (B) show the LSPR spectra of pure AuNCs (inner and outer plasmonic surfaces) and AuPtNCs (outer plasmonic surface only) after mixing with pure 4NP, pure BH, and a mixture of 4NP and BH at 0 min and after 30 min of mixing, respectively.[74] Two sequential blue shifts of the LSPR peak are observed on using the AuNC hollow nanocatalyst, one immediate and the other later on. The first shift is due to catalysis on the outer surface and the rate of the catalysis reaction being slow; the rate

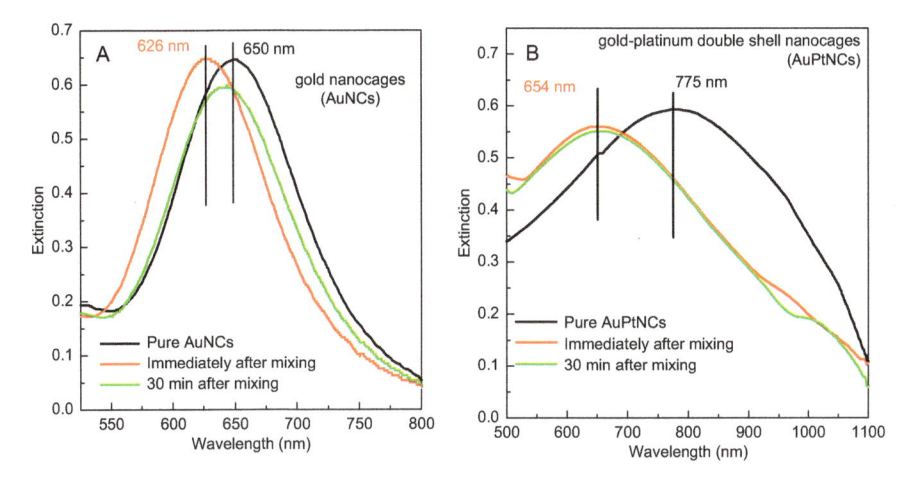

Figure 9.6 The localized surface plasmon resonance spectra of pure nanocatalysts (black), and immediately after mixing with borohydride and 4-nitrophenol (red), and 30 min after mixing with borohydride and 4-nitrophenol (green): (A) AuNCs, (B) AuPtNCs.

increases when the second blue shift occurs due to the reaction on the inner surface and the production of hydrogen gas.[74] A single rapid LSPR blue shift is observed when using a AuPtNC hollow nanocatalyst and the reaction begins at a slow rate, becoming faster after some time. This is consistent with the presence of one plasmonic surface for AuPtNCs. Based on these results, it can be deduced that the catalysis reaction on the inner surface is more efficient than the catalysis on the outer surface of the hollow nanocatalyst due to the cage nanoreactor effect.[74]

9.5 Proposed Mechanism for Nanocatalysis Based on Spectroscopic Studies

Nanocatalysts are small and in some cases, they approach the size of the reacting materials. It is useful to study the mechanism of nanocatalysis to determine whether it is heterogeneous (in which the reaction takes place on the surface of the nanocatalyst) or homogeneous (in which the reaction takes place *via* the formation of a complex resulting from the reaction between one of the reacting materials with the surface of the nanocatalyst, which reacts with the other reacting material in the bulk of the solution). In order to study the mechanism of nanocatalysis, the reduction reaction of hexacyanoferrate(III) (HCF III) by thiosulfate (TS) in the presence of platinum nanoparticles (PtNPs) was used as an example. Figure 9.7(A) shows the IR spectrum of the HCF III and TS reaction with molar concentration ratios of 0.5 : 0.05, 0.5 : 0.1, 0.5 : 0.5, 0.1 : 0.5 and 0.05 : 0.5, measured by attenuated total reflectance (ATR) IR *in situ* during mixing of HCF III with a PtNPs-TS mixture (solutions of HCF$_{III}$ and PtNCs-TS flowed separately and were mixed inside the ATR cell). The products of the HCF$_{III}$ and TS reaction are tetrathionate and hexacyanoferrate(II) (HCF$_{II}$);[102] the bands corresponding to the

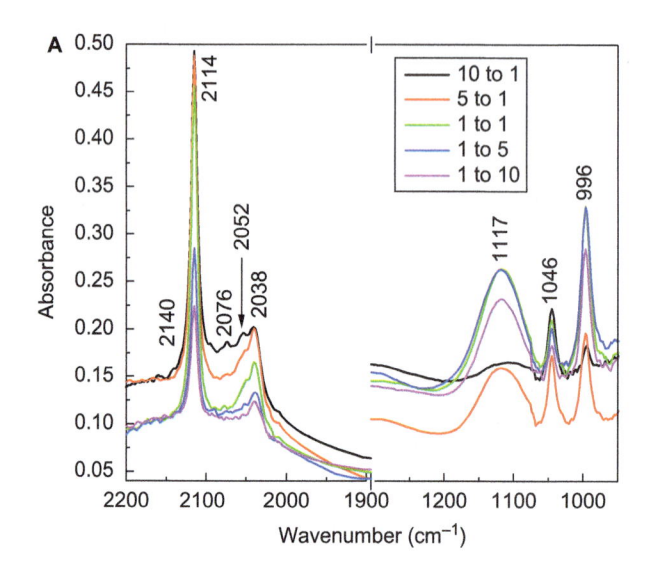

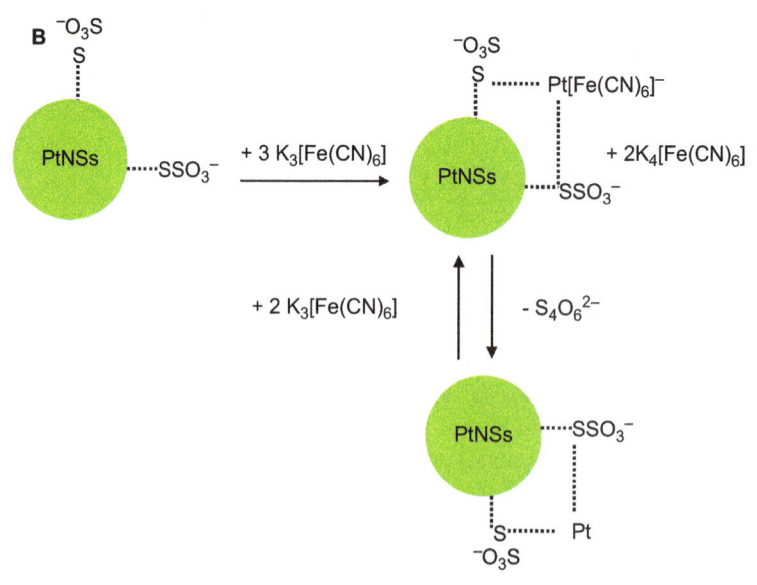

Figure 9.7 (A) The FTIR spectrum of the hexacyanoferrate(III) and thiosulfate reaction with concentration ratios of 0.5 M : 0.05 M, 0.5 M : 0.1 M, 0.5 M : 0.5 M, 0.1 M : 0.5 M and 0.05 M : 0.5 M at zero time (mixing the reacting materials inside the ATR cell during detection). (B) Schematic diagrams for the mechanism of the reaction of hexacyanoferrate(III) and thiosulfate catalyzed by PtNP.

reactant and the products are represented in the IR spectrum. The CN stretching bands for HCFIII and HCFII are at 2114 and 2038 cm^{-1}, respectively.[103] The band at 2052 cm^{-1} corresponds to the CN stretching of the Prussian blue analogue KPt^{II}[Fe^{III}(CN)$_6$] complex, while that at 2074 cm^{-1} corresponds to Pt$_{IV}$[Fe^{II}(CN)$_6$] complex formation.[103,104] The band at 2140 cm^{-1} is assigned as bridged cyanide.[94] The bands at 996 and 1117 cm^{-1} represent the symmetric and asymmetric SO stretching of free thiosulfate, respectively, while the adsorbed TS appears at 1046 cm^{-1}. Based on these results, the mechanism of HCFIII-TS catalyzed by PtNPs takes place through the adsorption of thiosulfate on the surface of the PtNP catalyst, and the Prussian blue analogue is formed during the reaction. Since TS is capable of forming a complex with platinum(II), the Prussian blue analogue will form on the surface of the platinum nanocatalyst.

The mechanism of the reaction proceeds by the reaction of two HCFIII molecules with a PtNP, leading to the oxidation of a platinum atom and formation of a Prussian blue analogue complex with another HCFIII molecule. This complex binds to the surface of the PtNP through two TS ions. The electron transfer takes place between the TS and the Prussian blue analogue, leading to the formation of tetrathionate and Pt atoms, which return to the surface leading to reshaping of the PtNS. The complete reaction is described by eqns (1)–(3), and the mechanism is summarized in Figure 9.7(B).

$$2K_3[Fe(CN)_6] + Pt(PtNPs) + 2K^+ \rightarrow Pt^{2+} + 2K_4[Fe(CN)_6] \tag{1}$$

$$K_3[Fe(CN)_6] + Pt^{2+} \rightarrow 2KPt^{II}[Fe^{III}(CN)_6] + 2K^+ \tag{2}$$

$$KPt[Fe(CN)_6] + 2S_2O_3^{2-} + 2K^+ \rightarrow K_3[Fe(CN)_6] + S_4O_6^{2-} + Pt \tag{3}$$

References

1. P. L. Freund and M. Spiro, *J. Phys. Chem.*, 1985, **89**, 1074.
2. R. Jin, G. Wu, Z. Li, C. A. Mirkin and G. C. Schatz, *J. Am. Chem. Soc.*, 2003, **125**, 1643.
3. A. J. Haes, L. Chang, W. L. Klein and R. P. Van Duyne, *J. Am. Chem. Soc.*, 2005, **127**, 2264.
4. C. E. H. Berger, T. A. M. Beumer, R. P. H. Kooyman and J. Greve, *Anal. Chem.*, 1998, **70**, 703.
5. I. Brigger, C. Dubernet and P. Couvreur, *Adv. Drug Delivery Rev.*, 2002, **54**, 631.
6. C. Aymonier, U. Schlotterbeck, L. Antonietti, P. Zacharias, R. Thomann, J. C. Tiller and S. Mecking, *Chem. Commun.*, 2002, 3018.
7. K. F. MacDonald, Z. L. Samson, M. I. Stockman and N. I. Zheludev, *Nat. Photonics*, 2009, **3**, 55.

8. M. Pohl, V. I. Belotelov, I. A. Akimov, S. Kasture, A. S. Vengurlekar, A. V. Gopal, A. K. Zvezdin, D. R. Yakovlev and M. Bayer, *Phys. Rev. B*, 2012, **85**, 081401.
9. V. V. Temnov, G. Armelles, U. Woggon, D. Guzatov, A. Cebollada, A. Garcia-Martin, J. M. Garcia-Martin, T. Thomay, A. Leitenstorfer and R. Bratschitsch, *Nat. Photonics*, 2010, **4**, 107.
10. V. V. Temnov, *Nat. Photonics*, 2012, **6**, 728.
11. I. H. El-Sayed, X. H. Huang and M. A. El-Sayed, *Nano Lett.*, 2005, **5**, 829.
12. C. Loo, A. Lowery, N. Halas, J. West and R. Drezek, *Nano Lett.*, 2005, **5**, 709.
13. P. Fortina, L. J. Kricka, D. J. Graves, J. Park, T. Hyslop, F. Tam, N. Halas, S. Surrey and S. A. Waldman, *Trends Biotechnol.*, 2007, **25**, 145.
14. M. D. Malinsky, K. L. Kelly, G. C. Schatz and R. P. Van Duyne, *J. Phys. Chem. B*, 2001, **105**, 2343.
15. U. Kreibig and M. Vollmer, *Optical Properties of Metal Clusters*, Springer Series in Materials Science 25, Berlin, 1995.
16. A. Tao, P. Sinsermsuksakul and P. D. Yang, *Angew. Chem., Int. Ed.*, 2006, **45**, 4597.
17. P. L. Freund and M. Spiro, *J. Phys. Chem.*, 1985, **89**, 1074.
18. S. Link and M. A. El-Sayed, *Int. Rev. Phys. Chem.*, 2000, **19**, 409.
19. P. K. Jain, X. Huang, I. H. El-Sayed and M. A. El-Sayed, *Acc. Chem. Res.*, 2008, **41**, 1578.
20. M. Grzelczak, J. Perez-Juste, P. Mulvaney and L. M. Liz-Marzan, *Chem. Soc. Rev.*, 2008, **37**, 1783.
21. J. Perez-Juste, I. Pastoriza-Santos, L. M. Liz-Marzan and P. Mulvaney, *Coord. Chem. Rev.*, 2005, **249**, 1870.
22. X. Huang, I. H. El-Sayed, W. Qian and M. A. El-Sayed, *J. Am. Chem. Soc.*, 2006, **128**, 2115.
23. D. Solis, B. Willingham, S. L. Nauert, L. S. Slaughter, J. Olson, P. Swanglap, A. Paul, W. S. Chang and S. Link, *Nano Lett.*, 2012, **12**, 1349.
24. R. Comparelli, E. Fanizza, M. L. Curri, P. D. Cozzoli, G. Mascolo, R. Passino and A. Agostiano, *Appl. Catal., B*, 2005, **55**, 81.
25. Y. G. Sun and Y. N. Xia, *Science*, 2002, **298**, 2176.
26. R. Jin, Y. C. Cao, E. Hao, G. S. Metraux, G. C. Schatz and C. A. Mirkin, *Nature*, 2003, **425**, 487.
27. S. J. Oldenburg, R. D. Averitt, S. L. Westcott and N. J. Halas, *Chem. Phys. Lett.*, 1998, **288**, 243.
28. A. M. Schwartzberg, T. Y. Olson, C. E. Talley and J. Z. Zhang, *J. Phys. Chem. B*, 2006, **110**, 19935.
29. M. A. Mahmoud and M. A. El-Sayed, *Nano Lett.*, 2009, **9**, 3025.
30. N. R. Jana, L. Gearheart and C. J. Murphy, *J. Phys. Chem. B*, 2001, **105**, 4065.
31. A. Tao, F. Kim, C. Hess, J. Goldberger, R. R. He, Y. G. Sun, Y. N. Xia and P. D. Yang, *Nano Lett.*, 2003, **3**, 1229.
32. J. Zhang, M. R. Langille and C. A. Mirkin, *Nano Lett.*, 2011, **11**, 2495.

33. C. L. Nehl, H. W. Liao and J. H. Hafner, *Nano Lett.*, 2006, **6**, 683.
34. F. Caruso, R. A. Caruso and H. Mohwald, *Science*, 1998, **282**, 1111.
35. Y. D. Yin, R. M. Rioux, C. K. Erdonmez, S. Hughes, G. A. Somorjai and A. P. Alivisatos, *Science*, 2004, **304**, 711.
36. N. Tian, Z.-Y. Zhou, S.-G. Sun, Y. Ding and Z. L. Wang, *Science*, 2007, **316**, 732.
37. C. Burda, X. Chen, R. Narayanan and M. A. El-Sayed, *Chem. Rev.*, 2005, **105**, 1025.
38. R. Narayanan and M. A. El-Sayed, *J. Phys. Chem. B*, 2004, **108**, 8572.
39. R. Narayanan and M. A. El-Sayed, *J. Am. Chem. Soc.*, 2004, **126**, 7194.
40. I. Lee, R. Morales, M. A. Albiter and F. Zaera, *Proc. Natl. Acad. Sci.*, 2008, **105**, 15241.
41. A. Kudo and Y. Miseki, *Chem. Soc. Rev.*, 2009, **38**, 253.
42. S. J. Kweskin, R. M. Rioux, S. E. Habas, K. Komvopoulos, P. Yang and G. A. Somorjai, *J. Phys. Chem. B*, 2006, **110**, 15920.
43. V. Subramanian, E. E. Wolf and P. V. Kamat, *J. Am. Chem. Soc.*, 2004, **126**, 4943.
44. A. Eppler, G. Rupprechter, L. Guczi and G. A. Somorjai, *J. Phys. Chem. B*, 1997, **101**, 9973.
45. P. L. Freund and M. Spiro, *J. Chem. Soc., Faraday Trans. 1*, 1986, **82**, 2277.
46. P. L. Freund and M. Spiro, *J. Chem. Soc., Faraday Trans. 1*, 1983, **79**, 481.
47. Y. Yamada, C. K. Tsung, W. Huang, Z. Y. Huo, S. E. Habas, T. Soejima, C. E. Aliaga, G. A. Somorjai and P. D. Yang, *Nat. Chem.*, 2011, **3**, 372.
48. I. Lee and F. Zaera, *J. Catal.*, 2010, **269**, 359.
49. R. Narayanan and M. A. El-Sayed, *Nano Lett.*, 2004, **4**, 1343.
50. R. Narayanan and M. A. El-Sayed, *J. Phys. Chem. B*, 2004, **108**, 8572.
51. R. Narayanan and M. A. El-Sayed, *J. Phys. Chem. B*, 2003, **107**, 12416.
52. Y. Borodko, S. E. Habas, M. Koebel, P. D. Yang, H. Frei and G. A. Somorjai, *J. Phys. Chem. B*, 2006, **110**, 23052.
53. R. Narayanan and M. A. El-Sayed, *J. Catal.*, 2005, **234**, 348.
54. M. Haruta, *Catal. Today*, 1997, **36**, 153.
55. J. Hain, M. Schrinner, Y. Lu and A. Pich, *Small*, 2008, **4**, 2016.
56. S. H. Cho, B. Q. Ma, S. T. Nguyen, J. T. Hupp and T. E. Albrecht-Schmitt, *Chem. Commun.*, 2006, 2563.
57. F. Bedioui, *Chem. Rev.*, 1995, **144**, 39.
58. C. A. Witham, W. Y. Huang, C. K. Tsung, J. N. Kuhn, G. A. Somorjai and F. D. Toste, *Nat. Chem.*, 2010, **2**, 36.
59. Y. G. Sun, B. Mayers and Y. N. Xia, *Adv. Mater.*, 2003, **15**, 641.
60. H. Song, F. Kim, S. Connor, G. A. Somorjai and P. D. Yang, *J. Phys. Chem. B*, 2005, **109**, 188.
61. A. M. Contreras, X. M. Yan, S. Kwon, J. Bokor and G. A. Somorjai, *Catal. Lett.*, 2006, **111**, 5.
62. A. M. Contreras, J. Grunes, X. M. Yan, A. Liddle and G. A. Somorjai, *Top. Catal.*, 2006, **39**, 123.
63. I. Lee and F. Zaera, *J. Am. Chem. Soc.*, 2005, **127**, 12174.

64. M. A. Mahmoud, C. E. Tabor, M. A. El-Sayed, Y. Ding and Z. L. Wang, *J. Am. Chem. Soc.*, 2008, **130**, 4590.
65. F. Tao, M. E. Grass, Y. W. Zhang, D. R. Butcher, J. R. Renzas, Z. Liu, J. Y. Chung, B. S. Mun, M. Salmeron and G. A. Somorjai, *Science*, 2008, **322**, 932.
66. J. M. Thomas, B. F. G. Johnson, R. Raja, G. Sankar and P. A. Midgley, *Acc. Chem. Res.*, 2003, **36**, 20.
67. M. A. Mahmoud, F. Saira and M. A. El-Sayed, *Nano Lett.*, 2010, **10**, 3764.
68. S. Alayoglu, A. U. Nilekar, M. Mavrikakis and B. Eichhorn, *Nat. Mater.*, 2008, **7**, 333.
69. R. Narayanan and M. A. El-Sayed, *J. Phys. Chem. B*, 2004, **108**, 5726.
70. M. Eddaoudi, J. Kim, N. Rosi, D. Vodak, J. Wachter, M. O'Keeffe and O. M. Yaghi, *Science*, 2002, **295**, 469.
71. M. Graeser, E. Pippel, A. Greiner and J. H. Wendorff, *Macromolecules*, 2007, **40**, 6032.
72. J. C. Park, J. U. Bang, J. Lee, C. H. Ko and H. Song, *J. Mater. Chem.*, 2010, **20**, 1239.
73. J. Zeng, Q. Zhang, J. Y. Chen and Y. N. Xia, *Nano Lett.*, 2010, **10**, 30.
74. M. A. Mahmoud and M. A. El-Sayed, *Nano Lett.*, 2011, **11**, 946.
75. J. E. Macdonald, M. Bar Sadan, L. Houben, I. Popov and U. Banin, *Nat. Mater.*, 2010, **9**, 810.
76. C. W. Yen, M. A. Mahmoud and M. A. El-Sayed, *J. Phys. Chem. A*, 2009, **113**, 4340.
77. M. A. Mahmoud, W. Qian and M. A. El-Sayed, *Nano Lett.*, 2011, **11**, 3285.
78. J. Snyder, I. McCue, K. Livi and J. Erlebacher, *J. Am. Chem. Soc.*, 2012, **134**, 8633.
79. M. Yadav, T. Akita, N. Tsumori and Q. Xu, *J. Mater. Chem.*, 2012, **22**, 12582.
80. J. W. Hong, S. W. Kang, B. S. Choi, D. Kim, S. B. Lee and S. W. Han, *ACS Nano*, 2012, **6**, 2410.
81. M. Faraday, *Philos. Trans. R. Soc. London*, 1847, **147**, 145.
82. R. Zsigmondy, *The Chemistry of Colloids*, John Wiley & Sons, Inc., New York, 1917.
83. H. J. Fan, M. Knez, R. Scholz, D. Hesse, K. Nielsch, M. Zacharias and U. Gosele, *Nano Lett.*, 2007, **7**, 993.
84. S. D. Berger, J. M. Gibson, R. M. Camarda, R. C. Farrow, H. A. Huggins, J. S. Kraus and J. A. Liddle, *J. Vac. Sci. Technol., B: Microelectron. Nanometer Struct.–Process., Meas., Phenom.*, 1991, **9**, 2996.
85. C. L. Haynes and R. P. Van Duyne, *J. Phys. Chem. B*, 2001, **105**, 5599.
86. J. H. Lee, M. A. Mahmoud, V. Sitterle, J. Sitterle and J. C. Meredith, *J. Am. Chem. Soc.*, 2009, **131**, 5048.
87. J. H. Lee, M. A. Mahmoud, V. B. Sitterle, J. J. Sitterle and J. C. Meredith, *Chem. Mat.*, 2009, **21**, 5654.
88. M. A. Mahmoud, B. Snyder and M. A. El-Sayed, *J. Phys. Chem. Lett.*, 2010, **1**, 28.
89. M. A. Mahmoud and M. A. El-Sayed, *J. Phys. Chem. C*, 2008, **112**, 14618.

90. R. Narayanan and M. A. El-Sayed, *J. Phys. Chem. B*, 2004, **108**, 5726.
91. M. A. Mahmoud and M. A. El-Sayed, *Langmuir*, 2012, **28**, 4051.
92. M. Graeser, E. Pippel, A. Greiner and J. H. Wendorff, *Macromolecules*, 2007, **40**, 6032.
93. J. C. Park, J. U. Bang, J. Lee, C. H. Ko and H. Song, *J. Mater. Chem.*, 2010, **20**, 1239.
94. A. Dawson and P. V. Kamat, *J. Phys. Chem. B*, 2001, **105**, 960.
95. A. Fujishima and K. Honda, *Nature*, 1972, **238**, 37.
96. J. Snyder, I. McCue, K. Livi and J. Erlebacher, *J. Am. Chem. Soc.*, 2012, **134**, 8633.
97. G. Weng, M. A. Mahmoud and M. A. El-Sayed, *J. Phys. Chem. C*, 2012, **116**, 24171.
98. M. A. Mahmoud and G. Weng, *Catal. Commun.*, 2013, **38**, 63.
99. J. W. Hong, S. W. Kang, B. S. Choi, D. Kim, S. B. Lee and S. W. Han, *ACS Nano*, 2012, **6**, 2410.
100. J. M. M. de la Hoz and P. B. Balbuena, *J. Phys. Chem. C*, 2011, **115**, 21324.
101. Y. G. Sun, B. Mayers and Y. N. Xia, *Adv. Mater.*, 2003, **15**, 641.
102. *Infrared and Raman Spectra of Inorganic and Coordination Compounds, Part A: Theory and Applications in Inorganic Chemistry*, ed. K. Nakamoto, John Wiley & Sons, Inc., New York, 5th edn, 1997.
103. M. A. Mahmoud and M. A. El-Sayed, *J. Phys. Chem. C*, 2007, **111**, 17180.
104. M. A. Mahmoud, *J. Catal.*, 2010, **274**, 215.

CHAPTER 10
Nanoreactor Catalysis

KYU BUM HAN, CURTIS TAKAGI AND AGNES OSTAFIN*

University of Utah Nano Institute, Department of Chemical Engineering, 36 S Wasatch Drive, Salt Lake City UT 84112
*Email: a.ostafin@utah.edu

10.1 Introduction

A catalyst is a substrate that accelerates chemical reactions by reducing their activation energy. The nanoscale catalyst has expanded the efficiencies of different industries *via* its greater active surface area, controllable physical morphology and properties (*via* chemical approaches), and improved stability. Examples of successful nanocatalysts include dendrimers, polymer microspheres and core–shell structures, microgels, liposomes, emulsions, micelles and carbon nanotubes. Many of these materials contain protected catalytic environments and so fall into the category of nanoreactor catalysts.

The nanocatalyst market has grown by an average 6.3% since 2004; the global market has risen from US$ 3.7 billion in 2004 to US$ 5.0 billion in 2009.[1] Petrochemicals, one of the major markets, is a US$ 742 million[2] global industry that focuses on obtaining premium quality fuel and improving selectivity control. Refineries previously used lead and benzene to increase the octane number of gasoline, but these materials were environmentally unfriendly. By adding nanoplatinum as a reforming catalyst with other metals, such as tin or rhenium, higher octane can be achieved without these additives. Zeolites are an early form of nanoreactor catalyst with customized cage-like structures ~1 nm in size. The method for customizing the structure was developed at Mobil in 1977. Methanol is reacted on aluminosilicate-based zeolite HZSM-5 to produce a mixture of hydrocarbons for gasoline blends (a mixture of branched alkanes and methylbenzenes).[3,4] Because the zeolite structure has holes of a certain size, only the desired molecules are allowed to react inside the cage.

These early nanocatalysts led to improved product quality and stability, faster reactions, and reduced costs. A reduced catalyst size increased the amount of surface available for reactions, and for some materials, changed the surface energy and nature of thermodynamics to promote reactions that ordinarily would not occur. Because the active surface was exposed, fouling and agglomeration of catalysts remained a possibility. Nanoreactor catalysts, such as zeolites, were aimed at controlling both the active surface area and the local environment dimension and chemical/physical properties. A smaller reaction environment changes the dynamics of molecule interactions with the catalysts, and can reduce unwanted reactions that lead to less effective catalytic power. Depending on the physical nature of the space, it is possible to control the molecular kinetics, specificity, and extend the operating range of the catalyst.

10.2 Steric and Structural Effects

10.2.1 Dendrimers

Maintaining well-dispersed catalyst in the reaction medium, then separating the catalyst from the final product is a particularly challenging task for nanocatalysts. To engineer methods of particle separation while ensuring phase cooperation to maximize catalytic efficiency, dendrimers, polymer coatings, and carbon nanotubes have been developed. These approaches were aimed at showing improvement in performance and stability.[5-8] However, either due to phase instability or molecular changes to the structure of the polymeric coating, the nanoparticles can still aggregate over time, resulting in a lowered surface area and a lower catalytic efficiency.

Dendrimers are a kind of highly-branched polymer whose completed size is generally larger than that of the metal nanoparticle catalyst. It contains many interior compartments that can act as a group of isolated nanoreactors for various chemical reactions.[9-15] Dendrimers can be customized to house or to grow metallic nanocatalysts *in situ* and result in a system that behaves as both a homogeneous and heterogeneous catalyst at the same time. Practically, the metal nanoparticle size can be restricted to maximize catalytic effect, while the entire unit can be more easily handled and separated from the product. The reaction interface between the catalyst and the reactant in the surrounding solvent is somewhat protected as well.

Dendrimers maintain the catalytic efficiency of metal nanoparticles, or complexes, by keeping them isolated from each other, and controlling the flow of reagents and product. For dendrimer systems, the metal nanoparticles are often created *in situ* by reduction chemistry in the dendrimer pockets where steric and electrostatic forces terminate the synthesizing reaction.[16] Catalyst growth can be initiated and stabilized through interactions between the π-electrons of the aromatic rings in the polymer, and vacant orbitals of the metal species.[17] Choosing the correct polymer functional groups, with ionic or hydrogen bonds, for the interior regions of the dendrimer, attracts and stabilizes the metal cation complexes in a relatively high concentration in the interior of the dendrimer. For example, Ooe et al. constructed a dendrimer system with interior amino groups that create a polar environment around a Pd complex, which leads to higher reactivity rates.[10] Careful manipulation of the dendrimer nanoreactors has often yielded higher catalytic efficiencies than the normal monomeric catalysts. This is because the catalytic site can force the reagents interacting with the catalysts into a specific configuration that is similar to their reaction transition state, in a phenomenon which is considered to be similar to biological enzyme mechanisms. This allows for increased selectivity in reactions and higher efficiencies, especially in reactions with two enantiomeric products.[12,18,19]

One of the most promising features of the dendrimeric catalyst nanoreactor systems is the customizability of the dendrimer molecules and, in effect, the properties of the reaction space and exterior periphery. The first factor that must be considered is the number of generations, or layers, that have been instilled in the structure. Dendrimers of high generations can be so sterically crowded at the edges that they act as filter membranes. Inside, the large open pores remain, but the exterior layers of the polymer itself can easily control access to these ports.[5,13,20,21] Control over the diffusion of species can be accomplished by keeping the dendrimer surface open with different functional groups. Simple steps, such as protonation or deprotonation of terminal groups, can open and close the pores at the surface, and these can be tuned to only allow the diffusion of specific reactants or products.[5,20,21] Very large pores allow for greater mass transfer, but at the same time the catalytic pockets can become saturated with reactants and products, and the diffusion kinetics are hindered.[22,23] However, if the pores are too small, then of course, there is not enough opportunity to bring in new reagents while releasing the product. This is why the pore sizes on the surface of the dendrimer need to be optimized, to optimize the catalytic properties.

Along with pore size, the peripheral molecules can also control the solubility properties of the particle. For example, it is possible to have a polar interior with a non-polar surface; this allows the system to incorporate itself into an organic environment but catalyze reactions with polar properties.[24,25] This property allows for unconventional designs and ability, and also it has improved the recoverability of the catalysts, leading to a greener industry with higher recyclability

10.2.2 Microgels

The microgel catalytic nanoreactor systems are similar to the dendrimers, except they are cross-linked linear polymer chains. They are generally more sensitive to chemicals, temperature, and hydration rates, than their dendrimer counterparts. Microgels are very stable and highly customizable.[26] Like the dendrimer systems, metallic nanoparticles are often embedded in the structures,[27–30] and access to these sites is controlled by other external factors. Microgels can expand and contract with temperature changes, or as they gain or lose water. This opens and closes the pores, allowing for selectively and responsiveness. For example, poly(N-isopropylacrylamide) (PNIPA) is soluble in water below 32 °C, due to the formation of hydrogen bonds; however, above this temperature the hydrogen bonds are disrupted, the water escapes and the gel collapses.[31–35] Pich et al. have demonstrated the capability of the microgel-supported catalyst by embedding gold nanoparticles into a microgel system. They found that the confined catalyst increased efficiency to 100% by preventing aggregation of the particles, and allowing for greater mass transport.[30] Because of the high water content that is possible in microgels, enzymes can also be placed inside rather than the metallic nanoparticle.[36,37] This has been demonstrated with glucose oxidase, which did not lose its catalytic activity when placed in a water-saturated microgel system. This discovery could have huge industrial implications where catalysis is needed, because natural enzymes are often the most efficient and selective catalysts, but their complexity makes it difficult to synthetically duplicate.

10.2.3 Polymer Core–shell Structures

The concept of a polymer core–shell structure is similar to the dendrimer. These systems typically consist of a solid core with a secondary polymer molecule that resides on the surface.[38] Often there are different environments throughout the structure to optimize the reaction space. Wei et al. have demonstrated that a molecular brush structure with a Pd core can perform well with the hydrophilic and hydrophobic substrates necessary for conducting Suzuki reactions.[39] The hydrophilic molecules stay within the regions of the polymer while the hydrophobic substrates are adsorbed deeper, near the Pd core. It is not necessary, however, to have a metallic core with the polymer brush. Lu and co-workers employed a nanoparticle embedding system, similar to that used in a dendrimer, to incorporate the catalyst into a nanoreactor with a polystyrene core and PNIPA shell.[40]

10.2.4 Hydrophobic–hydrophilic Structures: Micelle, Emulsion, and Liposome

The basic principle of these surfactant-supported structures is similar to others such as dendrimers, microgels, and carbon nanotubes. They have been commonly used as nanoreactors for increasing reaction rates. In systems where there are two different micellar structures, the micelles fuse together to create a nanoreactor environment with high concentrations of both reagents. The significant factor is that the smaller size of the micelles resulted in a higher number of mixed micelles and higher catalytic activity. The advantage of using micelles, emulsions and liposomes as catalysts is their recovery after the chemical product has formed. A water-soluble Rh/TPPTS catalyst, for example, was used and the hydrogenation of PB-*b*-PEO in mixed micelles was performed with an increase in efficiency.[41] Using the small interior space of the micelle, a simple metal salt (ammonium molybdate), was injected as an oxidation catalyst.

There are some advantages to the micelle catalyst, including an optimized interfacial area between two immiscible solvents, the enhanced solubility of hydrophobic substrates in an aqueous phase, and the conversion of a batch to continuous manufacturing processes.[42] Micellar support structures for metal catalytic nanoparticles reduce aggregation, which increases the surface area and also the catalytic potential. In addition, they also result in high solubility in solution, and may improve the selectivity of metal nanoparticle catalysts.[20,43–46] Encapsulators are most commonly spherical in shape, but they can also be fiber-like polymer. Polymer nanotubes with Pd–Rh nanoparticles embedded in them catalyze the hydrogenation reaction of double and triple bonds, control distribution of reagents, and also increase their stability.[47]

A maximum reaction rate is obtained at a certain emulsion size (~hundreds of nanometres). Generally, the particle size and reaction rate are inversely proportional due to the reduced surface area to volume ratio. Larger voids inside emulsions allow for nanoparticle placement with complexes, high selectivity and control of diffusion with different layers at the interface. Using computer simulations, the kinetics of the reaction were shown to rely on the surface activity of reactants and the potential barrier to the reaction.[48] It was theoretically demonstrated that the concentration of species inside the emulsion was higher than that of the bulk phase; consequently, the reaction rate was higher. Further, an optimal drop size was determined so that the reaction rate was maximized. As the reaction proceeded, the substrate concentration decreased while the catalyst concentration remained constant. The relationship between the maximum radius of the emulsion and the substrate concentration was inversely proportional.[49,50] This principle was also applied to forming ZrO_2, CeO_2, $LiMn_2O_4$, polythiophene nanoparticles, and $Ba_2YCu_3O_7$,[21,22] using crosslinked acrylic emulsion polymers. Further, heterogeneous Co(III)-salen catalyst cores were more efficient and had quicker reaction times while requiring less catalyst, than when in homogeneous form. The hydrophobic surface prevented water from penetrating the interior where it would interfere with the reaction and eventually stop it.

The reaction rate in a micelle can also be controlled by temperature. In a recent study, Wang *et al.* synthesized gold nanoparticles using a supporting micelle structure made of poly(*N*-isopropylacrylamide)-*b*-poly(4-vinyl pyridine) (PNIPAM-*b*-P4VP). This specific polymer combination is thermally responsive and allowed temperature controlled rates by altering the coordination kinetics between $AuCl_4^-$ and micelles.[41] Below the lower critical solution temperature (32 °C) of the micelle, the PNIPAM segments were hydrophilic, allowing certain reactants, such as *p*-aminophenol and NaBH4, to easily reach the catalyst surface (micelle). This resulted in an increased reaction rate, and a more efficient method to produce gold nanoparticles. After the particles were formed, the temperature was raised above 32 °C at which point the PNIPAM folded back onto the gold nanoparticles and hindered the reaction.

An alternative method for particle synthesis is using a reverse micelle. The basic principle of confining the reaction to increase the reaction rate is maintained, but hydrogen-bonding at the interface of the micelles is utilized to produce an enzymatic reaction. The catalytic efficiency was higher in the micelle system than in a bulk solution such as water. It was concluded that the hydrolysis reaction occurred at the interior interface, resulting in the conversion of 2-naphthyl acetate by α-chymotrypsin.[51]

The macroscale structure that results from the dispersion of micelles is the nanogel, which is a network of cross-linked polymeric particles similar to hydrogels. Since the gel phase has recently come into the spotlight, due to its potential applications in drug delivery and nanoreactors, the nanogel has been studied greatly along with its networks of dispersed micelles. The size of the nanogel was controlled by both pH and irradiation of UV light. The UV light-responsive nanogel was made by reversibly photo-cross-linking diblock copolymer micelles. Then the pH was reduced and this allowed for the collapse of the micellar solution into a more condensed gel state. Inside this nanogel, gold nanoparticles were synthesized[52] *via* atom transfer radical polymerization (ATRP). However, cross-linking the polymer branches creates a stabilized micelle, but changes the catalytic properties. Catalysts—biological and synthetic—were placed inside, and substrates were selectively passed through the micelle membrane. The cross-linking caused a reduced diffusive flux across the membrane resulting in decreased reaction efficiency.[53]

An accelerated reaction in a confined space is not only true of micelles, but also liposomes. Copolymer liposomes were made to house enzymes for greater concentration capacity, protection from proteases, and for multistage enzymatic reactions. ABA copolymer, made of hydrophilic and hydrophobic regions, spontaneously forms a liposome-like structure that has thicker and more rigid walls. Nardin *et al.* put an enzyme inside then inserted membrane channels to allow for adequate diffusion across the membrane. Due to the artificial nature of the system, the protein channels were slower but still functional in the synthetic membrane.[54]

Polymersomes like liposomes have widely been used as a catalyst support structure; they are formed by the self-assembly of amphiphilic block copolymers, which builds a bilayer. Since the polymersome system encapsulates species in a sterically small space, chemical reactions are accelerated due to a higher concentration of species. However, their demerit is a low permeability through the bilayer, which results in a high concentration of product in the interior region. So people have attempted to synthetically replicate the functions of natural enzymes, but natural enzymes are very complex. Attempts have been made to copy them by taking a molecule in the transition state and stamping it into a polymer; the idea being that the most energetic parts will be identified, and leave behind a stabilizing molecule on the mold polymer. However, this seems very difficult and isn't as effective as some metal nanoparticle based catalysts.[55] In order to fix this problem, channel proteins or proton pumps were placed in the bilayer so that substrate molecules could easily diffuse in and out.[56,57] Further, an easier synthesis method has been developed by incorporating an additional block copolymer, which intrinsically provided the porous bilayer. Later, researchers improved the polymersomes as multifunctional nanocatalysts, which allowed biochemical reactions not only inside the polymersomes but also on the surface.[58-60]

10.3 Absorbing Nanocatalyst Surface

10.3.1 Micelle and Emulsion

Since reactant concentrations are higher at interfaces in emulsions than in the bulk phase, the interface of a droplet in an emulsion is a special medium for an accelerated reaction. The locally high concentration of molecules accelerates the reactions in which substrates and catalysts can be concentrated in micelles. In a modified approach, small polymer particles are used to accumulate hydrophobic substrates. The reaction rate of poly(N-vinylcaprolactam-co-1-vinylimidazole) and poly(N-isopropylacrylamide-co-1-vinylimidazole) was increased due to concentrated substrate at the interface of aggregates, and obeyed theoretical Michaelis–Menten kinetics.[61] The reaction rate of p-nitrophenyl diphenyl phosphate increased the amount of positively charged surfactant such as cetyltrimethylammonium chloride (CTACl).[62,63] Further, horseradish peroxidase (HRP) was attached on the surface of polymersomes to convert 2,2'-azinobis(3-ethylbenzothiazoline-6-sulfonic acid) (ABTA) into ABTS$^+$.[64]

10.3.2 Carbon Nanotubes

In addition to the polymer-based support systems for catalysis, carbon nanotubes (CNT) have demonstrated their ability to increase catalytic efficiency. These increases in efficiency can be due to improved protection of the catalyst species, similar to the effects described for polymer systems earlier, but additionally, the properties of the CNT can act as a stabilizing force for reducing activation energies and increasing reaction rates.[65-67] Of course, CNTs have unique mechanical properties, but they are also unique in their electron configuration.[68,69] Depending on the orientation of the CNT, the electronic structure changes and, consequently, there are different interactions with the catalysts inside SWCNTs.[70] Li et al. suggested that the electrons of bis(cyclopentadienyl) cobalt and bis(ethyl-cyclopentadienyl) cobalt nanoparticles encapsulated in CNTs are more readily transferred between the particle and interior surface than when on the outer surface.[71,72] For the control of metal nanoparticle catalysts, the chemical and catalytic properties of atoms at corners or edges is different. So the way the CNT (also applies to polymer-based structures) attaches may change the catalytic properties.[73] The walls of the CNT are polarized, which may seem odd for a network of carbon–carbon single bonds, but it is this dipolar interaction between the CNT interior and the embedded catalytic species that can stabilize a transition state and promote more efficient product formation. The Menshutkin reaction, an SN2 mechanism, showed significant increase in reaction rate when conducted within a CNT, evidenced by a reduction in the endothermicity and reaction energy barrier.[74]

10.4 Conclusion

The nanoscale catalyst has been developed within the last couple of decades, in the hopes of making industrial processes cheaper, greener, and more efficient. The nanospace inside nanoreactor structures accelerated chemical and biochemical reactions due to increased concentrations; it also improved the selectivity of reactions due to the different mechanisms of reagent isolation and selection. However, it is not necessary for these catalysts to be confined to produce higher efficiency rates; this can also be achieved by increasing the surface area or by stabilizing the catalytic species, such as in dendrimers and other stabilization procedures. More complex systems, where a cascading effect that utilizes two or more catalysts, are currently being developed, mainly for biological scenarios where an enzymatic chain is desirable. The scale at which it is now possible to isolate and control environments allows for greater efficiency with less chemical waste, thus allowing industries to expand their capability while maintaining the state of the environment. Carbon nanotubes, dendrimers, polymers, and emulsion systems, all work to exploit the differences that occur at small scales. If successful, the way certain industries operate can change dramatically, and the desire to control various chemical reactions can be realized.

References

1. *Nanotechnology: nanocatalysts*, BCC Inc. Research Market Forecasting, 2009.
2. *Global Nanocatalysts Market to Reach $6.0 Billion by 2015, According to New Report by Global Industry Analysts, Inc.*, in Nanotechnology Now, 2009 (cited 2011 17, Dec.); available from: http://www.nanotech-now.com/news.cgi?story_id = 32763.
3. C. Chang and A. Silvestri, *J. Catal.*, 1977, **47**, 249–259.
4. C. Chang, *J. Catal. Rev.*, 1983, **25**, 1–118.
5. M. Zhao and R. M. Crooks, *Angew. Chem., Int. Ed.*, 1999, **38**(3), 364–366.
6. M. Zhao and R. M. Crooks, *Adv. Mater.*, 1999, **11**(3), 217–220.
7. Y. Niu, L. Yeung and R. Crooks, *J. Am. Chem. Soc.*, 2001, **123**, 6840–6846.

8. L. Balogh and D. Tomalia, *J. Am. Chem. Soc.*, 1998, **120**, 7355–7356.
9. Y. M. Chung and H. K. Rhee, *Catal. Lett.*, 2003, **85**(3–4), 159–164.
10. M. Ooe, M. Murata, T. Mizugaki, K. Ebitani and K. Kaneda, *J. Am. Chem. Soc.*, 2004, **126**(6), 1604–1605.
11. D. De Groot, B. de Waal, J. Reek, A. Schenning, P. Kamer, E. Meijer and P. van Leeuwen, *J. Am. Chem. Soc.*, 2001, **123**(35), 8453–8458.
12. L. K. Yeung and R. M. Crooks, *Nano Lett.*, 2001, **1**(1), 14–17.
13. J. F. G. A. Jansen, E. M. M. De Brabander-van Den Berg and E. W. Meijer, *Science*, 1994, **266**(5188), 1226–1229.
14. A. W. Bosman, H. M. Janssen and E. W. Meijer, *Chem. Rev.*, 1999, **99**(7), 1665–1688.
15. R. Breinbauer and E. Jacobsen, *Angew. Chem., Int. Ed.*, 2000, **39**, 3604–3607.
16. R. M. Crooks, M. Zhao, L. Sun, V. Chechik and L. Yeung, *Acc. Chem. Res.*, 2001, **34**(3), 181–190.
17. R. Akiyama and S. Kobayashi, *Chem. Rev.*, 2009, **109**(2), 594–642.
18. P. Bhyrappa, J. Young, J. Moore and K. Suslick, *J. Am. Chem. Soc.*, 1996, **118**, 5708–5711.
19. H. W. I. Peerlings and E. W. Meijer, *Chem.-Eur. J.*, 1997, **3**(10), 1563–1570.
20. M. Zhao, H. Tokuhisa and R. M. Crooks, *Angew. Chem., Int. Ed. Engl.*, 1997, **36**(23), 2596–2598.
21. M. Wells and R. M. Crooks, *J. Am. Chem. Soc.*, 1996, **118**(16), 3988–3989.
22. A. Pashornik, Polymer-Bound Reagents and Catalysis, in *Modern Synthetic Methods*, ed. R. Scheffold, Sauerlander, Aarau, 1976, pp. 113–168.
23. C. Mak and H. Chow, *Macromolecules*, 1997, **30**, 1228–1230.
24. M. Kimura, M. Kato, T. Muto, K. Hanabusa and H. Shirai, *Macromolecules*, 2000, **33**, 1117–1119.
25. M. Piotti, F. Rivera, R. Bond, C. Hawker and J. Frechet, *J. Am. Chem. Soc.*, 1999, **121**, 9471–9472.
26. J. Zhang, S. Xu and E. Kumacheva, *Adv. Mater.*, 2005, **17**(19), 2336–2340.
27. M. Antonietti, F. Grohn, J. Hartmann and L. Bronstein, *Angew. Chem., Int. Ed. Engl.*, 1997, **36**(19), 2080–2083.
28. N. T. Whilton, B. Berton, L. Bronstein, H. Hentze and M. Antonietti, *Adv. Mater.*, 1999, **11**(12), 1014–1018.
29. A. Biffis, *J. Mol. Catal., A*, 2001, **165**(1–2), 303–307.
30. A. Pich, A. Karak, Y. Lu, A. Ghosh and H. Adler, *J. Nanosci. Nanotechnol.*, 2006, **6**(12), 3763–3769.
31. H. G. Schild, *Prog. Polym. Sci.*, 1992, **17**(2), 163–249.
32. C. Wu and X. Wang, *Phys. Rev. Lett.*, 1998, **80**(18), 4092–4094.
33. T. Hellweg, C. Dewhurst, W. Eimer and K. Kratz, *Langmuir*, 2004, **20**(11), 4330–4335.
34. Y. Mei, Y. Lu, F. Polzer, M. Ballauff and M. Drechsler, *Chem. Mater.*, 2007, **19**(5), 1062–1069.
35. Y. Lu, Yu, M. Drechsler and M. Ballauff, *Macromol. Symp.*, 2007, **254**, 97–102.

36. J. R. Retama, B. Lopez-Ruiz and E. Lopez-Cabarcos, *Biomaterials*, 2003, **24**(17), 2965–2973.
37. S. Lu and K. S. Anseth, *Macromolecules*, 2000, **33**(7), 2509–2515.
38. H. Yang, L. Zhang, P. Wang, Q. Yang and C. Li, *Green Chem.*, 2009, **11**(2), 257–264.
39. G. Wei, W. Zhang, W. Fei, Y. Wang and M. Zhang, *J. Phys. Chem., C*, 2008, **112**(29), 10827–10832.
40. Y. Lu, Y. Mei, M. Drechsler and M. Ballauff, *Angew. Chem., Int. Ed.*, 2006, **45**(5), 813–816.
41. V. Kotzabasakis, E. Georgopoulou, M. Pitsikalis, N. Hadjichristidis and G. Papadogianakis, *J. Mol. Catal., A*, 2005, **231**(1–2), 93–101.
42. S. C. Tsang, N. Zhang, L. Fellas and A. Steele, *Catal. Today*, 2000, **61**(1), 29–36.
43. H. Bönnemann and R. M. Richards, *Eur. J. Inorg. Chem.*, 2001, **10**, 2455–2480.
44. T. Teranishi, M. Hosoe, T. Tanaka and M. Miyake, *J. Phys. Chem., B*, 1999, **103**(19), 3818–3827.
45. C. L. Lee, C. C. Wan and Y. Y. Wang, *Adv. Funct. Mater.*, 2001, **11**(5), 344–347.
46. X. Zuo, H. Liu and M. Liu, *Tetrahedron Lett.*, 1998, **39**(14), 1941–1944.
47. M. Graeser, E. Pippel, A. Greiner and J. Wendorff, *Macromolecules*, 2007, **40**(17), 6032–6039.
48. P. Ablyazov, V. Vasilevskaya and A. R. Khokhlov, *Colloid J*, 2007, **69**, 265–271.
49. V. V. Vasilevskaya, A. A. Aerov and A. R. Khokhlov, *Dokl. Phys. Chem.*, 2004, **398**(4–6), 258–261.
50. V. V. Vasilevskaya, A. A. Aerov and A. R. Khokhlov, *Colloid Polym. Sci.*, 2006, **284**(5), 459–467.
51. F. Moyano, R. Falcone, M. Dario, J. Mejuto, J. Silber and N. Correa, *Chem.-Eur. J*, 2010, **16**(29), 8887–8893.
52. Q. Jin, G. Liu and J. Ji, *Eur. Polym. J.*, 2010, **46**, 2120–2128.
53. Y. Liu, Y. Wang, Y. Wang, J. Lu, V. Pinon and M. Weck, *J. Am. Chem. Soc.*, 2011, **133**(36), 14260–14263.
54. C. Nardin, S. Thoeni, J. Widmer, M. Winterhalter and W. Meier, *Chem. Commun.*, 2000, 1433–1434.
55. A. Katz and M. E. Davis, *Nature*, 2000, **403**(6767), 286–289.
56. H. Choi and C. Montemagno, *Nano Lett.*, 2005, **5**, 2538–2542.
57. M. Nallani, S. Benito, O. Onaca, A. Graff, M. Lindemann, M. Winterhalter, S. Wolfgang and U. Schwaneberg, *Biotechnology*, 2006, **123**, 50–59.
58. H. de Hoog, D. Vriezema, M. Nallani, S. Kuiper, J. Cornelissen, A. Rowan and R. Nolte, *Soft Matter*, 2008, **4**, 1003–1010.
59. D. Vriezema, J. Hoogboom, K. Velonia, K. Takazawa, P. Christianen, J. Maan, A. Rowan and R. Nolte, *Angew. Chem., Int. Ed.*, 2003, **42**, 772–776.
60. D. Vriezema, A. Kros, R. De Gelder, J. Cornelissen, A. Rowan and R. Nolte, *Macromolecules*, 2004, **37**, 4736–4739.

61. I. Okhapkin, L. Bronstein, E. Makhaeva, V. Matveeva, E. Sulman, M. Sulman and A. Khokhlov, *Macromolecules*, 2004, **37**, 7879-7883.
62. R. Moss, K. Alwis and G. Bizzigotti, *J. Am. Chem. Soc.*, 1983, **105**, 681-682.
63. P. Hammond, J. Forster, C. Lieske and H. Durst, *J. Am. Chem. Soc.*, 1989, **111**, 7860-7866.
64. S. van Dongen, M. Nallani, J. Cornelissen, R. Nolte and J. van Hest, *Chem.-Eur. J.*, 2009, **15**, 1107-1114.
65. X. Pan, Z. Fan, W. Chen, Y. Ding, H. Luo and X. Bao, *Nat. Mater.*, 2007, **6**, 507-511.
66. R. Abbaslou, M. Reza, A. Tavassoli, J. Soltan and A. Dalai, *Appl. Catal., A*, 2009, **367**, 47-52.
67. R. Abbaslou, M. Reza, A. Tavassoli, J. Soltan and A. Dalai, *Angew. Chem., Int. Ed.*, 2011, **50**(21), 4913-4917.
68. H. Shiozawa, T. Pichler, A. Grueneis, R. Pfeiffer, H. Kuzmany, Z. Liu, K. Suenaga and H. Kataura, *Adv. Mater.*, 2008, **20**, 1443-1449.
69. J. Lee, H. Kim, S. Kahng, G. Kim, Y. Son, J. Ihm, H. Kato, Z. Wang, T. Okazaki and H. Shinohara, *et al.*, *Nature*, 2002, **415**, 1005-1008.
70. P. Ayala, R. Kitaura, R. Nakanishi, H. Shiozawa, D. Ogawa, P. Hoffmann, H. Shinohara and T. Pichler, *Phys. Rev. B: Condens. Matter Mater. Phys.*, 2011, **83**(8), 085407.
71. L. Li, A. Khlobystov, J. Wiltshire, G. Briggs and R. Nicholas, *Nat. Mater.*, 2005, **4**, 481-485.
72. P. Kondratyuk and J. Yates, *Acc. Chem. Res.*, 2007, **40**, 995-1004.
73. U. Heiz, A. Sanchez, S. Abbet and W.-D. Schneider, *Chem. Phys.*, 2000, **262**(1), 189-200.
74. M. D. Halls and H. Schlegal, *J. Phys. Chem., B*, 2002, **106**, 1921-1925.

CHAPTER 11
Nanoparticle Mediated Clock Reaction: a Redox Phenomenon

TARASANKAR PAL* AND CHAITI RAY

Department of Chemistry, IIT Kharagpur, Kharagpur, West Bengal 721302, India
*Email: tpal@chem.iitkgp.ernet.in

In 1803 Berthollet introduced the concept of a 'reversible chemical reaction'. He questioned the deposition of sodium carbonate in one of the lakes in Egypt and got the idea of a reversible reaction. In general, the hardness of water due to $CaCl_2$ is removed by Na_2CO_3. Soluble calcium ions are expelled from a body of water as the insoluble precipitate $CaCO_3$ and eventually, the water becomes soft. However, in the lake, the reverse reaction was observed. Naturally occurring limestone, $CaCO_3$, was dissolved in the lake water, bearing a high concentration of common salt, NaCl. Excess NaCl present in the lake water reversed the familiar reaction. Thus Berthollet, for the first time, proposed that there exists an equilibrium between the reactants and products.

The familiar reaction is:

$$Na_2CO_3 + CaCl_2 \rightleftharpoons 2NaCl + CaCO_3 \downarrow$$

This removes the hardness and a possible reversal of the reaction takes place in lakes. This is an unique and natural consequence.

Theoretically, all reactions are now considered to be reversible, even if the reaction is a redox one. However, if the free energy changes of a reaction are large, the equilibrium constant, K, becomes large and the reaction is considered to be irreversible. In an extreme case where one of the products escapes as a volatile gas, then the reaction becomes irreversible.

There are certain classes of reactions that serve as an example of non-equilibrium thermodynamics. The reactions are theoretically important and they show that chemical reactions do not have to be dominated by equilibrium thermodynamic behavior. A reactions in which the concentration of one or more components changes periodically, or a certain property of the reactant changes after a predicted time, is known as a 'clock reaction'. The clock reaction is a popular example of a redox reaction in which dramatic color changes engross the viewers. One of the best known of these is the Landolt clock reaction between sulfite and excess iodate, which dates back to 1886. After that, dozens of interesting redox reactions have been documented in the literature as a clock reaction. Quite often, these reactions are simply crowd-pleasing demonstrations.

11.1 History

There are many reports of clock reactions in the field of chemistry describing the reversible pathway of a process that takes place periodically. The following are popular demonstrations.

11.1.1 Iodine Clock Reaction

In 1886, Hans Heinrich Landolt reported the clock reaction between bisulfite and excess iodate known as the iodine clock reaction. Here, two colorless solutions are mixed and left for some time. After a short time lag, the colorless solution gradually turns dark blue. The reaction starts with an easily available common reagent such as potassium iodate and an acidic solution of sodium bisulfite (acidified with sulfuric acid) in the presence of starch.

In this protocol, an iodide ion is generated by the following slow reaction between the iodate and bisulfite:

$$IO_3^- + 3HSO_3^- \rightarrow I^- + 3HSO_4^-$$

This is the rate determining step. Excess iodate will oxidize the iodide generated in the last step to form iodine:

$$IO_3^- + 5I^- + 6H^+ \rightarrow 3I_2 + 3H_2O$$

However, the iodine is reduced immediately back to iodide by the bisulfite:

$$I_2 + HSO_3^- + H_2O \rightarrow 2I^- + HSO_4^- + 2H^+$$

When the bisulfite is fully consumed, the iodine will survive to form the dark blue complex with starch.

The solution will repeatedly cycle from colorless to blue and back to colorless, until the reagents are depleted.[1]

11.1.2 B–Z Reaction

The Belousov–Zhabotinsky (BZ) reaction is one of a class of oscillating chemical reactions. The only common element in these oscillating systems is the inclusion of bromine and an acid.[2] The discovery of the phenomenon is attributed to Boris Belousov who mixed a solution of potassium bromate, cerium(IV) sulfate, and citric acid in dilute sulfuric acid. The solution oscillates between a yellow solution and a colorless solution as the ratio of the concentration of cerium(IV) to cerium(III) ions alternates; this is due to the cerium(IV) ions being reduced by malonic acid to cerium(III) ions, which are then oxidized back to cerium(IV) ions by bromate(V) ions. Later, in 1964, Anatoly Zhabotinsky rediscovered this reaction sequence.[3] In this case, ferroin, a complex of phenanthroline and iron, was used as a common indicator. Mainly, this is a cerium-catalyzed oxidation of malonic acid by a bromate ion in the presence of dilute sulfuric acid. The overall equation is given:

$$3CH_2(CO_2H)_2 + 4BrO_3^- \rightarrow 4Br^- + 9CO_2 + 6H_2O$$

This process involves following three solutions: solution A contains 0.23 M $KBrO_3$, solution B is a mixture of 0.31 M malonic acid and 0.059 M KBr and solution C consists of 0.019 M cerium(IV) ammonium nitrate and 2.7 M sulfuric acid.

This oscillatory cycle can be qualitatively described in the following way: when a sufficiently high Ce(IV) concentration is present in the system, Br^- will be produced rapidly and its concentration will also be high. As a result, autocatalytic oxidation of Ce(III) is completely inhibited, and the [Ce(IV)] decreases due to its reduction by malonic acid. Then, the concentration of Br^- decreases along with that of [Ce(IV)]. When [Ce(IV)] reaches its lower threshold, the bromide ion concentration drops abruptly. The rapid autocatalytic oxidation starts and rises [Ce(IV)]. When [Ce(IV)] reaches its higher threshold, [Br^-] increases sharply and inhibits the autocatalytic oxidation of Ce^{3+}. The cycle then repeats (Figure 11.1).[4–8]

The reaction turns from green to blue, purple, and red, which arise due to the oxidation and reduction of iron and cerium complexes. When [Ce(IV)] increases in the system, it oxidizes the iron of ferroin from the red Fe(II) to the blue Fe(III). Ce(III) is colorless but Ce(IV) is yellow, which turns green in the presence of blue Fe(III).

11.1.3 Bray–Liebhafsky Reaction

William C. Bray described the first oscillating reaction in a stirred solution for the first time in 1921. Here, the role of iodate was investigated in the catalytic conversion of H_2O_2 to oxygen and water. Further, his student Herman A. Liebhafsky investigated the involvement of free radicals in a nonradical step and the modified reaction is known as the Bray–Liebhafsky reaction.[9,10]

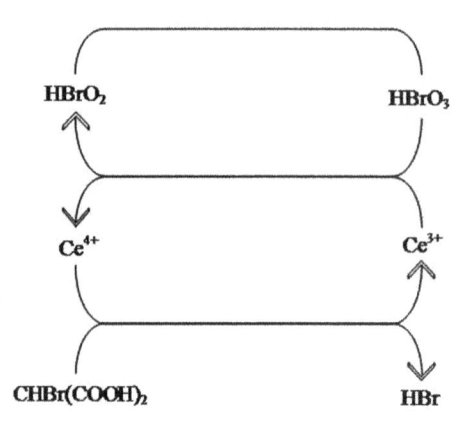

Figure 11.1 Schematic representation of a B-Z reaction.

Fundamentally, in this reaction, the redox potential value of H_2O_2 enables the simultaneous oxidation of iodine to iodate:

$$5H_2O_2 + I_2 \rightarrow 2IO_3^- + 2H^+ + 4H_2O$$

And the opposite reduction reaction:

$$5H_2O_2 + 2IO_3^- + 2H^+ \rightarrow I_2 + 5O_2 + 6H_2O$$

The net reaction is:

$$2H_2O_2 \rightarrow 2H_2O + O_2$$

But the presence of a catalyst and IO_3^- are essential.

11.1.4 Briggs–Rauscher Reaction

In 1972, the Briggs–Rauscher reaction was discovered by replacing the bromate in the B–Z reaction with iodate and hydrogen peroxide. Optionally, starch is used as an indicator for the abrupt change in iodide ion concentration. This reaction is very important in demonstrating the purpose of an oscillating chemical reaction in which a colorless solution is gradually transformed to an amber solution and is suddenly changed into a dark blue solution.

Initially, the aqueous solution consists of hydrogen peroxide, an iodate, divalent manganese (Mn^{2+}) as a catalyst, a strong chemically unreactive acid such as sulfuric acid (H_2SO_4) or perchloric acid ($HClO_4$), and an organic compound such as malonic acid with an active hydrogen atom attached to carbon which will slowly reduce free iodine (I_2) to iodide (I^-). The reaction shows habitual cyclic changes, both gradual and sudden, which are visibly slow changes in the intensity of color, interrupted by abrupt changes in color. This demonstrates that a difficult arrangement of slow and fast reactions is taking place simultaneously. In the slow process, free iodine is consumed by malonic acid involving an intermediate species, iodate.

The second step involves a fast auto-catalytic process containing manganese and free radical intermediates, which converts hydrogen peroxide and iodate to free iodine and oxygen. The slow generation of the amber solution is due to the formation of iodine through the second step. After completion of the second step, a sudden increase in the concentration of iodide gives a dark blue color in the presence of starch. But because the first step continues, the blue color is slowly faded out.[11] This reaction is generally used as an assay procedure for antioxidants in foodstuffs.

11.1.5 The Blue Bottle Experiment

A chemical reaction in a closed bottle containing an aqueous solution of glucose, sodium hydroxide, methylene blue and some air that transforms a colorless reduced product to the blue colored methylene blue after gentle shaking is described as the 'blue bottle' reaction. The blue color of the methylene blue is decolorized due to the reduction of methylene blue to leucomethylene blue by glucose. Glucose itself is oxidized to gluconic acid, which is converted into sodium gluconate under the alkaline conditions. When the bottle is shaken in air, the oxygen in the air oxidizes the leucomethylene blue to methylene blue. After the solution comes to rest, glucose again reduces methylene blue and the color of the solution disappears again. This reaction is first order with respect to glucose, sodium hydroxide, methylene blue and zero order with respect to oxygen in the air. If the solution is heated stepwise from 10 °C to a maximum of 50 °C, the color change process requires less time. This incident agrees with the rules of chemical kinetics, with the increase in temperature resulting in an increase in reaction rate by a factor of two to four.[12]

11.2 Recent Work

Recently, many reports have appeared in print that circumvent reversible redox reactions. Again, one such reaction may be described taking coinage metals (Cu, Ag and Au) into consideration. The nobility of coinage metals varies from copper through to gold. Their redox chemistry also differs markedly. We know that gold is the noblest amongst these three coinage metals. The same feature, *i.e.* nobility, of these three metals truly exists while they are taken to their nanodimension (10^{-9} metres). Silver occupies an intermediate position in the 11th group of the periodic table and it has redox properties in between copper and gold. Consider that silver particles with much smaller particles in solution would exhibit a yellow color with an intense band in the 380–400 nm range in an absorption spectrum. This band is attributed to a collective oscillation of the electrons in the particles, with a periodic change in electron density at the surface (surface Plasmon absorption) of the particles. Copper, silver and gold nanoparticles show a rich Plasmon band in the visible range, unlike the other noble metal hydrosols. Platinum, palladium and other metallic nanoparticles show featureless absorption spectra in the visible region. Thus, a crowd-pleasing demonstration can easily be achieved from coinage metals, and the best of these is the silver nanoparticle redox reaction. The Plasmon absorption, *i.e.* the coherent oscillation of the valence electron with the impinging electromagnetic radiation interacting with metal nanoparticle surfaces, is easily demonstrated taking silver into consideration. The absorption maximum due to stabilized metal nanoparticles varies with particle size, shape and also with the dielectric medium that stabilizes the nanoparticles. The metal nanoparticles become progressively reactive as the size decreases. Henglein described a 'push–pull' reduction method for the preparation of plausibly monodisperse silver particles in aqueous solution.[13] Slow particle growth is achieved in this method by simultaneously reducing Ag^+ ions and partially oxidizing Ag particles. In particular, a narrow surface Plasmon band at 382 nm was generated for the formation of Ag(0) particles in the solution. In this investigation, the particles used were prepared by the fast reduction of Ag^+ ions and they had a broader Plasmon absorption band. Figure 11.2 shows the absorption spectrum before and after addition of two concentrations of iodide to the silver sol. It can be seen that the absorption band is broadened and slightly red-shifted.

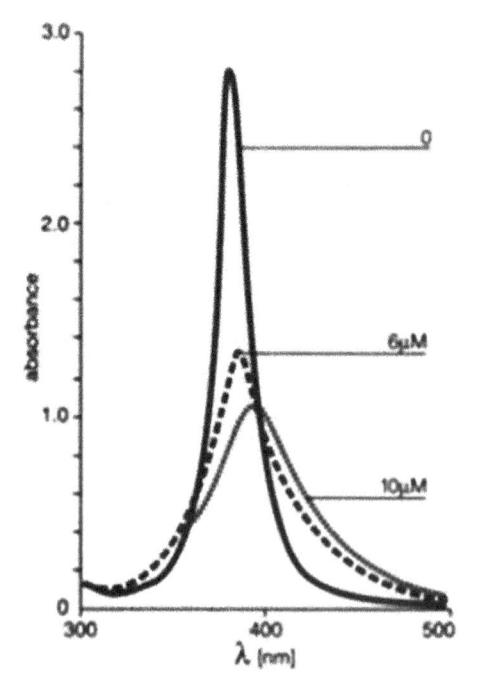

Figure 11.2 Absorption spectrum before and after the addition of various concentrations of KI. At [KI] = 10 pM, full coverage of the surface of the particles was reached, as a further increase in the KI concentration did not lead to additional changes in the shape of the absorption band; the CTTS absorption band of free I- appeared at [KI] > 10 pM.

Below a critical size, the coinage metal nanoparticles also show featureless absorption curves in the visible region like platinum, palladium *etc*. These particles (sub-nanometre particles) are called metallic clusters.[14] It is very difficult to deal with them in solution because of their pronounced reactivity, identification and stabilization problems. It is also a difficult task to obtain catalytic activity from these sub-nanometre particles because the easy access to their surfaces is lost once they are stabilized by a capping agent.

Metal hydrosol bears a negative charge, which is the manifestation of the capping agent present on the nanoparticle's surface. A capping agent may be replaced by a place exchange reaction with another suitable capping agent. Still, they may remain as charged particles. Interestingly, the capping agent may be replaced by neutral long chain amines, thiols, proteins *etc*. The metal nanoparticles remain stable and retain their identity with these capping agents. Then the particles are said to be sterically stabilized and the phenomenon is known as 'steric stabilization' (Figure 11.3).

To elaborate on the fascinating reversible reaction, dilute $AgNO_3$ solution (10^{-4}–10^{-5} M) may be placed in a test tube as a representative coinage metal salt. Dilute ice cold $NaBH_4$ (0.01 M) has to be introduced into the test tube with vigorous stirring. A transparent and stable yellow silver hydrosol would be obtained.[15] The sol particles as usual would bear a negative charge, which makes the silver particles stable due to Coulomb's interaction (Figure 11.4).

The maximum absorption of the solution would appear at ~ 400 nm. Excess BH_4^- would remain as adsorbed ions, *i.e.* the adsorbate BH_4^- on the silver nanoparticles. Before the decomposition of BH_4^- ions, if some surfactant molecules are introduced, a dramatic reversible redox reaction may be observed. Upon shaking the yellow solution, the color completely disappears and then reappears, keeping the solution for some time. The whole operation may be summarized as a crowd-pleasing demonstration: $NaBH_4$ reduces $AgNO_3$ into small silver nanoparticles. The particles are in the nanodimension (<50 nm) and they impart a yellow color (Plasmon absorption) to the solution. Excess BH_4^-, more diffusive than NO_3^- or OH^-, would adhere onto the silver nanoparticle surfaces. This would shift the Fermi level towards a more negative region (Figure 11.5).

Thus oxygen in the air would easily oxidize the silver particles' surface. Surfactant in the reaction medium would dissolve out the oxide layer

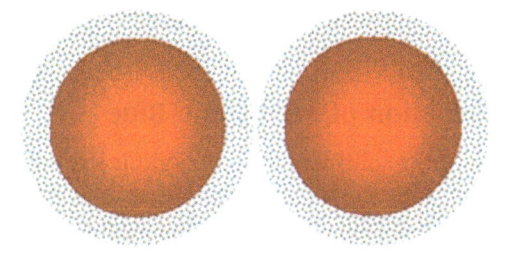

Figure 11.3 Steric stabilization.

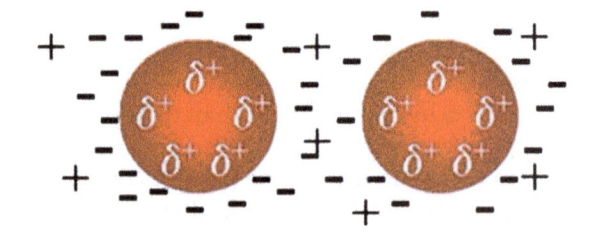

Figure 11.4 Electrostatic stabilization.

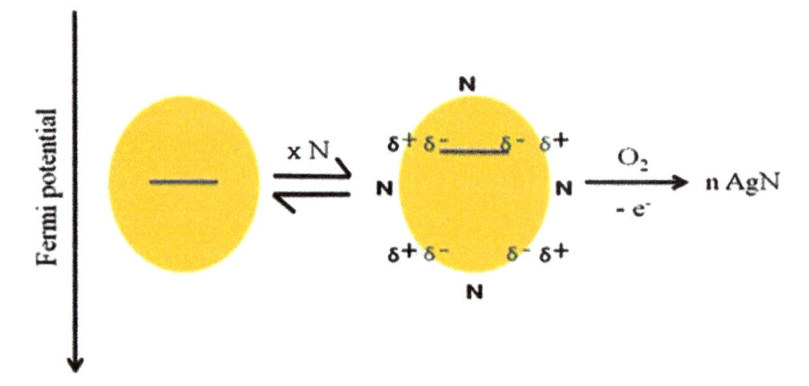

Figure 11.5 Schematic representation of a shift in Fermi potential after adsorption of a nucleophile (N).

successively from the silver particles' surface. The oxidation and surfactant-assisted removal of the oxide layer would take place after the oxidation of the silver surface by air, but in the presence of a strong nucleophile, such as BH_4^- (in the present case) or a cyanide ion (in the other). As a result of oxidation, the silver nanoparticles would be smaller (more reactive) and finally would dissolve/disappear upon shaking. Oxygen in the air assists the oxidation. Now, the unstirred colorless solution with argento cyanide, still with excess $NaBH_4$, would again evolve silver nanoparticles. This reversible cycle would go on and on, just shaking and standing for oxidation (disappearance of the yellow color) and reduction (re-appearance of the yellow color), respectively, for a crowd-pleasing demonstration (Figure 11.6). This can even be demonstrated with Au(III) chloride.[16]

11.2.1 Clock Reaction of Methylene Blue

Methylene blue is a water-soluble, cationic, organic dye with a thiazine group in its basic skeleton. Blue colored methylene blue is easily reduced to the colorless leucomethylene blue by a reducing agent. The reduced form is prone to oxidation by oxygen in the air after gentle shaking and is again transformed into methylene blue. This visual color change for the oxidation

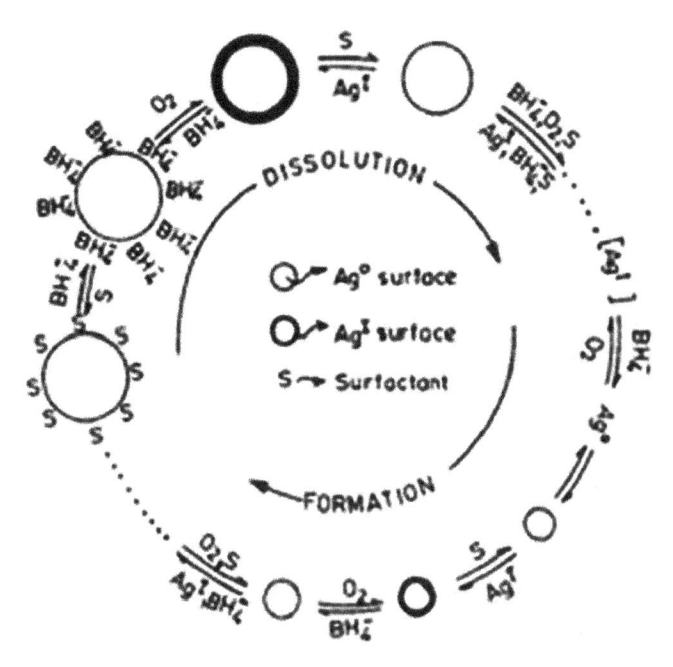

Figure 11.6 Mechanism of the reversible formation and dissolution of silver nanoparticles.

and reduction processes makes the clock reaction of methylene blue important in chemistry and biological systems.

Snehalatha et al.[17] demonstrated the clock reaction of methylene blue using ascorbic acid as a reducing agent. Here, a reduction was performed in the presence of dilute sulfuric acid. It was observed that with an increase in acid concentration, the time of reaction decreased, that is, the rate of the reduction gradually increased. The clear and visible color change made the reaction simple and so the mechanism of the reaction is easy to explain.

But in the absence of a catalyst, the reduction process becomes time consuming. So, a methylene blue-ascorbate ion redox reaction has been catalyzed by a micellar system in which catalysis occurred by the binding of the substrates onto the micellar surface by hydrophobic and electrostatic interactions. The rate of the reaction increases with an increase in the possibility of encounters. In this case, the effect of various salting-in and salting-out agents, size of reverse micellar water pool, and hydrophobicity of the dye were studied and no change in reaction rate was found. The reversible electron-transfer reactions for the MB-AA- and MB-BH$_4$- systems become facile in the presence of coinage metal sol particles as catalysts (Figure 11.7).[18]

Additionally, it has been shown that Au nanocrystals act as an effective catalyst for the reduction of nitrobenzene to aniline and methylene blue to leucomethylene blue in the presence of NaBH$_4$ and N$_2$H$_4$, respectively, as reducing agents (Figure 11.8).[19]

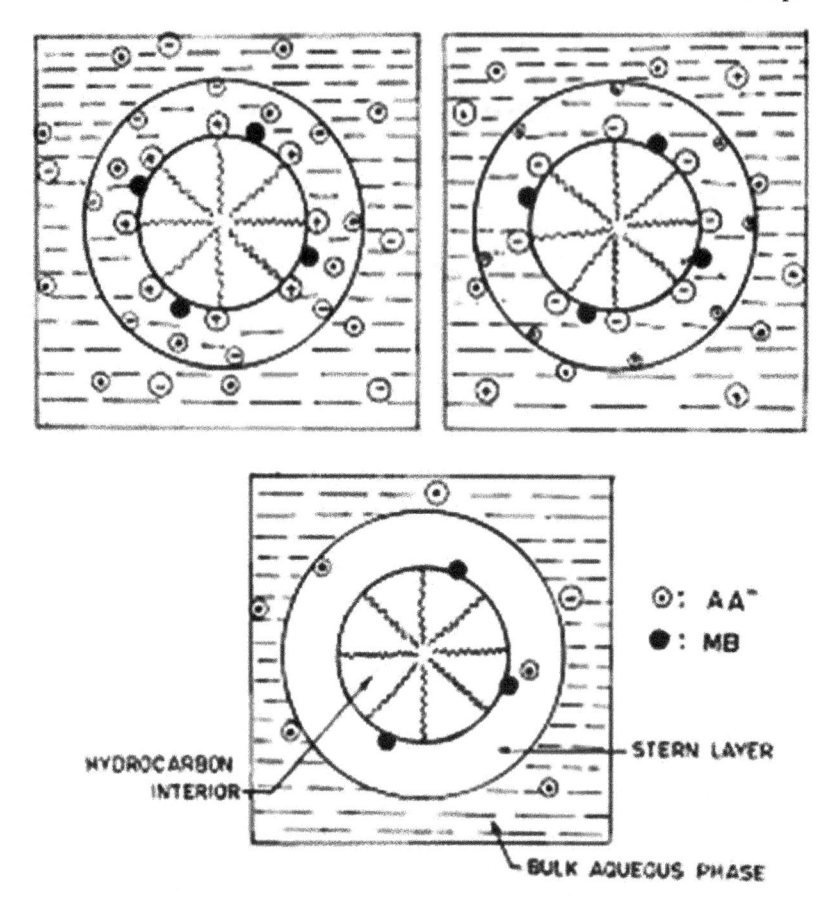

Figure 11.7 Model of the reaction conditions prevailing in a (a) cationic micelle, (b) anionic micelle, and (c) neutral micelle.

Only a few noble metals in the elemental form are present in our Earth's crust. However, most metals combine with oxygen and form a binary compound—an oxide. Oxides are robust and have proved to be excellent catalysts. Proper tuning of an oxide structure might develop a good semiconductor. Then the semiconductor may be useful for scavenging solar light. The bottle neck with most oxide materials is their tendency to be activated under ultraviolet light. Thus, there is a necessity for tuning the size, shape and/or doping the oxides to make them visible light photocatalysts. The other aspect of oxide materials is their usefulness in chemical reactions. In the present context, it may be shown that metallic oxides also exhibit reversible redox reactions.

The popular Fehling test for aldehyde detection and quantification results in precipitating cuprous oxide, Cu_2O. An ore—cuprite or red copper oxide, Cu_2O—occurs in nature. This is an unique monovalent simple compound of copper. In the bulk stage, red Cu_2O gets stabilized as a robust crystal.

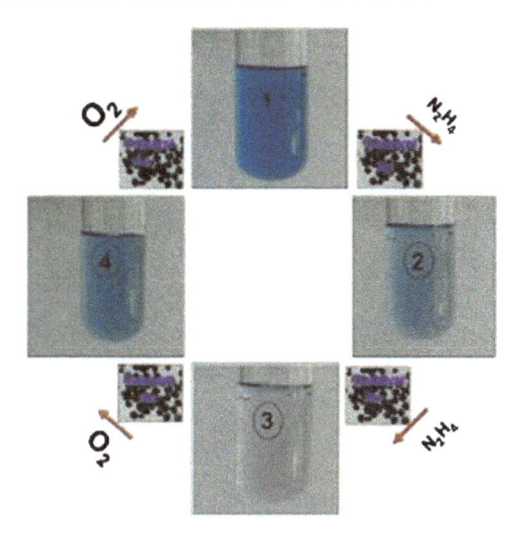

Figure 11.8 Digital images of the 'decolorization–colorization' reaction of MB (*i.e.*, color fading and regeneration of MB) in the presence of an Au catalyst.

One can restrict the growth of Cu_2O in solution and in the nanodimension, Cu_2O remains as a yellow solid, which has been exploited as a catalyst to exhibit reversible clock reactions.[3] One challenge is to take bulk Cu_2O to a nanoregime and to stabilize it in solution. Cu_2O in its nanoregime would act as a photocatalyst. We report a new synthetic protocol to obtain monodisperse, stable, exclusively cubic Cu_2O nanoparticles under surfactant-free conditions and its catalytic action in the methylene blue (MB)-hydrazine reaction in an aqueous medium. The blue color of the dye, MB, faded away upon the addition of hydrazine, producing colorless leucomethylene blue (LMB), indicating the progress of the redox reaction. The rate of this redox reaction has been found to be enhanced in the presence of the nanocatalyst, Cu_2O. The success of the reaction demonstrates a simple 'clock reaction'. An oscillation between a blue MB color and colorless solution due to the formation of LMB is observed on periodic shaking. This oscillation continues for over 15 cycles. Studies on the effect of bulk Cu_2O and nanoparticles of CuO and Cu(0) have not been successful for the demonstration of the 'clock reaction'. Thus, the importance of Cu_2O nanoparticles in the clock reaction is established beyond doubt. The electrochemical studies indicate that nano-Cu_2O shows a couple of redox peaks, which correspond to the redox Cu(II)–Cu(I) system. Kinetic studies authenticate a first-order reaction mechanism. Further, quantum chemical calculations reveal that the nanoparticles reduce the activation energy by $\sim$17 kcal mol^{-1}, thereby making the reaction 2.4×10^7 times faster compared to the gas phase (Figure 11.9).[20]

Truncated Cu_2O cubes have been successfully synthesized using a stable Cu(II)-EDTA precursor and glucose as the reducing agent under alkaline conditions and five minutes of microwave irradiation. Interestingly, in the

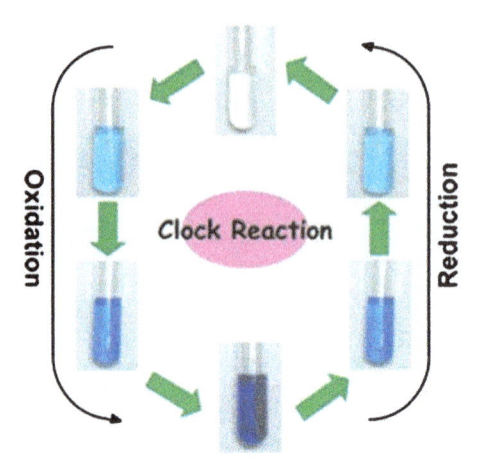

Figure 11.9 Schematic representation of the clock reaction (*i.e.* color fading and regeneration of MB).

presence of Cu$_2$O particles, the blue-colored methylene blue (MB) in aqueous acidic (pH ≈ 1.0) solution is successively bleached to a colorless leucoMB (LMB). The redox reaction generates a colorless solution that easily reverts back to blue in air. The reversible color change, *i.e.* oscillation between a blue MB solution and a colorless LMB solution, happens to be a periodic phenomenon for more than 50 cycles and is reproducibly demonstrated as a simple 'clock reaction'. Under acidic (pH ≈ 3.0) conditions, redox phenomena favor zero-order kinetics. Dilute H$_2$SO$_4$ has been proven to be the best choice to provide a passive reaction medium, and the undisturbed reaction mixture showed oscillatory behavior even after one month (Figure 11.10).[21]

A clock reaction of methylene blue by sodium borohydride in the presence of selenium nanowire with a metalloid as a catalyst has been carried out. With the addition of borohydride solution in methylene blue solution containing a nanocatalyst, the color of the solution disappeared, confirming the formation of leucomethylene blue. In air, after gentle shaking of the mixture, the colorless solution turned blue, indicating air oxidation of leucomethylene blue. This amazing reversible color alteration went on for 30 to 50 cycles depending upon the quantity of catalyst added. The same procedure was followed for bulk selenium, but it failed to show the clock reaction. Again, sodium borohydride solution could not bleach the color of methylene blue in the absence of a nanocatalyst. This reduction reaction undergoes two processes. In the first step, the rate of the reduction is slow. This step is designated as the 'induction period', which occurred due to the surface oxidation of the selenium nanowire in an aqueous medium. This reduced the catalytic property of the catalyst, after which the rate became slow. Then, the reducing agent donated an electron to the catalyst's surface; after that, the catalyst was capable of transferring electrons to methylene blue. After this, the electron transfer rate of the reduction increased gradually. Noble metal deposited selenium nanowire enhances the rate of the clock

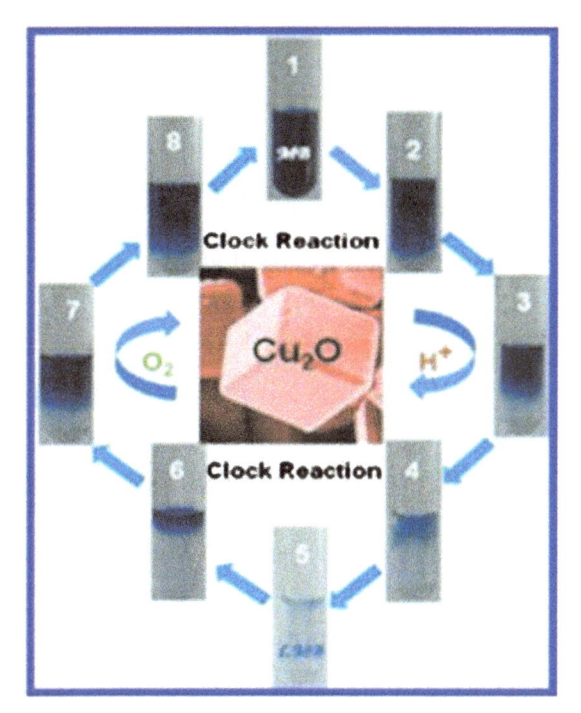

Figure 11.10 Schematic representation of the clock reaction involving MB and Cu$_2$O in acidic pH.

reaction compared to selenium nanowire due to the high conducting property of the noble metal.

11.3 Mechanistic Approach

In the clock reaction, a reducing agent reacts with an oxidizing compound in the presence of a solid catalyst surface. The reaction on the catalyst's surface proceeds *via* two steps: adsorption of one or more reactants on the catalyst's surface and product formation. One of the steps is the rate determining step. To explain the above phenomenon, two mechanisms are proposed: the Eley–Rideal mechanism and the Langmuir–Hinshelwood mechanism.

11.3.1 Eley–Rideal Mechanism

Here, one of the reactants (A), generally a reducing agent, is adsorbed on the catalyst's surface and reacts with another reactant (B) present in the system. Then the newly formed product desorbs from the catalyst's surface (Figure 11.11).[22]

$$A + S\,(\text{support}) \rightleftharpoons AS$$

$$AS + B \rightarrow \text{Product (P)}$$

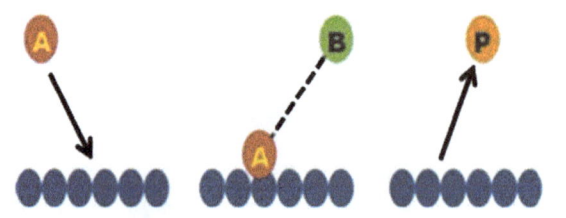

Figure 11.11 Representation of the Eley–Rideal mechanism.

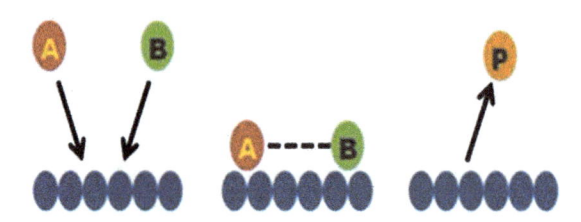

Figure 11.12 Representation of the Langmuir–Hinshelwood mechanism.

11.3.2 Langmuir–Hinshelwood Mechanism

In this case, reducing agent (A) adsorbed on the catalysts's surface, reacts with the surface and transfers surface-hydrogen species to the catalyst's surface. Again, another reactant (B) is also adsorbed on the catalyst's surface. Both steps are assumed to be fast and also termed as Langmuir isotherm. Now, reactant B reacts with the adsorbed surface-hydrogen species and is reduced. After completion of the reaction, the new product apart from catalyst and again the vacant catalyst surface is ready for repetitive reduction cycles (Figure 11.12).[23]

$$A + \text{Support (S)} \rightleftharpoons AS$$

$$B + \text{Support (S)} \rightleftharpoons BS$$

$$AS + BS \rightarrow \text{Product (P)}$$

11.4 Applications

The clock reaction of methylene blue is used in various applications such as:

11.4.1 Water Purification

A plasma jet is used for water purification. A plasma jet produced in water using a submerged ac excited electrode in a coaxial dielectric barrier discharge configuration was studied. The set-up contains an oxidation–reduction indicator methylene blue dye to show the clearing of the water visually and also spectrophotometrically.[24]

11.4.2 Memory Facilitation by Methylene Blue and Brain Oxygen Consumption

Incorporation of methylene blue improves memory in prevention avoidance and appetitive tasks. It helps to restore impaired spatial memory by behaving as an inhibitor of cytochrome oxidase. Methylene blue also develops the memory by escalating the utilization of oxygen present in brain. In this case, the reduced form, leucomethylene blue, is used as an indicator for brain oxygen and it increased the oxygen consumption, whereas methylene blue is involved in cytochrome C oxidation measurement.[25]

11.4.3 Novel UV-activated Colorimetric Oxygen Indicator

Oxygen plays an important and significant role in analytical chemistry and in a variety of other areas. So, it is essential to detect and measure the presence of oxygen in a gaseous phase or dissolved in solution, because oxygen is an important feature in environmental analysis, patient monitoring and biotechnology. Methylene blue is used as an indicator for oxygen detection where MB is reduced by triethanolamine using UV irradiation in the presence of TiO_2.[26]

11.5 Conclusion

In this chapter, we have described various types of clock reaction, which is mainly controlled by a reversible redox mediated reaction. In most cases, the oxidizing agent is a dye or colored molecule, which is reduced by a general reducing agent; in this context, two reaction mechanisms are discussed in the preceding section. To make the reduction process facile, various nanocatalysts are used. An investigation into the clock reaction with new catalysts or dye molecules will be very fascinating.

References

1. H. Landolt, *Ber. Dtsch. Chem. Ges.*, 1886, **19**, 1317.
2. B. P. Belousov, *Collection of Abstracts on Radiation Medicine*, 1959, **147**, 145.
3. A. M. Zhabotinsky, *Biophysics*, 1964, **9**, 306.
4. R. J. Field, E. Koros and R. M. Noyes, *J. Am. Chem. Soc.*, 1972, **94**, 8649.
5. R. J. Field and R. M. Noyes, *J. Am. Chem. Soc.*, 1974, **96**, 2001.
6. I. R. Epstein and K. Showalter, *J. Phys. Chem.*, 1996, **100**, 13132.
7. Z. Nagy-Ungvarai, J. J. Tyson and B. Hess, *J. Phys. Chem.*, 1989, **93**, 707.
8. V. K. Vanag, L. F. Yang, M. Dolnik, A. M. Zhabotinsky and I. R. Epstein, *Nature*, 2000, **406**, 389.
9. W. C. Bray, *J. Am. Chem. Soc.*, 1921, **43**, 1262.
10. W. C. Bray and H. A. Liebhafsky, *J. Am. Chem. Soc.*, 1931, **53**, 38.
11. T. S. Briggs and W. C. Rauscher, *J. Chem. Educ.*, 1973, **50**, 496.

12. L. Anderson, S. M. Wittkopp, C. J. Painter, J. J. Liegel, R. Schreiner, J. A. Bell and B. Z. Shakhashiri, *J. Chem. Educ.*, 2012, **89**, 1425.
13. M. Gutierrez and A. Henglein, *J. Phys. Chem.*, 1993, **97**, 11368.
14. H. Xu and K. S. Suslick, *Adv. Mater.*, 2010, **22**, 1078.
15. J. A. Creighton, C. G. Blatchford and M. G. Albrecht, *J. Chem. Soc., Faraday Trans. 2*, 1979, **75**, 790.
16. T. Pal, T. K. Sau and N. R. Jana, *Langmuir*, 1997, **13**, 1481.
17. T. Snehalatha, K. C. Rajanna and P. K. Saiprakash, *J. Chem. Educ.*, 1997, **74**, 228.
18. T. Pal, S. De, N. R. Jana, N. Pradhan, R. Mandal, A. Pal, A. E. Beezer and J. C. Mitchell, *Langmuir*, 1998, **14**, 4724.
19. A. K. Sinha, M. Basu, S. Sarkar, M. Pradhan and T. Pal, *J. Colloid Interface Sci.*, 2013, **398**, 13.
20. S. Pande, S. Jana, S. Basu, A. K. Sinha, A. Datta and T. Pal, *J. Phys. Chem. C*, 2008, **112**, 3619.
21. M. Basu, A. K. Sinha, M. Pradhan, S. Sarkar, A. Pal, C. Mondal and T. Pal, *J. Phys. Chem. C*, 2012, **116**, 25741.
22. Y. Khalavka, J. Becker and C. Sonnichsen, *J. Am. Chem. Soc.*, 2009, **131**, 1871.
23. S. Wunder, F. Polzer, Y. Lu, Y. Mei and M. Ballauff, *J. Phys. Chem. C*, 2010, **114**, 8814.
24. J. E. Foster, B. Weatherford, E. Gillman and B. Yee, *Plasma Sources Sci. Technol.*, 2010, **19**, 1.
25. P. D. Riha, A. K. Bruchey, D. J. Echevarria and F. Gonzalez-Lima, *Eur. J. Pharmacol.*, 2005, **511**, 151.
26. S. Lee, M. Sheridan and A. Mills, *Chem. Mater.*, 2005, **17**, 2744.

CHAPTER 12

Theoretical Insights into Metal Nanocatalysts

PING LIU

Chemistry Department, Brookhaven National Laboratory, Upton, NY 11973, United States of America
Email: pingliu3@bnl.gov

12.1 Introduction

Catalysis has been essential to life and as a frontier in science. Transportation fuels, lubricants, chlorine-free refrigerants, high-strength polymers, stain-resistant fibers, drugs, and many thousands of other products required by modern societies would not be possible without the existence of catalysts.[1] Most catalysts consist of nanometre-sized particles dispersed on a high-surface-area support. Metal nanocatalysts have a high surface-to-volume ratio and very active surface atoms, and are very attractive for a wide variety of reactions, compared to bulk catalysts.[2-6] The nature of the active phase(s) in these metal/support catalysts and the reaction mechanism are still unclear. What are the effects of size, shape and support on the catalytic properties? It is extremely difficult to obtain such necessary details from current experimental techniques, which has been an obstacle in understanding their structure-dependent catalytic activity and selectivity and therefore hinders further catalyst development.

Over the past few decades, advances in methodology, software and the power of supercomputers have made computational approaches, specifically density functional theory (DFT), capable of providing qualitative, and in many cases quantitative, insights into catalysis.[2,3,7-18] In this chapter, we will

---

RSC Catalysis Series No. 17
Metal Nanoparticles for Catalysis: Advances and Applications
Edited by Franklin Tao
© The Royal Society of Chemistry 2014
Published by the Royal Society of Chemistry, www.rsc.org

report our recent DFT studies on metal and supported metal nanocatalysts. In section 2, DFT and DFT codes that we used will briefly be introduced. Section 3 will address the DFT studies of catalysis on unsupported metal nanocatalysts. The effect of support on the catalytic behavior of metal nanocatalysts will be discussed in section 4. As you will see in this chapter, theoretical studies play a very important role. It describes catalytic reactions on nanocatalysts with the detail and accuracy required for computational results to compare with experiments in a meaningful way. More importantly, it gives insights into the size and shape effects of nanocatalysts on catalyst activity and stability. Such understanding can lead to the rational design of better catalysts.

12.2 Computational Method

Depending on the capability of describing the time scale of chemical reactions and the length-scale of materials, the computational method ranges from quantum mechanics, molecular dynamics, kinetic theory and statistical mechanics to continuum theory. The accuracy increases from continuum theory to quantum mechanics, while the capable time scale decreases from real time to pico-second and the length decreases from macro-scale to the atomic scale. So far, DFT has been widely employed for many catalysis applications.

DFT was used to attempt to solve the Schrödinger equation, where the kinetic and potential energies are transformed into the Hamiltonian and act upon the wavefunction to generate the evolution of the wavefunction in time and space.[19-21] It is derived directly from theoretical principles with no experimental data, providing accurate and reliable theoretical results with reasonable computational costs. DFT permits more realistic models of the intrinsic reaction than can be afforded by other quantum mechanical methods and is likely the one that offers a compromise between accuracy and computational cost as the system increases in size.

The DFT calculations reported below were carried out using DMol3,[22,23] CASTEP,[24] and VASP.[25-27] These codes utilize numerical (DMol3) or plane-wave (CASTEP, VASP) basis sets, and allow DFT calculations at the GGA level with the PW91, PBE or RPBE functional. DMol3 can be used to study molecules, nanoparticles, and periodic surfaces with the same level of accuracy.[22,23]

12.3 Metal Nanocatalysts

Size and shape have been considered as the main parameters to explain the different catalytic behaviors of metal nanoparticles from their parent bulk materials.[28,29] With the variation in size and shape, the intrinsic catalytic properties of the material may change due to a combination of localized coordination effects, surface site distribution, surface relaxation and quantum size effects.[30,31] However, even the most careful experimental measurements provide difficulty in deconvoluting these effects. Thus, in spite of the profound influence that these changes can have on catalytic properties, they remain relatively poorly understood.

12.3.1 Copper Nanocatalysts for Water–Gas Shift Reactions: the Importance of Low-coordinated Sites

The water–gas shift (WGS, $CO + H_2O \rightarrow CO_2 + H_2$) reaction is primarily used to increase the H_2 content as well as reducing the CO concentration in synthesis gas and is an essential part of a hydrogen plant.[32] Current industrial catalysts (Zn-Al-Cu oxides) are pyrophoric and require complex activation steps before usage.[33] Recent studies show that the disadvantages of the industrial catalysts can be overcome by the use of Cu nanoparticles supported on an oxide (CeO_2, TiO_2, MoO_2).[7,10,11,34–38] Various effects have been proposed to be of fundamental importance to the catalytic activity of the supported nanoparticles, including size, shape, the metal oxidation state, the degree of reduction of the oxide support and/or the metal–oxide interactions. There is a clear need for a better understanding of the underlying mechanism of the WGS over metal and metal/oxide catalysts for optimizing their performance.

We employed DFT-GGA calculations to investigate the WGS on Cu surfaces and nanoparticles, aiming to understand the role of Cu in the WGS reaction.[3] To model a Cu nanocatalyst, a Cu_{29} model (Figure 12.1) is adopted with a size of ∼1.2 nm. It exhibits a pyramidal structure formed by the interconnection of the (111) and (100) faces of the bulk metals, which have been observed on CeO_2(111) with scanning tunneling microscopy.[3] A distortion from a C_{4v} pyramid to a C_{2v} boat-shaped structure was observed for Cu_{29} after full geometry optimization. This gives a high number of atoms exposed in the (111) face, the most stable orientation for Cu. We checked that the distorted geometry does not have any imaginary frequency and therefore corresponds to a local minimum. Thus, Cu_{29} can be taken as a reasonable models for unsupported nanocatalysts.

Figure 12.1 shows the lowest energy pathway for the WGS reaction on Cu_{29}. The catalytic cycle starts with a bare Cu_{29} nanoparticle and gas-phase CO and H_2O, which is followed by CO and H_2O adsorption, H_2O dissociation to H and OH, CO oxidation with OH to form carboxyl HOCO species and HOCO dissociation into CO_2 and H. The cycle is closed with CO_2 and H_2 desorbed from the surface and the bare Cu_{29}. The micro-kinetic modeling based on the DFT calculations shows that the overall WGS rate is faster on Cu_{29} than Cu(100). A similar trend is also observed for Au. The predicted trend in WGS rate for Cu, Au nanoparticles and surfaces agrees well with the experimental measurements on model catalysts (Figure 12.2).

According to the DFT calculations, H_2O dissociation is the most difficult step along the WGS reaction path for both surface and nanoparticle, which slows down the overall WGS rate on Cu and Au. The enhanced WGS activity

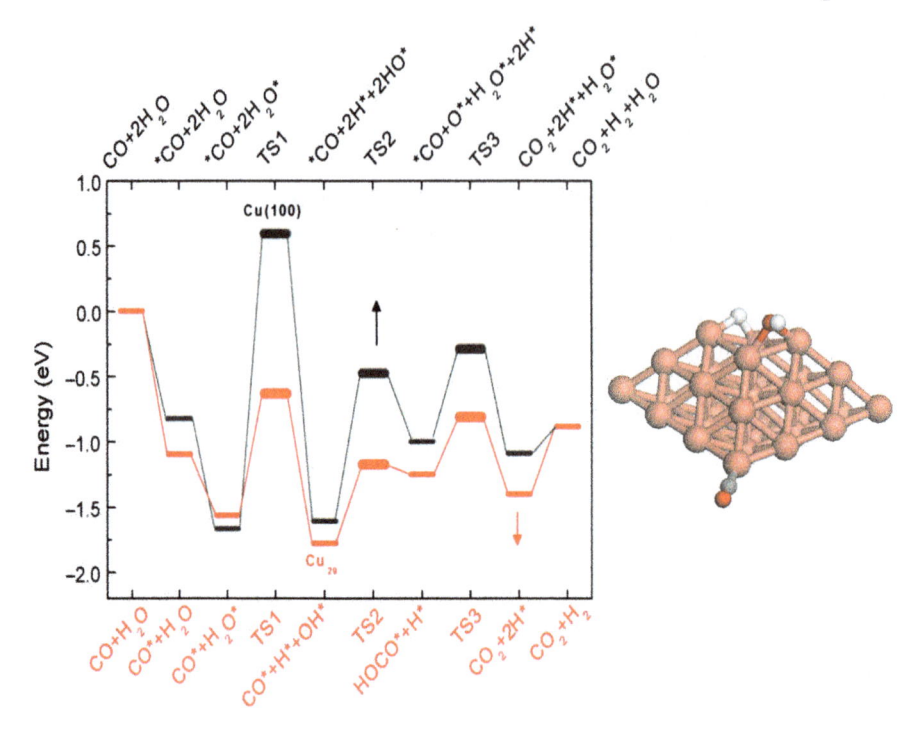

Figure 12.1 DFT calculated energy profiles for the WGS on Cu_{29} (red) and Cu(100) (black). Thin bar: intermediate; thick bar: transition state. Inset: CO, H and OH adsorbed on Cu_{29}. Brown: Cu; red: O; grey: C; white: H. Adsorbed species are denoted by asterisks (*).
The figure is reproduced from ref. 3 and ref. 37.

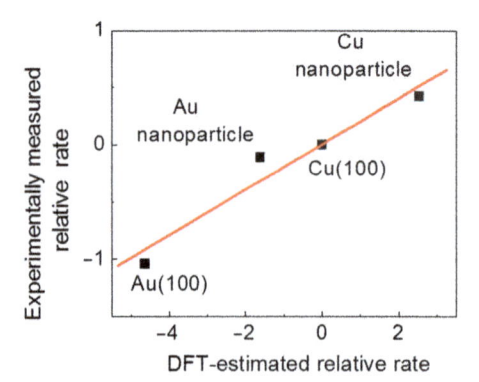

Figure 12.2 Relative WGS rate with respect to Cu(100).
The experimental data are from ref. 2.

of the nanoparticles is associated with the low-coordinated corner and the edge sites of the nanocatalysts, which are more active than the sites present on the flat surfaces. In addition, the fluxionality of the particles also contributes to strengthening the bindings of the intermediates and transition states with the catalyst (Figure 12.1). Given that, the bottleneck water dissociation becomes more facile when going from extended surfaces to nanoparticles, which facilitates the overall WGS reaction.

The importance of low-coordinated sites of metal nanocatalysts was also observed for other catalytic processes: methanol synthesis,[39] CO oxidation,[40] ammonia synthesis,[41] methanation[42] *etc.* Turnover rates and selectivity of catalytic reactions can be influenced strongly by the size and shape of nanoparticles, as their surface structures and active sites can be tailored at the molecular level. For these reactions, the bond breaking of reactants or reaction intermediates is the most difficult step that hinders the overall conversion. Metal nanocatalysts with high surface areas can provide the sites with the enhanced ability to chemisorb and activate bond cleavage as the coordination number is reduced at corners and step edges. Consequently, the reaction pathways leading to product yield and selectivity are in turn affected. In this case, the catalytic activity is promoted when the size of catalysts decreases to the nanoscale.

However, for the reaction that is hindered by the bond formation to produce products or the removal of the products from the surface, the strengthened interaction associated with the low-coordinated sites of metal nanoparticles is likely to slow down the overall conversion. As you will see in the following, the adsorbates may either stay on the surface as spectators and poison the low-coordinated sites, or destabilize these sites and therefore the nanocatalysts. In both cases, catalyst degradation with time and decreased activity are expected. That is, nano-sized catalysts may not always be the choice of catalyst to achieve high performance.

12.3.2 Metal (Core)–Platinum Shell Nanocatalysts for Oxygen Reduction Reactions in Fuel Cells: the Essential Role of Surface Contraction

Fuel cells are now widely accepted as the next step in power generation for many applications including commercial power, automotive transportation and portable electronic devices. Pt undoubtedly is the best elemental catalyst in fuel cells. However, there are two main drawbacks that hinder the widespread use of fuel cells. One is the high cost of Pt; the other is the sluggish kinetics, in particular for the cathodic reaction at low temperatures, *i.e.* the oxygen reduction reaction (ORR), even on Pt.[43,44] Large efforts have been spent searching for non-Pt electrocatalysts to lower the cost while maintaining high ORR activity; however, the inadequate activity and stability remain a major obstacle.[45,46] Recent studies show that core–shell nanocatalysts consisting of a single-layered shell of Pt atoms on cores of metal- or alloy- nanoparticles are able to achieve higher activity and stability for ORR compared to pure Pt.[6,47–52] The monolayer concept facilitates an ultra-low Pt content, high Pt utilization, and the possibility of tuning the activity of the Pt$_{ML}$ shell through interactions with the supporting nanoparticle cores.

To understand the enhanced activity and stability of the core–shell nanocatalysts under the ORR condition, we developed a simplified model to describe the nanoparticles in real size, which allows us to gain a better understanding of the effects of size and shape on the stability and activity under the ORR conditions. In our calculations, assuming that the shape of the nanoparticle is symmetric, we constructed a semi-sphere-like nanoparticle model (Figure 12.3) to reduce the computational cost, considering half of the nanoparticle should be representative for the whole. Different from the slab model,[53–59] relaxations were allowed in both out-of-plane and in-plane directions for the atoms, while only the Z directions of the atoms on the bottom layer were fixed to keep the symmetry of the particle. As shown, this simplified model is able to capture the surface contraction associated with the particle size and shape with an affordable cost, which was not included in the previous slab model,[53–57] and affects the ORR activity and stability significantly. As you will see, using this particle model, our DFT calculations are capable of describing the experimental results on well-defined core–shell nanoparticles well. Note that for the ORR, the low-coordinated sites of Pt-based nanoparticles are not active and are blocked by the O-containing species due to the strong O–Pt interaction. Accordingly, only the sites on the terrace, which bind O moderately, are considered for describing the ORR activity.

(A) Size Effect

To better understand the factors determining the ORR specific activity on well-defined Pd(core)–Pt(shell) nanoparticles, we carried out DFT calculations using the semi-sphere-like nanoparticle model as shown in Figure 12.3.[47] In agreement with the experimental findings, our DFT calculations for a Pt nanoparticle with a size of 1.8 nm demonstrate a considerable decrease in average in-plane spacing on the (111) terraces compared to that on the extended Pt(111) surface (−4.2%).[6] While this nanosize-induced

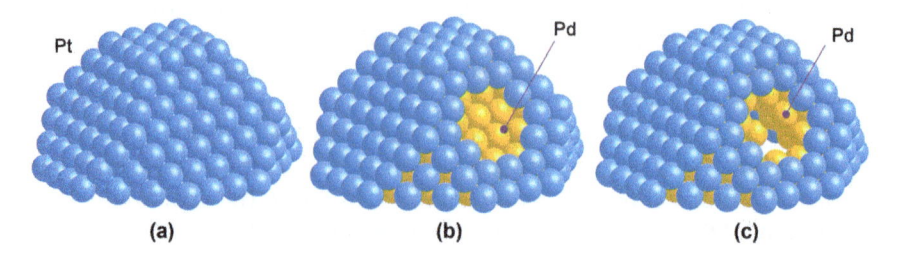

Figure 12.3 Schematic illustration of surface models for (a) pure Pt, (b) solid Pd core and Pt$_{ML}$ shell (Pd$_c$Pt$_1$), and (c) partially hollow Pd core and Pt$_{ML}$ shell (hPd$_1$Pt$_1$). "c", "s", and "1" correspond to core, shell, and 1 monolayer (ML), respectively.
The figure is reproduced from ref. 47.

Theoretical Insights into Metal Nanocatalysts 225

in-plane contraction decreases with increasing particle size, the corresponding effect on the binding energy for oxygen (BE-O), a descriptor for the ORR activity,[56,60,61] cannot be ignored for particles of at least up to 5 nm. For Pd(core)–Pt(shell) nanocatalysts, the size and lattice mismatch (−0.9%) jointly determine the strain that affects the BE-O on the (111) facets.[6] Considering the fact that the intrinsic ORR activity is largely determined by the BE-O,[53–55] one can see that the particle size affects not only the fraction of active sites at the surface, but also the turnover frequency (TOF) per active site. The product of the two determines the specific activity. A previous study of the ORR on transition metal alloys suggested that the best ORR electrocatalysts may bind oxygen more weakly, by about 0.2 eV,[55] than Pt(111) does, *i.e.* corresponding to −3.89 eV in our calculations (dashed line in Figure 12.4). This value is reached on 1.8-nm Pt particles solely by the nanosize-induced surface contraction. Larger particle sizes for optimal BE-O were found using the 3.4-nm Pd_CPt_3 and 4.4-nm Pd_CPt_2 models, where a reduced nanosize effect is compensated by the increased influence of the Pd core. Therefore, the enhancements in specific activity are largely attributed to the effect of nanosize-induced surface contraction on facet-dependent BE-O. In accordance with the experiment, the DFT calculations suggest that the moderately compressed (111) facet is most conducive to the ORR on small nanoparticles. Our results indicate the importance of concerted structure and component optimization for enhancing core–shell nanocatalysts' activity and stability.

DFT calculations using the particle model were also performed to elucidate the observed high stability of Pd(core)–Pt(shell) nanoparticles.[47] Three nanoparticle systems with different sizes were considered in our calculations (Figure 12.3): pure Pt, a Pt monolayer shell with a solid Pd core (Pd_CPt_1), and a Pt monolayer shell with only a 1 ML Pd inner core (hPd_1Pt_1). The extended

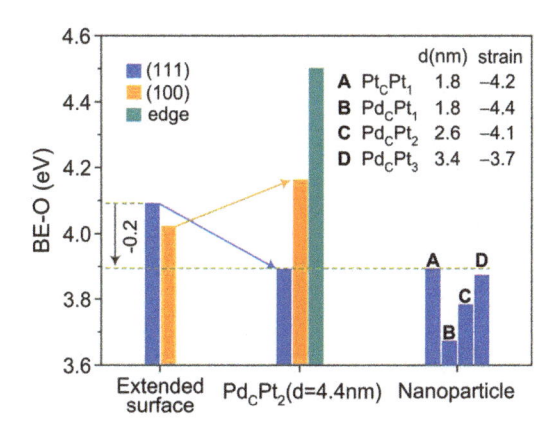

Figure 12.4 Binding energy for oxygen on different sites calculated using the slab model for extended surfaces (as a reference), and the nanoparticle model for various core–shell nanoparticles.
The figure is reproduced from ref. 6.

surfaces were also considered as an extreme case to model big nanoparticles. DFT calculations show that the relative shift in 1 ML Pt dissolution potentials for Pt and Pd (core)–Pt (shell) nanoparticles depends strongly on the particle size (Figure 12.5). For both Pt and Pd (core)–Pt (shell) nanocatalysts, the big particles display the lower surface contraction and therefore a higher dissolution potential than the small ones. The trend in dissolution potential from Pt, Pd$_C$Pt$_1$ to hPd$_1$Pt$_1$ also varies with the size. For small particles (<~3.5 nm), Pt shows the strongest binding to the Pt shell and the highest dissolution potential among all. When the particle size is bigger than ~3.5 nm, hPd$_1$Pt$_1$ stabilizes the Pt shell the most (Figure 12.5). Our calculations agree reasonably well with the experiment, which observes a higher stability of the Pd (core)–Pt (shell) with a size of 3.6 nm than Pt and partial dissolution of Pd under the ORR condition. The theoretical results imply that the particle size, the surface contraction, and the bonding activity of the inner core layer directly interact with the shell interplay and play an essential role in tuning the stability of the core–shell catalysts.

(B) Shape Effect

The shape of nanoparticles is also important. To understand the effect of shape on the ORR activity of Pd(core)–Pt(shell) nanocatalysts, both sphere-like (SP) and tetrahedral (TH) shapes were considered.[52] As shown in Figure 12.6(a), to construct plausible surface models, we initially compared the stability of Pd$_{TH}$ and Pd$_{SP}$ nanoparticles as a function of particle sizes. The results show that at the same size the Pd$_{SP}$ nanoparticles are more stable than the Pd$_{TH}$ ones, while the Pd$_{TH}$ consists of a much smaller number of atoms than Pd$_{SP}$ (*i.e.* 220 *versus* 550 atoms of ~2.5 nm particles, respectively). As shown in Figure 12.6(b), BE-O on Pt$_{ML}$Pd$_{TH}$ is weaker than that on

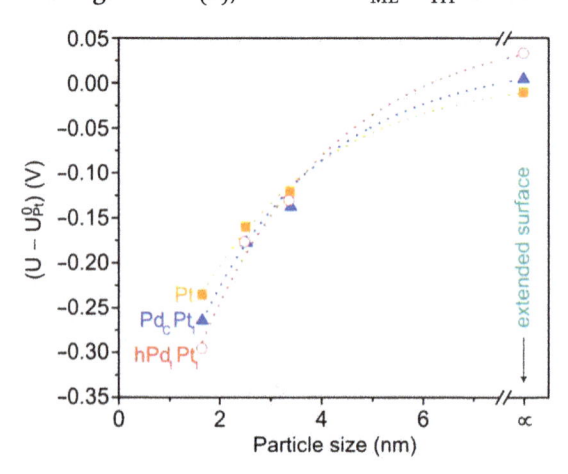

Figure 12.5 DFT-predicted dissolution potentials of a 1 ML Pt shell from nanoparticles and extended surfaces of Pt, Pd$_C$Pt$_1$, and hPd$_1$Pt$_1$ as a function of particle sizes.
The figure is reproduced from ref. 47.

Theoretical Insights into Metal Nanocatalysts 227

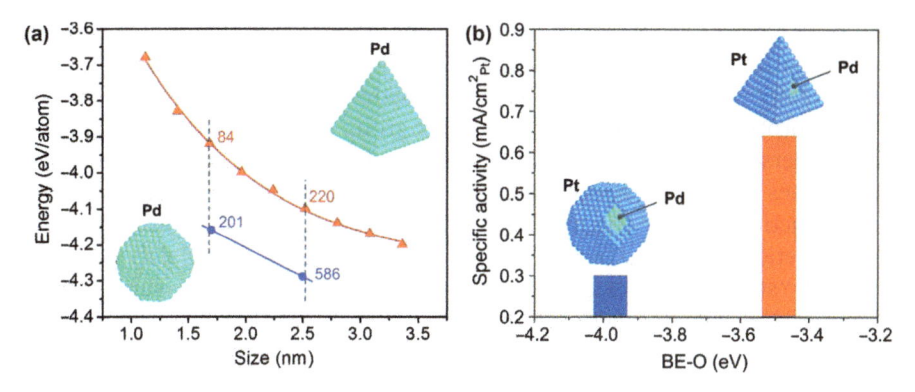

Figure 12.6 Comparison of stability of the Pd$_{TH}$ (red triangle) and Pd$_{SP}$ (blue circle) nanoparticles using DFT calculations. The numbers labeled in the figure correspond to the number of atoms of which the ~1.7 nm and ~2.5 nm nanoparticles consist respectively. (b) The Pt specific activity against BE-O on Pt$_1$Pd$_{TH}$ and Pt$_1$Pd$_{SP}$ nanoparticles. For the binding energy of oxygen calculations, ~1.7 nm nanoparticles were used. The specific activities of the electrocatalysts were from experimental measurements.
The figure is reproduced from ref. 52.

Pt$_{ML}$Pd$_{SP}$ (~0.5 eV). This agrees well with experimental observations, showing a significant increase in ORR activity on moving from Pt$_1$Pd$_{TH}$ to Pt$_1$Pd$_{SP}$ (Figure 12.6(b)). The weakened O–Pt interaction can be attributed to the surface contraction. Pt$_1$Pd$_{TH}$ introduces more surface contraction (−5.51%) compared to that of Pt$_1$Pd$_{SP}$ and Pt (−3.51% and −3.04%), resulting in the lower-lying Pt d-band and therefore the weaker O adsorption is observed.

Overall, for small nanoparticles (<5 nm), both size and shape have a significant effect on the stability and ORR activity of the catalysts. As shown above, the semi-sphere-like nanoparticle model we developed is able to capture these features with an affordable computational cost, and describe the experiments on the well-defined model systems well. Both the geometric (the contraction in the Pt shell introduced by the core) and electronic effects (the modification in the electronic structure of the Pt shell atoms associated with the core–shell interaction) are found to play a key role in promoting the ORR activity and stability of core–shell nanocatalysts. The theoretical understanding of the promoting effects provides new insights into the rational design of better Pt-based ORR nanocatalysts.

12.4 Supported Metal Nanocatalysts

Section 12.3 addresses the great promise of metal nanoparticles for heterogeneous catalysis and electrocatalysis. However, the poor thermal stability of nano-sized particles limits their use to low temperature conditions and constitutes one of the key hurdles on the way to industrial application.[62] A conventional approach towards stabilizing metal nanoparticles is to use a support. Now, the question is: what role does the support play in the catalytic performance of a catalyst? Tauster *et al.* used the term "strong metal/support interaction" (SMSI) to describe the drastic changes that occur in the chemisorption properties of Pt and other group VIII (8–10) metals when they are dispersed on the surfaces of titanium oxide.[63,64] The spreading of TiO_x aggregates on the surfaces of the supported metals could be responsible for the reduction in their chemical and catalytic activity.[64,65] By combining DFT and kinetic modelling, our theoretical studies show a completely different type of "strong metal/support interaction", where supports substantially enhance the catalytic activity of metal nanoparticles for many reactions, *e.g.* the WGS reaction,[10,35,66] desulfurization,[8,67] CO oxidation[68] and CO_2 activation.[69] In the following, we will show two examples to illustrate the promoting effect of supports.

Associated with SMSI, supports can not only stablize the small metal nanoparticles that are unstable in the gas phase, they can also produce large electronic perturbations for the supported small particles and significantly enhance the catalytic activity. This is the case for the WGS reaction on Pt/CeO_2.[66] It is observed experimentally that upon adding Pt to $CeO_2(111)$, there is a continuous increase in the catalytic activity until a maximum is reached at a coverage of ~0.2 ML. Comparing with the STM results reported for the $Pt/CeO_2(111)$,[70] one can conclude that the maximum activity corresponds to catalysts that contain Pt particles with a diameter smaller than 1.7 nm and a height below 0.4 nm. According to our DFT calculations, Pt(111) has problems in adsorbing and dissociating the water molecule (Figure 12.7).

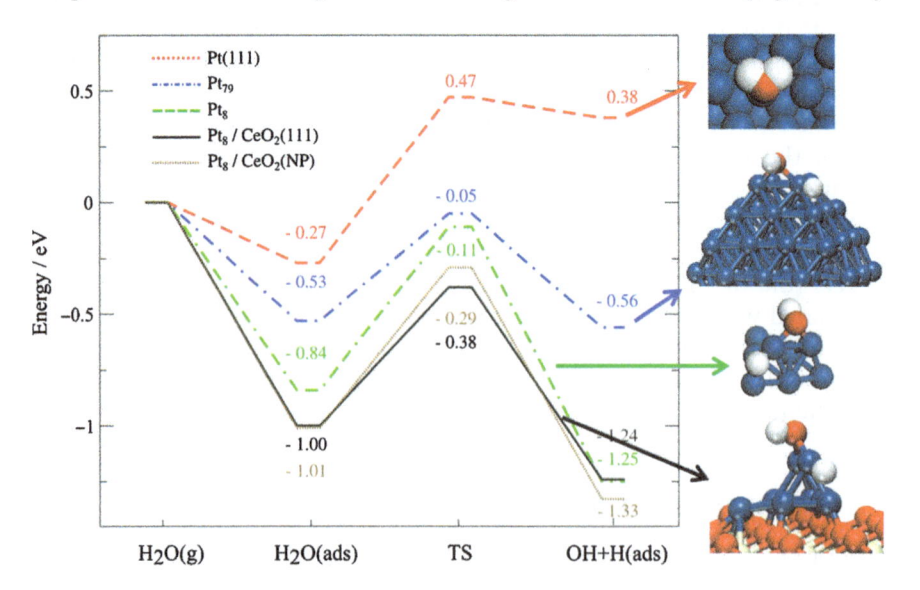

Figure 12.7 DFT calculated energy profile and the geometries for the dissociation on Pt(111), Pt_{79} and Pt_8 clusters, and Pt_8 on CeO_2. Blue: Pt; white: H; red: O; cream: Ce.
The figure is reproduced from ref. 66.

Theoretical Insights into Metal Nanocatalysts

Water dissociation is a rate-limiting step for the WGS reaction on Pt(111) and therefore it is not a good WGS catalyst.[71] The exothermicity of the rate-limiting H_2O dissociation increases when going from Pt(111) to free Pt_{79} and even more on free Pt_8, where the resulting OH can react with CO to eventually yield H_2 and CO_2 in the WGS. Similarly, as in the case of Cu shown in Section 12.3, the corner atoms present in the Pt_{79} and Pt_8 nanocatalysts (Figure 12.7) facilitate the cleavage of O–H bonds, and the smaller the Pt particles, the easier the dissociation of water.[66] However, in the gas phase, the particles as small as Pt_8 are not stable and are not practical as catalysts. When Pt_8 is deposited on a CeO_2(111) surface or on a $Ce_{40}O_{80}$ nanoparticle, the DFT calculations show an enhancement in the stablity and ability of Pt_8 for the cleavage of O–H bonds (Figure 12.7). The key to the chemical activity of Pt_8/CeO_2(111) is in the electronic perturbations induced by ceria on Pt_8, which is observed using DFT and X-ray photoemission spectroscopy (XPS). Thus, the ceria-supported Pt_8 appears as a fluxional system that can turn the geometry and charge distribution to better accommodate the adsorbates. Small nanoparticles of Pt in contact with CeO_2 can perform all the steps, adsorption of H_2O and CO, H_2O dissociation and CO oxidation, of the WGS reaction, making this system more active than Cu/CeO_2, Au/CeO_2 or Ni/CeO_2.[66]

Supports can also participate in the reaction directly by catalysing some of the elementary steps, which are highly activated on the metal nanoparticles, and therefore facilitate the overall conversion. For the WGS reaction, the experiments show that Au/TiO_2(110) is clearly a better catalyst than each parent system as well as the commercial catalyst Cu/ZnO.[10] To understand the role of TiO_2 support, the WGS on each component and the interface of the composite catalyst are investigated using DFT. As demonstrated in Section 12.3, water dissociation is a rate-limiting step for the WGS reaction on a Au nanoparticle. TiO_2(110) bonds H_2O more strongly than Au, being active for the first and the second O–H scissions of H_2O; however, due to the strong $O-TiO_2$ interaction, the CO oxidation by the O from H_2O dissociation to form CO_2 is rather difficult (Figure 12.8). That is, neither Au nor TiO_2(110) is good enough to catalyze the WGS reaction. Au bonds the H_2O too weakly to adsorb and dissociate H_2O, while TiO_2 is too active to produce CO_2 from CO oxidation. When the two components are combined to form the Au/TiO_2 catalyst, the calculations indicate that it works as a bifunctional catalyst. Both components participate in the reaction directly and work in a cooperative way. According to energetics, for the H_2O dissociation, Au/TiO_2 behaves differently from Au, but as facile as TiO_2(110). For the CO oxidation, on other hand, Au/TiO_2 acts more like Au, rather than TiO_2, where no high activation barrier is observed (Figure 12.8). As a consequence, the WGS reaction runs faster on Au/TiO_2 than on each parent individual, in agreement with experimental observations.[10] According to the geometries, CO prefers the Au sites; H_2O adsorbs and dissociates on TiO_2 and the rest of the steps occur at the Au/TiO_2 interface. The TiO_2 support helps to release the bottleneck H_2O dissociation on Au and therefore facilitates the overall WGS

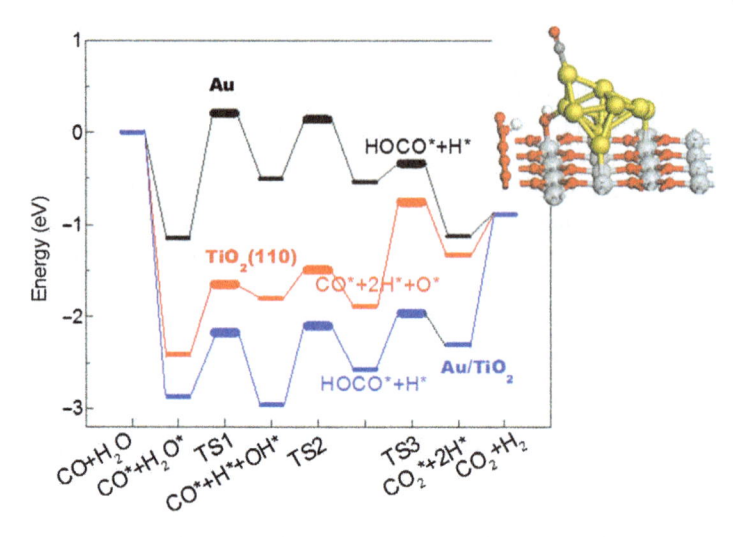

Figure 12.8 Potential energy diagram for the WGS on Au$_{29}$, TiO$_2$(110) and Au$_{10}$/TiO$_2$(110). The thin bars represent the intermediates; thick bars correspond to the transition states (TS). Inset: corresponding structures for the intermediates and TSs involved in the WGS at the Au/TiO$_2$ interface. Large yellow spheres: Au; large gray spheres: Ti; small red spheres: O; small white spheres: H; small dark grey spheres: C. Adsorbed species are denoted by asterisks (*).
The figures are reproduced from ref. 10.

reaction. Our results show that the relatively inert gold nanoparticles can become very good catalysts when combined with an over-reactive TiO$_2$ surface.

Overall, supports can play an essential role in promoting the catalytic performance of metal nanoparticles *via* SMSI. The support is able to stabilize the small metal nanoparticles, which are catalytically active, by hindering diffusion and coalescence under the reaction conditions. In addition, it can modify the electronic structure of the supported small particles and therefore promote their catalytic activities. Finally, it can participate in the reaction directly by facilitating the steps that are activated on the supported particles and therefore facilitate the overall reaction. Such understanding is the key for the further development of metal/support catalysts.

12.5 Conclusions

The powerful new computing infrastructures are starting to close the gap between theoretical and experimental studies in the field of nanocatalysis. Our theoretical studies on unsupported and supported metal nanocatalysts provide a molecular-scale picture of the mechanisms underlying catalytic activity. Firstly, the low-coordinated sites at step edges are essential for metal nanocatalysts, which behave differently from those on terraces. These low-coordinated sites can be the only active sites for the reaction and influence turnover rates and selectivity of catalytic reactions. Secondly, for small nanoparticles (<5 nm), tremendous surface contractions are introduced depending on particle size and shape, which have a significant effect on the stability and activity of the metal nanocatalysts. Finally, supports can play an essential role for promoting the catalytic performance of metal nanocatalysts *via* SMSI by stabilizing the supported particles, modifying the electronic structure of the supported small particles and/or participating in the reaction directly by facilitating the steps that are activated on the supported metal particles. Such a fundamental understanding would allow us to accurately predict the performance of a catalyst exclusively based on computer models, without the need for extensive trial-and-error experimental screening.

Acknowledgements

The research is supported by the US Department of Energy, Division of Chemical Sciences, under contract DEAC02-98CH10886.

References

1. A. T. Bell, *Science*, 2003, **299**, 1688.
2. J. A. Rodriguez, P. Liu, J. Hrbek, J. Evans and M. Perez, *Angew. Chem., Int. Ed.*, 2007, **46**, 1329.
3. P. Liu and J. A. Rodriguez, *J. Chem. Phys.*, 2007, **126**, 164705.
4. L. Barrio, P. Liu, J. A. Rodriguez, J. M. Campos-Martin and J. L. G. Fierro, *J. Phys. Chem. C*, 2007, **111**, 19001.
5. L. Barrio, P. Liu, J. A. Rodriguez, J. M. Campos-Martin and J. L. G. Fierro, *J. Chem. Phys.*, 2006, **125**, 164715.
6. J. X. Wang, H. Inada, L. Wu, Y. Zhu, Y. Choi, P. Liu, W.-P. Zhou and R. R. Adzic, *J. Am. Chem. Soc.*, 2009, **131**, 17298.
7. J. A. Rodriguez, S. Ma, P. Liu, J. Hrbek, J. Evans and M. Pérez, *Science*, 2007, **318**, 1757.
8. J. A. Rodriguez, P. Liu, J. F. Viñes, F. Illas, Y. Takahashi and K. Nakamura, *Angew. Chem., Int. Ed.*, 2008, **47**, 6685.
9. A. Kowal, M. Li, M. Shao, K. Sasaki, M. B. Vukmirovic, J. Zhang, N. S. Marinkovic, P. Liu, A. I. Frenkel and R. R. Adzic, *Nat. Mater.*, 2009, **8**, 325.
10. J. A. Rodríguez, J. Evans, J. Graciani, J. Park, P. Liu, J. Hrbek and J. F. Sanz, *J. Phys. Chem. C*, 2009, **113**, 7364.
11. J. B. Park, J. Graciani, J. Evans, D. Stacchiola, S. Ma, P. Liu, A. Nambu, J. F. Sanz, J. Hrbek and J. A. Rodriguez, *Proc. Natl. Acad. Sci.*, 2009, **106**, 4975.
12. J. A. Rodriguez, P. Liu, Y. Takahashi, K. Nakamura, F. Viñes and F. Illas, *J. Am. Chem. Soc.*, 2009, **131**, 8595.
13. J. X. Wang, H. Inada, L. Wu, Y. Zhu, Y. Choi, P. Liu, W. P. Zhou and R. R. Adzic, *J. Am. Chem. Soc.*, 2009, **131**, 17298.

14. P. Liu, J. A. Rodriguez, Y. Takahashi and K. Nakamura, *J. Catal.*, 2009, **262**, 294.
15. P. Liu, A. Logadottir and J. K. Nørskov, *Electrochim. Acta*, 2003, **48**, 3731.
16. P. Liu and J. A. Rodriguez, *J. Phys. Chem. B*, 2006, **110**, 19418.
17. P. Liu, *J. Chem. Phys.*, 2010, **133**, 204705.
18. A. B. Vidal and P. Liu, *Phys. Chem. Chem. Phys.*, 2012, **14**, 16626.
19. P. Hohenberg and W. Kohn, *Phys. Rev.*, 1964, **136**, B864.
20. W. Kohn and L. J. Sham, *Phys. Rev.*, 1965, **140**, A1133.
21. L. H. Thomas, *Math. Proc. Cambridge Philos. Soc.*, 1927, **23**, 542.
22. B. Delley, *J. Chem. Phys.*, 2000, **113**, 7756.
23. B. Delley, *J. Chem. Phys.*, 1990, **92**, 508.
24. T. Arias, M. C. Payne and J. D. Joannopoulos, *Phys. Rev. Lett.*, 1992, 69.
25. G. Kresse and J. Furthmüller, *Phys. Rev. B*, 1996, **54**, 11169.
26. G. Kresse and J. Hafner, *Phys. Rev. B*, 1993, **47**, 558.
27. G. Kresse and J. Furthmüller, *Comput. Mater. Sci.*, 1996, **6**, 15.
28. M. S. Chen and D. W. Goodman, *Science*, 2004, **306**, 252.
29. M. Haruta, *CATTECH*, 2002, **6**, 102.
30. K. An and G. A. Somorjai, *ChemCatChem*, 2012, **4**, 1512.
31. I. V. Yudanov, A. Genest, S. Schauermann, H.-J. Freund and N. Rösch, *Nano Lett.*, 2012, **12**, 2134.
32. *Catalyst Handbook*, ed. M. V. Twigg, Wolfe, England, 1989.
33. R. Burch, *Phys. Chem. Chem. Phys.*, 2006, **8**, 5483.
34. Q. Fu, H. Saltsburg and M. Flytzani-Stephanopoulos, *Science*, 2003, **301**, 935.
35. J. B. Park, J. Graciani, J. Evans, D. Stacchiola, S. D. Senanayake, L. Barrio, P. Liu, J. F. Sanz, J. Hrbek and J. A. Rodriguez, *J. Am. Chem. Soc.*, 2010, **132**, 356.
36. J. A. Rodriguez, J. Graciani, J. Evans, J. B. Park, F. Yang, D. Stacchiola, S. D. Senanayake, S. Ma, M. Pérez, P. Liu, J. F. Sanz and J. Hrbek, *Angew. Chem., Int. Ed.*, 2009, **48**, 8047.
37. J. A. Rodriguez, P. Liu, X. Wang, W. Wen, J. Hanson, J. Hrbek, M. Pérez and J. Evans, *Catal. Today*, 2009, **143**, 45.
38. J. A. Rodriguez, P. Liu, J. Hrbek, M. Pérez and J. Evans, *J. Mol. Catal. A: Chem.*, 2008, **281**, 59.
39. Y. Yang, J. Evans, J. A. Rodriguez, M. G. White and P. Liu, *Phys. Chem. Chem. Phys.*, 2010, **12**, 9909.
40. H. Falsig, B. Hvolbæk, I. S. Kristensen, T. Jiang, T. Bligaard, C. H. Christensen and J. K. Nørskov, *Angew. Chem., Int. Ed.*, 2008, **47**, 4835.
41. K. Honkala, A. Hellman, I. N. Remediakis, A. Logadottir, A. Carlsson, S. Dahl, C. H. Christensen and J. K. Nørskov, *Science*, 2005, **307**, 555.
42. H. S. Bengaard, J. K. Nørskov, J. Sehested, B. S. Clausen, L. P. Nielsen, A. M. Molenbroek and J. R. Rostrup-Nielsen, *J. Catal.*, 2002, **209**, 365.

43. M. R. Tarasevich, A. Sadkowski and E. Yeager, in *Comprehensive Treatise of Electrochemistry*, ed. B. E. Conway, J. O. M. Bockris, E. Yeager, S. U. M. Khan and R. E. White, Plenum Press, New York, 1983, vol. 7, p. 301.
44. R. R. Adzic, in *Electrocatalysis*, ed. J. Lipkowski and P. N. Ross, Wiley, New York, 1998, p. 197.
45. H. A. Gasteiger, S. S. Kocha, B. Sompalli and F. T. Wagner, *Appl. Catal., B*, 2005, **56**, 9.
46. H. A. Gasteiger and N. M. Marković, *Science*, 2009, **324**, 48.
47. K. Sasaki, H. Naohara, Y. Cai, Y. M. Choi, P. Liu, M. B. Vukmirovic, J. X. Wang and R. R. Adzic, *Angew. Chem., Int. Ed.*, 2010, **49**, 8602.
48. J. X. Wang, C. Ma, Y. M. Choi, D. Su, Y. M. Zhu, P. Liu, R. Si, M. B. Vukmirovic, Y. Zhang and R. R. Adzic, *J. Am. Chem. Soc.*, 2011, **133**, 13551.
49. K. A. Kuttiyiel, K. Sasaki, Y. Choi, D. Su, P. Liu and R. R. Adzic, *Nano Lett.*, 2012, **12**, 6266.
50. K. A. Kuttiyiel, K. Sasaki, Y. Choi, D. Su, P. Liu and R. R. Adzic, *Energy Environ. Sci.*, 2012, **5**, 5297.
51. K. Sasaki, H. Naohara, Y. Choi, Y. Cai, W.-F. Chen, P. Liu and R. R. Adzic, *Nat. Commun.*, 2012, **3**, 1115.
52. K. Gong, Y. Choi, M. B. Vukmirovic, P. Liu, C. Ma, D. Su and R. R. Adzic, *Z. Phys. Chem.*, 2012, **226**, 1025.
53. J. K. Nørskov, J. Rossmeisl, A. Logadottir, L. Lindqvist, J. R. Kitchin, T. Bligaard and H. Jónsson, *J. Phys. Chem. B*, 2004, **108**, 17886.
54. A. U. Nilekar and M. Mavrikakis, *Surf. Sci.*, 2008, **602**, L89.
55. V. Stamenkovic, B. S. Mun, K. J. J. Mayrhofer, P. N. Ross, N. M. Markovic, J. Rossmeisl, J. Greeley and J. K. Nørskov, *Angew. Chem., Int. Ed.*, 2006, **118**, 2963.
56. M. Shao, P. Liu and R. R. Adzic, *J. Phys. Chem. B*, 2007, **111**, 6772.
57. M. Shao, T. Huang, P. Liu, J. Zhang, K. Sasaki, V. B. Vukmirovic and R. R. Adzic, *Langmuir*, 2006, **22**, 10409.
58. J. L. Zhang, M. B. Vukmirovic, Y. Xu, M. Mavrikakis and R. R. Adzic, *Angew. Chem., Int. Ed.*, 2005, **44**, 2132.
59. Y. Ma and P. B. Balbuena, *J. Phys. Chem. C*, 2008, 112.
60. J. K. Nørskov, J. Rossmeisl, A. Logadottir, L. Lindqvist, J. Kitchin and T. Bligaard, *J. Phys. Chem. B*, 2004, **108**, 17886.
61. A. U. Nilekar and M. Mavrikakis, *Surf. Sci.*, 2008, **602**, L89.
62. A. Cao, R. Lu and G. Veser, *Phys. Chem. Chem. Phys.*, 2010, **12**, 13499.
63. S. J. Tauster, S. C. Fung and R. L. Garten, *J. Am. Chem. Soc.*, 1978, **100**, 170.
64. S. J. Tauster, *Acc. Chem. Res.*, 1987, **20**, 389.
65. F. Pesty, H.-P. Steinrück and T. E. Madey, *Surf. Sci.*, 1995, **339**, 83.
66. A. Bruix, J. A. Rodriguez, P. J. Ramírez, S. D. Senanayake, J. Evans, J. B. Park, D. Stacchiola, P. Liu, J. Hrbek and F. Illas, *J. Am. Chem. Soc.*, 2012, **134**, 8968.
67. J. A. Rodriguez, P. Liu, Y. Takahashi, K. Nakamura, F. Viñes and F. Illas, *J. Am. Chem. Soc.*, 2009, **131**, 8595.

68. J. A. Rodriguez, P. Liu, Y. Takahashi, F. Vines, L. Feria, E. Florez, K. Nakamura and F. Illas, *Catal. Today*, 2011, **166**, 2.
69. A. B. Vidal, L. Feria, J. Evans, Y. Takahashi, P. Liu, K. Nakamura, F. Illas and J. A. Rodriguez, *J. Phys. Chem. Lett.*, 2012, **3**, 2275.
70. Y. Zhou, J. M. Perket and J. Zhou, *J. Phys. Chem. C*, 2010, **114**, 11853.
71. D. W. Flaherty, W.-Y. Yu, Z. D. Pozun, G. Henkelman and C. B. Mullins, *J. Catal.*, 2011, **282**, 278.

CHAPTER 13

Porous Cryptomelane-type Manganese Oxide Octahedral Molecular Sieves (OMS-2); Synthesis, Characterization and Applications in Catalysis

SAMINDA DHARMARATHNA AND S. L. SUIB*

Department of Chemistry, University of Connecticut, 55, North Eagleville Road, Storrs, Connecticut, 06269, USA
*Email: steven.suib@uconn.edu

13.1 Introduction

Manganese oxide octahedral molecular sieves (OMS) have attracted a large amount of attention from researchers in recent years as they are excellent catalysts. Natural OMS, or cryptomelane-type manganese oxide, occurs in deep sea beds as nodules.[1] The motivation to synthesize this mineral was triggered by the synthesis of a todorokite-type 3×3 manganese oxide molecular sieve by Shen *et al.* in 1993, which was named OMS-1.[2] Then, in 1994, DeGuzman *et al.* synthesized a cryptomelane-type material with a 2×2 tunnel structure, as shown in Figure 13.1, using MnO_4^- and Mn^{2+}; the material was highly porous and showed semiconducting properties.[3] After this discovery, many researchers around the world became very interested in this material. Many new avenues of research have opened up since then. As an example,

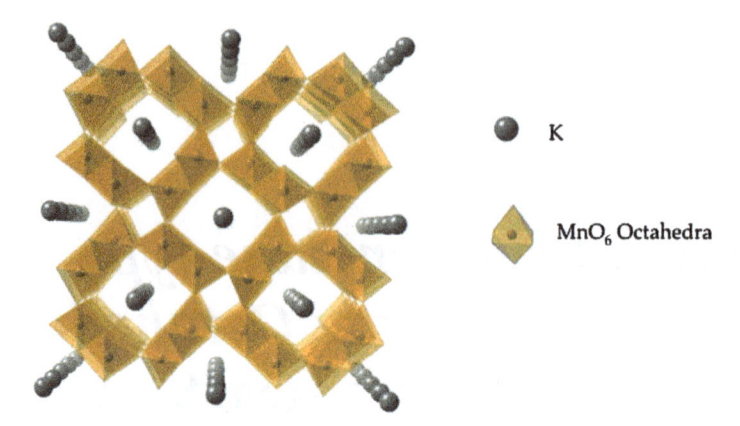

Figure 13.1 Tunnel-type crystal structure of OMS-2 with 2×2 edge shared MnO_6 octahedra (orange) and K^+ (black) in the center of the tunnel as the counter ion. Tunnels have dimensions 4.6 Å×4.6 Å.

biocompatibility has been tested by implanting OMS-2 in a rat brain, and when functionalized, these OMS-2 phases can even be used in medical applications.[4] Synthetic cryptomelane-type OMS-2 has a formula of $KMn_8O_{16} \cdot nH_2O$ and has an average oxidation state of 3.8. This is an indication of mixed valency in the framework, and it was further proven by electrochemical experiments that these materials possess mixed Mn^{4+}, Mn^{3+}, and Mn^{2+} valencies.[5] Flexibility in tuning the particle size, surface area, morphology, doping the structure with a variety of metals, and easy regeneration using air or oxygen, are highly advantageous in the area of catalysis. The porous tunnel-type structure and mixed valency in the framework play a significant role in controlling the selectivity and conversion of various different types of redox reactions. Recent examples of molecular trafficking inside the tunnels and shape selectivity of the catalyst towards substrates suggest a general reaction mechanism involving active tunnel sites.[6–9] This review will emphasize the synthesis of cryptomelane-type OMS-2 materials using different techniques and the progress of synthetic routes to alter the morphology, synthesis time, and particle size. It will also discuss catalytic applications in different types of reactions including oxidation, dehydration, epoxidation, and degradation.

13.2 Synthesis and Morphology Control

Since the initial synthesis of OMS-2, many other synthetic routes were reported in the literature. These novel methods offer control over particle size, morphology and the amount of defects in the material, by controlling reaction time, temperature, pH, and precursor used. Here, we will discuss convectional, microwave-assisted methods, solvent free, and ultrasound routes to synthesize OMS-2 materials.

Porous Cryptomelane-type Manganese Oxide Octahedral Molecular Sieves (OMS-2)   237

Hydrothermal processing of materials has been used as a facile and environmentally benign synthetic approach for years. This is one of the most versatile solution-based strategies to synthesize nanomaterials, due to the ease of control of the reaction time, temperature, precursor, product separation and simple setup.[10,11]

Typical fabrication of OMS-2 using any synthetic technique involves oxidation of Mn^{2+} with a suitable oxidant such as MnO_4^-, $Cr_2O_7^{2-}$, $S_2O_8^{2-}$, H_2O_2, or ozone.[12–16] Hydrothermal techniques in particular have been identified in the literature as an effective way to control the morphology and particle size of the OMS-2 materials. Figure 13.2 shows field emission scanning electron micrographs (FESEM) of OMS-2 nanomaterials, fabricated using hydrothermal techniques. Typical fiber-like morphology was obtained by using a MnO_4^-/Mn^{2+} redox couple under acidic conditions (Figure 13.2(a)).[3,12] However, when the oxidant was changed to $Cr_2O_7^{2-}$, a self-assembled three-dimensional dendritic structure was formed (Figure 13.2(b,c)).[13] A spontaneous free-standing membrane was formed using $S_2O_8^{2-}$ as the oxidant. When the fibers grew to sufficient length (~100 µm), they were arranged in an interlocking fashion, Figure 13.2(d), to yield a free-standing membrane, which was used in sustainable energy and environmental applications. A recent study by Yuan *et al.* showed that these membranes can be used in oil spill cleanup by selective adsorption of oil.[17,18] Compared to fibers of several hundreds of micrometres obtained using conventional hydrothermal methods, a microwave-assisted

Figure 13.2   Morphology of OMS-2 synthesized using different precursors under hydrothermal conditions. (a) MnO_4^-/Mn^{2+},[12] (b, c) $Cr_2O_7^{2-}/Mn^{2+}$,[13] (d, e) $S_2O_8^{2-}/Mn^{2+}$,[14,19] and ozone/Mn^{2+}.[16]
Copyright permission granted by the American Chemical Society and Wiley-VCH Verlag GmbH & Co. KGaA, Weinheim.

hydrothermal technique was used to obtain shorter fibers (∼100–200 nm), Figure 13.2(e). This OMS-2 material showed a higher surface area (129 ± 1 m² g⁻¹) compared to that of a material synthesized using conventional hydrothermal techniques (44 ± 1 m² g⁻¹).[17,19] A recent study by Galindo et al. yielded uniform hollow spheres of OMS-2 when ozone was used as an oxidant (Figure 13.2(f)).[16]

Further, particle size and surface area were easily controlled by altering the means of energy supply during the synthesis. For instance, Figure 13.3(a) shows materials synthesized by conventional reflux methods, with fiber lengths of ∼500–1000 nm and surface areas of ∼90 ± 1 m² g⁻¹.[20] A considerable increase in surface area (∼156 ± 1 m² g⁻¹) and decrease in fiber length (∼50–100 nm) were obtained by Ding et al. during a solvent-free synthesis of OMS-2.[21] Nyutu et al. successfully used a microwave-assisted reflux method, coupled with various organic co-solvents, to control the particle size of OMS-2 and they also obtained a higher surface area (∼227 ± 1 m² g⁻¹).[22] A recent study by Dharmarathna et al. focused on the use of ultrasound to obtain a self-assembled three-dimensional dendritic structure without the use of Cr. The materials obtained also showed a very high surface area (∼288 ± 1 m² g⁻¹) and catalytic activity compared to commercial MnO₂.[23]

Another unique feature of these OMS-2 materials is that they consist of edge and vertex shared Mn octahedral units (MnO₆$^{x-}$), which is uncommon for many oxides.[1] This is an advantage, since many transition metal cations

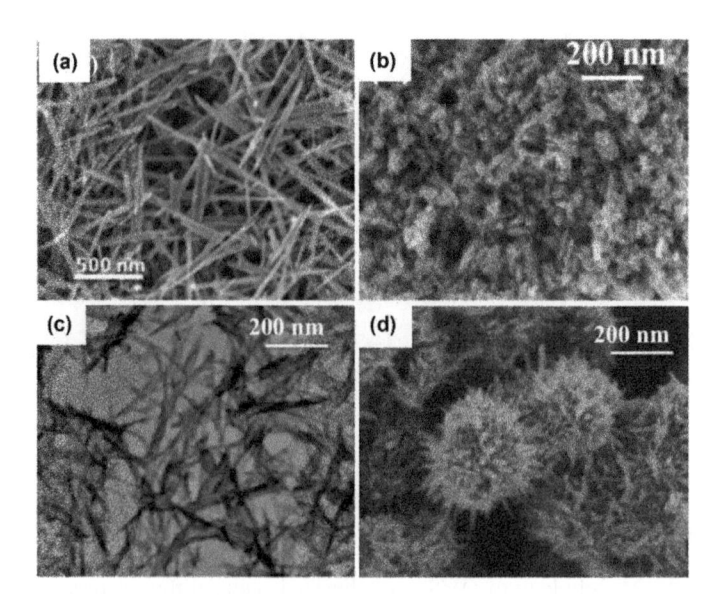

Figure 13.3 Particle size control of OMS-2 using different synthetic techniques. (a) Reflux,[23] (b) solvent free,[21] (c) microwave-assisted reflux,[22] and (d) ultrasound-assisted synthesis.[23]
Copyright permission granted by the American Chemical Society.

can be accommodated in an octahedral environment and various transition metal cations can be doped into the mixed valent OMS-2 structure as a solid solution. Many reports in the literature show single and multiple framework substitutions of transition metal cations such as Ce^{4+}, Co^{2+}, Cu^{+2}, Cr^{3+}, Fe^{3+}, In^{3+}, Mo^{6+}, Ni^{2+}, W^{6+}, and Zn^{2+}.[3,24–32] Introduction of foreign ions into the framework causes distortions and stress in the structure. These imperfections alter the acidity, porosity and adsorption properties of the material, hence they become more active in catalysis as compared to pristine materials.[33–37]

This extreme flexibility in starting materials, synthesis technique, and control over morphology, porosity, particle size, and metal doping offer OMS-2 a huge advantage in catalysis. The next part of this summary will focus on the effect of these properties on a wide variety of catalytic reactions, where OMS-2 was used as the catalyst.

13.3 Catalysis

13.3.1 Selective Oxidation and Fine Chemical Synthesis

Selective catalysis has gained a great deal of research interest due to the many economical as well as environmental advantages of the process. OMS-2 is an excellent catalyst that has the proven ability to catalyze both product and substrate selective reactions.[6,7,38] OMS-2 has shown the ability to selectively oxidize many organic functional groups including alcohols, alkenes, thiols, and amines.[7,9,33–43] Son et al. presented a selective alcohol oxidation to an aldehyde without any over oxidation to carboxylic acids, Scheme 13.1.[38] The mixed valent Mn active sites for the selective oxidation were located inside the tunnels, due to the lower activity observed towards larger substrates.

A more elaborate mechanistic study on alcohol oxidation was done using isotope labeled oxygen (^{18}O) by Makwana et al.[9] Selective aerobic benzyl alcohol oxidation was carried out as the model reaction, in which 100% conversion as well as 100% selectivity was achieved towards the aldehyde. A further study confirmed the active participation of structural oxygen during oxidation on OMS-2 catalysts, which were later replenished by atmospheric oxygen. The kinetic data also suggested the presence of a redox cycle between Mn^{2+} and Mn^{4+} that drives the oxidation of the substrate, Scheme 13.2. All of the above observations led to the conclusion that the

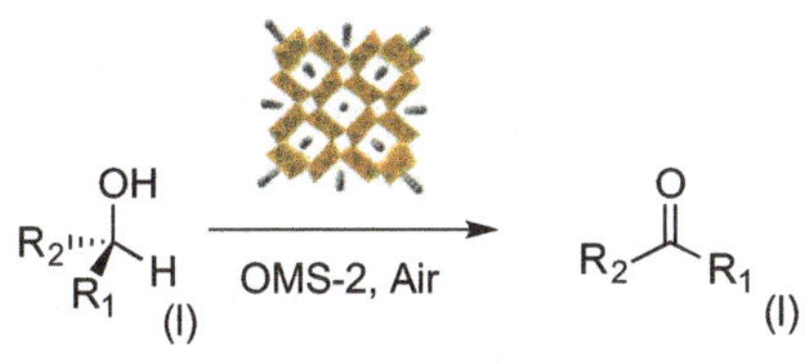

Scheme 13.1

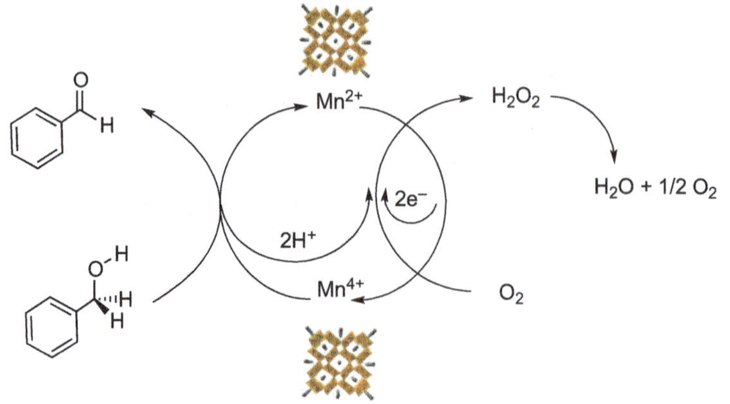

Scheme 13.2

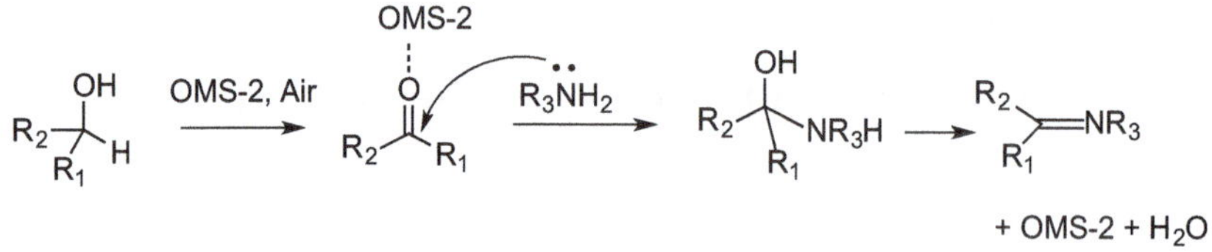

Scheme 13.3

Mars van Krevelen mechanism is active on OMS-2 materials during oxidations.[1,9,20]

Sithambaram et al. further utilized this excellent catalytic activity and selectivity of OMS-2 towards alcohol oxidation in tandem catalysis.[20] The team synthesized imines directly from alcohols and amines. The reaction progresses through a concerted fashion, Scheme 13.3, where selective oxidation of alcohols to aldehydes through a Mars van Krevelen mechanism occurs followed by nucleophilic attack from the amine to the carbonyl carbon producing an imine bond. Conversion and selectivity towards the imine of up to 100% was observed and also the reaction was found to be valid for a wide variety of alkyl and aryl substrates.

Advancement of this tandem imine formation led to another study by Sithambaram et al. on the synthesis of quinoxalines, Scheme 13.4, which are nitrogen-containing heterocyclic compounds, essential intermediates in the fine chemical industry.[44]

This reported process was highly efficient (100% yield in 1 h), had good substrate generality, and used mild reaction conditions. The catalyst, OMS-2, was also recycled after reaction without major structural changes or deactivation.

Another transformation important in fine chemical synthesis catalyzed by OMS-2 is hydrocarbon oxidation, including both alkanes and alkenes. Ghosh et al. reported the liquid-phase epoxidation of olefins using porous OMS-2

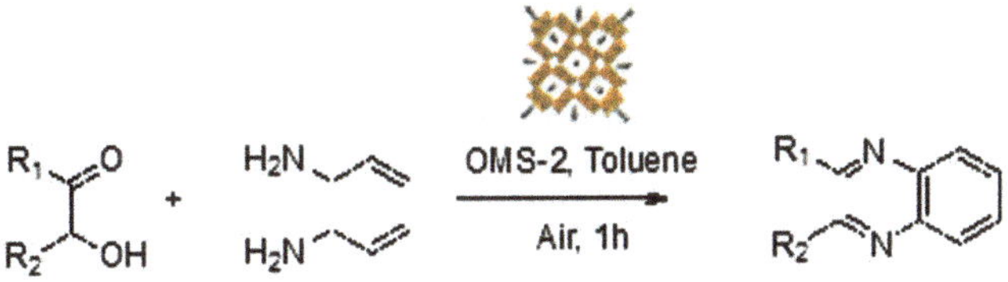

Scheme 13.4

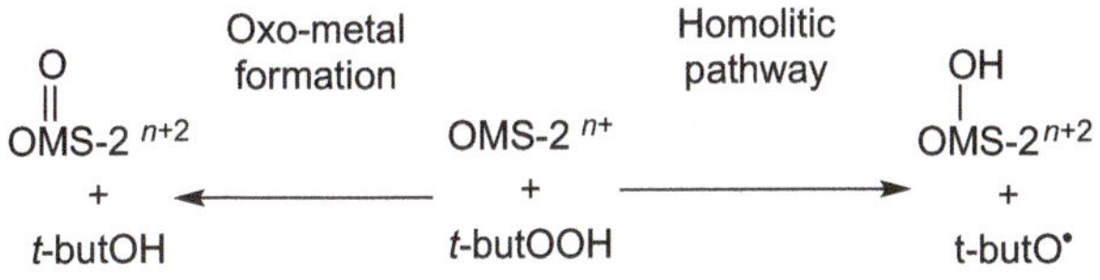

Scheme 13.5

and TBHP (*tertiary*-butyl hydroperoxide), Scheme 13.5.[45] The reaction gave good conversion with up to 100% selectivity towards epoxide at low temperature (60 °C). Significant conversion was obtained at temperatures as low as room temperature and catalyst amounts as low as 0.3 mol%. In order to understand the reaction mechanism, the reaction was carried out in the presence of radical scavengers/inhibitors. Two different radical scavengers/inhibitors were used in two separate reactions. When inhibitor 4-Oxo TEMPO (4-oxo-2,2,6,6-tetramethylpiperidine-1-oxyl) was used, the conversion decreased from 59 to 37% and in the presence of radical inhibitor IONOL (2,6-di-*tert*-butyl-4-methylphenol), the conversion of cyclooctene was reduced from 59% to 22.5%.

The results confirmed that the epoxidation occurs on OMS-2 surfaces *via* two co-operative mechanisms: oxo-metal formation, and a homolytic pathway due to the presence of Mn^{2+}, Mn^{3+}, and Mn^{4+} oxidation states (Scheme 13.5).[45] A follow up to this study was done by Ghosh *et al.* for a deeper understanding of the involvement of the Lewis and Brønsted acid sites and also the effect of the preparation method of OMS-2 catalysts, which we discussed previously in this chapter. OMS-2 materials synthesized by five different synthetic techniques were used in the study, namely reflux (OMS-2$_R$), hydrothermal (OMS-2$_{HY}$), high temperature (OMS-2$_{HT}$), solvent free (OMS-2$_{SF}$), and microwave (OMS-2$_{MW}$). The results of this study show that the acidity of the material together with porosity play essential roles. The mesopore volume of OMS-2 synthesized using different methods ranged from 0.48–0.027 cm^3 g^{-1}. The results also showed that the optimum mesopore volume for the styrene epoxidation was 0.29 cm^3 g^{-1}, which corresponds to OMS-2$_R$.

13.3.2 C–H Activation

The petroleum industry serves as the primary source of raw materials for the fine chemical industry; functionalization of these raw chemicals, especially alkanes, requires very high energy due to the inherent inertness of the C–H bond.[46] Hence, a catalytic process for the activation of the C–H bond is highly desirable. There are many examples of OMS-2 materials in the literature for the activation of C–H bonds and functionalized compounds such as 9-fluorene,[42] cyclohexane,[34] and, in some instances, toxic chemicals such as BTX (benzene, toluene, and xylene) were totally oxidized to CO_2.[7] These functionalized compounds have applications in the energy, medical, and environmental fields.[34,42] Scheme 13.6 shows a study by Opembe et al., where 9H-fluorene was oxidized to 9-fluorenone.[42]

The process developed by Opembe et al. used air as an oxidant, unlike the other processes, which used peroxide, oxygen, and other activation methods, which are harmful.[47–49] Excellent conversions (up to 99%) were observed with 100% selectivity towards 9-fluorenone. Furthermore, the reaction mechanism and kinetics were studied by isotope-labeled [D_{10}]-fluorene. The results show the reaction occurs via a late transition state in which C–H bond cleavage is the rate-determining step. Further, the mechanism was proposed to be a free radical mechanism, as shown in Scheme 13.6.

Recently, Kumar et al. developed a catalytic route for the oxidation of cyclohexane using OMS-2 as a catalyst.[34] This process is highly important in industry because the products from cyclohexane oxidation—cyclohexanol and cyclohexanone—are starting materials for Nylon and other value-added chemicals.[50–54] In this study, the acidity of the OMS-2 catalyst was altered by using an ion-exchange procedure involving concentrated nitric acid. The high

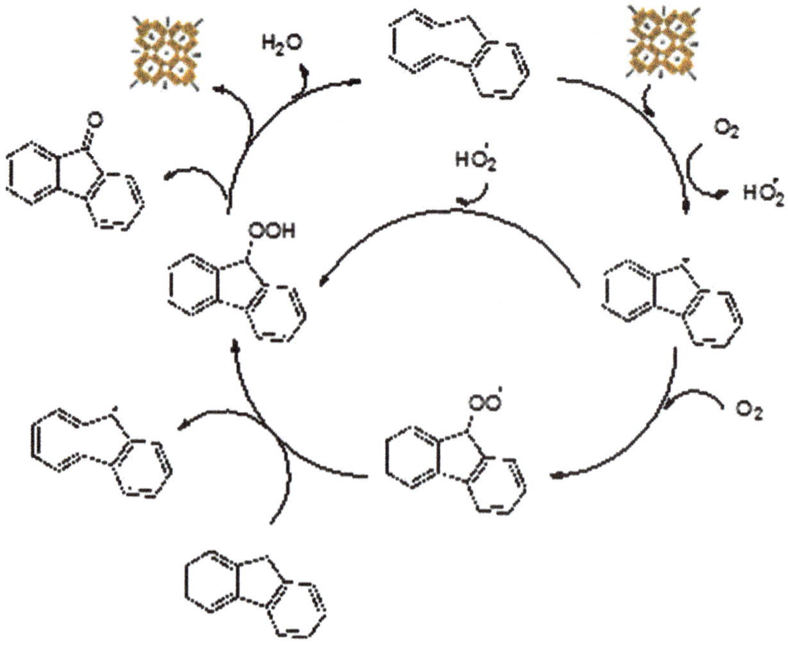

Scheme 13.6

H^+ concentration in HNO_3 drives the exchange of K^+ ions in the tunnels of OMS-2 with H^+ ions in solution. The results show that after four exchange cycles in concentrated HNO_3, 64% of K^+ in the structure of OMS-2 was exchanged with H^+, and this ion-exchanged catalyst was named as H-K-OMS-2. The ion-exchanged H-K-OMS-2 showed a remarkable increase in conversion (59.9%) compared to K-OMS-2 (25.3%). The selectivity towards cyclohexanone was also increased up to 56.3%, where K-OMS-2 gave only 24.1%.

13.3.3 CO₂ Activation

Utilization of CO_2—the most abundant C source in nature—as a raw material for fine chemical synthesis has gained immense research attention.[55-58] However, due to the high inertness or the stability of CO_2 molecules, a great amount of energy input is required to activate CO_2 in order to be utilized in a chemical reaction.[56,59]

Hu et al. developed a OMS-2 catalyzed electrochemical and thermal hybrid activation of CO_2 (Scheme 13.7) to produce paraformaldehyde, a useful chemical in the polymer and pesticide industry.[56] In this work, OMS-2 was used as a support for Pt nanoclusters, deposited using chemical vapor deposition (CVD). The metal/metal oxide interface was activated by means of a DC current, and the reaction proceeds through electron and ionic conduction from metal to metal oxide.

Since OMS-2 is a good conductor for electrons as well as ions and also possesses mixed valency,[1,5,9,60] when OMS-2 was used as a support, the reaction temperature was significantly reduced (350–450 °C) as compared to calcia-stabilized zirconia (600–900 °C).

Another catalytic transformation on OMS-2 utilizing CO_2 is ethane dehydrogenation to ethylene as reported by Jin et al.[61] In this study, OMS-2 yields a higher amount of ethane (up to 66%) compared to a conventional Cr-ZSM-5-type catalyst (36%). CO_2 utilization was also very effective when

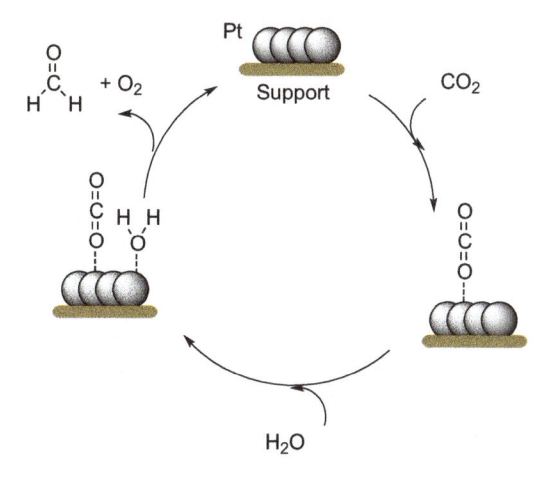

Scheme 13.7

using OMS-2 (50%) as compared to chromium catalysts (17%). The reaction mechanism was proposed to consist of two steps: (1) catalytic dehydrogenation of ethane and (2) a reverse water–gas shift reaction (RWGS). Carbon dioxide gas plays two important roles in the mechanism.

$$C_2H_{6(g)} \longrightarrow C_2H_{4(g)} + H_{2(g)} \qquad (1)$$

$$H_{2(g)} + CO_{2(g)} \longrightarrow H_2O_{(g)} + CO_{(g)} \qquad (2)$$

$$\overline{C_2H_{6(g)} + CO_{2(g)} \longrightarrow C_2H_{4(g)} + CO_{(g)} + H_2O_{(g)}}$$

Since $CO_{2(g)}$ utilizes $H_{2(g)}$ produced from reaction (1) in a RWGS reaction, that prevents further reaction between $H_{2(g)}$ and $C_2H_{6(g)}$ to give unwanted byproduct $CH_{4(g)}$. $CO_{2(g)}$ also prevents catalyst deactivation by coke formation, by acting as an oxidizing agent on the OMS-2 surface.

13.3.4 Environmental and Green Chemistry

Increasing concerns about environmentally friendly and sustainable processes have driven heterogeneous catalytic research towards improving air, water, and soil quality in recent years.[62–67] OMS-2 has shown environmental remediation ability in many instances, such as removal of organic dyes, phenol, and oil from water, as well as oxidation of carbon monoxide and toxic VOCs (volatile organic compounds).[7,17,36,68,69] Organic dyes such as methylene blue are highly toxic to living beings.[70–73] Therefore, removal of these dyes by means of a green process is highly desirable in our modern world. Green decomposition of organic dyes by using OMS-2 as a catalyst has been reported by Sriskandakumar et al.[36] This study shows that OMS-2 and metal-doped OMS-2 are highly active in degrading methylene blue (MB), an organic dye. Among metal-doped OMS-2 materials, Mo-doped catalysts showed the highest decomposition activity (98%). Lower pH and higher temperatures favored the decomposition reaction.

Another recent advancement in wastewater treatment using OMS-2 is adsorption and removal of phenol as reported by Hu et al.[68] Phenol is a polar compound that is highly miscible with water, and is a byproduct mostly from the petroleum, pharmaceutical, pesticide, and paper industries.[68,74–79] Hu et al. used Cu^{2+}, Co^{3+} and, Ce^{4+} doped OMS-2 materials in this study. During the phenol removal, phenol was first adsorbed onto the catalyst to form a phenolic polymer; this phenolic polymer then underwent oxidation to $CO_{2(g)}$. The results show that the Co, Ce, and Cu doped OMS-2 materials gave 99.8, 98.9, and 97.9% conversion of phenol respectively. Co-OMS-2 showed higher removal efficiency 3.7 kg phenol/h.kg catalyst, than the conventional activated carbon (AC) 0.002 kg phenol/h.kg AC.

The recent British Petroleum *Deepwater Horizon* explosion and oil spill in the Gulf of Mexico triggered researchers to develop porous materials that can be used in oil spill cleanup.[80] Due to the adverse effects of oil on aquatic life and ecology, spillage of any extent needs immediate attention and cleanup. Yuan et al. have reported oil removal via adsorption onto a modified super hydrophobic OMS-2 membrane.[17] A porous OMS-2 membrane was synthesized according to literature reported methods[81] and then the membrane was further modified using a polydimethyl siloxane (PDMS) polymer.

The resulting material showed an exceptional wettability profile, which can be altered to be super hydrophobic or super hydrophilic by changing the temperature. The removal efficiency of OMS-2 was 20× the weight of the membrane, which is much higher than the conventional adsorbents (Figure 13.4). This OMS-2 membrane can be produced easily to treat a

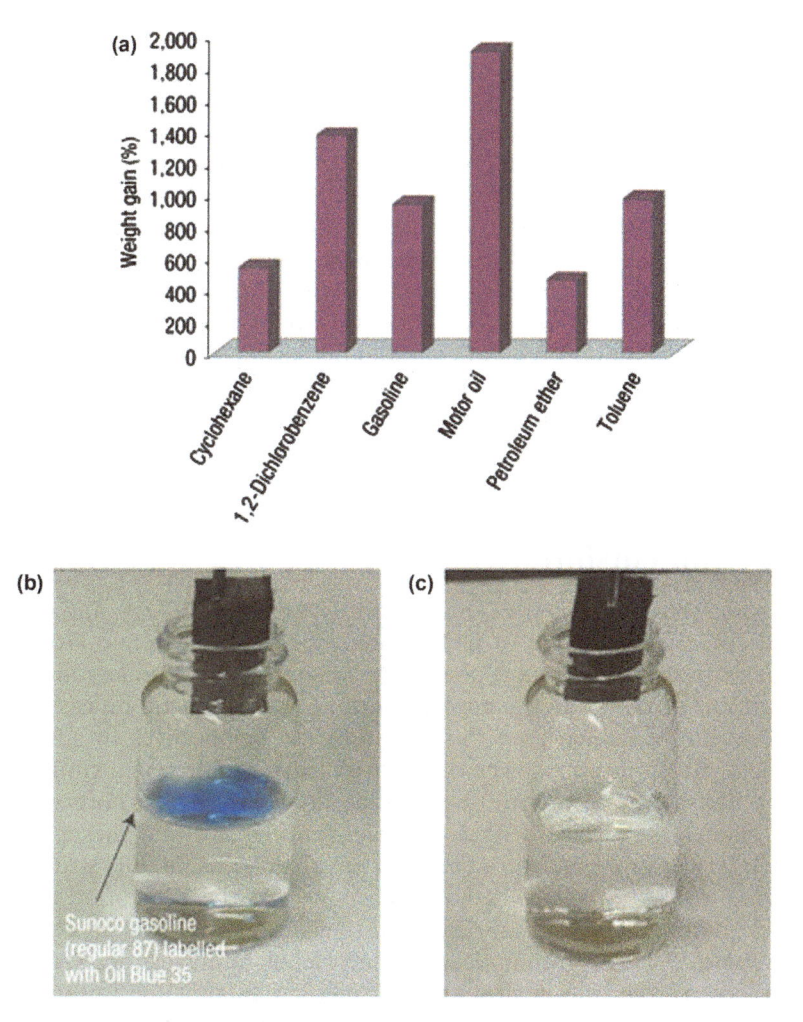

Figure 13.4 (a) Absorption capacity of polymer-modified OMS-2 membrane. (b) Colored gasoline on water before adsorption. (c) After removal of gasoline by adsorption onto OMS-2 membrane.
Copyright permission granted by the Nature Publishing Group.

large-scale oil spill and also easily recovered. Therefore, OMS-2 materials are highly desirable in large-scale industrial applications for wastewater treatment.

Air quality has declined at a rapid rate in recent years as a result of heavy use of fossil fuels, petroleum-based chemicals, and global industrialization.[82,83] VOCs (volatile organic compounds), carbon monoxide, nitrogen oxides, and particulate matter generation in the lower atmosphere are the primary cause of smog, which relates to high rates in cardiovascular disease in human beings.[82-84] Genuino et al.[7] have reported a catalytic route to oxidize BTX (benzene, toluene, and xylene), a group of VOCs that are highly toxic, to CO_2.[7] The reactivity of the substrates studied was xylenes < ethylbenzene < benzene < toluene. The process developed was highly selective; other byproducts, such as CO or partially oxidized VOC, were not observed. OMS-2 showed higher activity than commercial MnO_2. Temperature programed desorption (TPD) studies of xylenes (*ortho*, *meta*, and *para*-xylene) showed peak maxima in temperature in the order of *para* < *meta* < *ortho*-xylene suggesting a shape selectivity in the OMS-2 catalysts. The higher catalytic activity of the OMS-2 catalyst was due to the higher hydrophobicity (0.91), pore volume (0.46 cm^3 g^{-1}), pore diameter (20.1 nm), and higher oxygen mobility. This study further suggests the potential application of OMS-2 in air and soil purification from soil–vapor extraction, automobile exhaust treatment, and gasoline-stripping of soil. Another example of OMS-2 in air purification is low temperature $CO_{(g)}$ oxidation. Xia et al. reported a catalytic carbon monoxide oxidation route using Ag^+, Co^{2+}, and Cu^{2+} doped OMS-2.[69] Doped catalysts were very stable under a reducing atmosphere for extended time on stream.

13.4 Conclusion

This review summarizes the synthesis, structure, morphology, and catalytic activity of a unique class of OMS materials—OMS-2 systems. From their initial synthesis in 1993 until now, OMS-2 materials have undergone a large amount of research throughout the world. Novel synthetic processes have been developed, which gave this material new perspectives in adsorption and catalytic applications. The different synthetic routes yield similar crystalline phases (cryptomelane) with different surface areas, pore volumes, pore diameters, surface defects, porosity (both micro and mesoporosity), particle sizes, and catalytic activities. Metal dopants in the structure of OMS-2 increased the catalytic activities as well as the adsorption capacity. OMS-2 materials can be fabricated as powders as well as free-standing membranes for different applications such as oil spill cleanup. The OMS-2 catalyst is a non-toxic environmentally benign catalyst for fine chemical synthesis, oxidation, environmental remediation, and C–H and CO_2 activation.

Future studies of OMS-2 materials will involve desulfurization, tandem processes, and their use in more environmental applications such as active materials in particulate filters. Research will also continue along the avenues of formation of novel morphologies in thin OMS-2 films for electrocatalysis.

Acknowledgments

We acknowledge the Geosciences and Biosciences Division, Office of the Basic Energy Sciences, Office of Science, and U.S. Department of Energy for support of our research on manganese octahedral molecular sieves and other porous materials, under grant DE-FG02-86ER13622-A000.

References

1. S. L. Suib, *J. Mater. Chem.*, 2008, **18**, 1623–1631.
2. Y. F. Shen, R. P. Zerger, R. N. DeGuzman, S. L. Suib, D. I. McCurdy and C. L. O'Young, *Science*, 1993, **260**, 511–515.
3. R. N. DeGuzman, Y. Shen, E. J. Neth, S. L. Suib, C. O'Young, S. Levine and J. M. Newsam, *Chem. Mater.*, 1994, **6**, 815–821.
4. T. López, E. Ortiz, M. Alvarez, J. Manjarrez, M. Montes, P. Navarro and J. A. Odriozola, *Mater. Chem. Phys.*, 2010, **120**, 518–525.
5. R. N. De Guzman, Y. F. Shen, B. R. Shaw, S. L. Suib and C. L. O'Young, *Chem. Mater.*, 1993, **5**, 1395–1400.
6. L. Espinal, W. Wong-Ng, J. A. Kaduk, A. J. Allen, C. R. Snyder, C. Chiu, D. W. Siderius, L. Li, E. Cockayne, A. E. Espinal and S. L. Suib, *J. Am. Chem. Soc.*, 2012, **134**, 7944–7951.
7. H. C. Genuino, S. Dharmarathna, E. C. Njagi, M. C. Mei and S. L. Suib, *J. Phys. Chem. C*, 2012, **116**, 12066–12078.
8. J. Li, R. Wang and J. Hao, *J. Phys. Chem. C*, 2010, **114**, 10544–10550.
9. V. D. Makwana, S. Young-Chan, A. R. Howell and S. L. Suib, *J. Catal.*, 2002, **210**, 46–52.
10. M. Yoshimura and K. Byrappa, *J. Mater. Sci.*, 2008, **43**, 2085–2103.
11. D. Wang, T. Xie and Y. Li, *Nano Res.*, 2009, **2**, 30–46.
12. G. Qiu, H. Huang, S. Dharmarathna, E. Benbow, L. Stafford and S. L. Suib, *Chem. Mater.*, 2011, **23**, 3892–3901.
13. J. Yuan, W. Li, S. Gomez and S. L. Suib, *J. Am. Chem. Soc.*, 2005, **127**, 14184–14185.
14. J. Yuan, K. Laubernds, J. Villegas, S. Gomez and S. L. Suib, *Adv. Mater.*, 2004, **16**, 1729–1732.
15. J. C. Villegas, L. J. Garces, S. Gomez, J. P. Durand and S. L. Suib, *Chem. Mater.*, 2005, **17**, 1910–1918.
16. H. M. Galindo, Y. Carvajal, C. Njagi, R. A. Ristau and S. L. Suib, *Langmuir*, 2010, **26**, 13677–13683.
17. J. Yuan, X. Liu, O. Akbulut, J. Hu, S. L. Suib, J. Kong and F. Stellacci, *Nat. Nanotechnol.*, 2008, **3**, 332–336.
18. X. Li, B. Hu, S. L. Suib, Y. Lei and B. Li, *J. Power Sources*, 2010, **195**, 2586–2591.

19. H. Huang, S. Sithambaram, C. Chen, C. K. Kithongo, L. Xu, A. Iyer, H. F. Garces and S. L. Suib, *Chem. Mater.*, 2010, **22**, 3664–3669.
20. S. Sithambaram, R. Kumar, Y. Son and S. L. Suib, *J. Catal.*, 2008, **253**, 269–277.
21. Y. Ding, X. Shen, S. Sithambaram, S. Gomez, R. Kumar, V. Crisostomo, S. L. Suib and M. Aindow, *Chem. Mater.*, 2005, **17**, 5382–5389.
22. E. K. Nyutu, C. Chen, S. Sithambaram, V. Crisostomo and S. L. Suib, *J. Phys. Chem. C*, 2008, **112**, 6786–6793.
23. S. Dharmarathna, C. K. King'ondu, W. Pedrick, L. Pahalagedara and S. L. Suib, *Chem. Mater.*, 2012, **24**, 705–712.
24. J. Cai, J. Liu, W. Willis and S. L. Suib, *Chem. Mater.*, 2001, **13**, 2413–2422.
25. C. Calvert, R. Joesten, K. Ngala, J. Villegas, A. Morey, X. Shen and S. L. Suib, *Chem. Mater.*, 2008, **20**, 6382–6388.
26. X. Chen, Y. Shen, S. L. Suib and C. O'Young, *J. Catal.*, 2001, **197**, 292–302.
27. A. Hashem, H. Mohamed, A. Bahloul, A. Eid and C. Julien, *Ionics*, 2008, **14**, 7–14.
28. R. Jothiramalingam, B. Viswanathan and T. Varadarajan, *Mater. Chem. Phys.*, 2006, **100**, 257–261.
29. R. Jothiramalingam, B. Viswanathan and T. Varadarajan, *J. Mol. Catal. A: Chem.*, 2006, **252**, 49–55.
30. C. K. Kingóndu, N. Opembe, C. Chen, K. Ngala, H. Huang, A. Iyer, H. Garcés and S. L. Suib, *Adv. Funct. Mater.*, 2011, **21**, 312–323.
31. A. Onda, A. Hara, K. Kajiyoshi and K. Yanagisawa, *Appl. Catal., A*, 2007, **321**, 71–78.
32. R. Wang and J. Li, *Catal. Lett.*, 2009, **131**, 500–505.
33. A. Iyer, H. Galindo, S. Sithambaram, C. Kingóndu, C. Chen and S. L. Suib, *Appl. Catal., A*, 2010, **375**, 295–302.
34. R. Kumar, S. Sithambaram and S. L. Suib, *J. Catal.*, 2009, **262**, 304–313.
35. S. Sithambaram, L. Xu, C. Chen, Y. Ding, R. Kumar, C. Calvert and S. L. Suib, *Catal. Today*, 2009, **140**, 162–168.
36. T. Sriskandakumar, N. Opembe, C. Chen, A. Morey, C. Kingóndu and S. L. Suib, *J. Phys. Chem. A*, 2009, **113**, 1523–1530.
37. H. Huang, C. Chen, L. Xu, H. Genuino, G. Martinez, H. Garces, L. Jin, C. Kithongo and S. L. Suib, *Chem. Commun.*, 2010, **46**, 5945–5947.
38. Y. Son, V. Makwana, A. R. Howell and S. L. Suib, *Angew. Chem., Int. Ed.*, 2001, **113**, 4410–4413.
39. S. Dharmarathna, C. Kingóndu, L. Pahalagedara, C. Kuo, Y. Zhang and S. L. Suib, *Appl. Catal., B*, 2014, **147**, 124–131.
40. S. Sithambaram, E. K. Nyutu and S. L. Suib, *Appl. Catal., A*, 2008, **348**, 214–220.
41. N. Opembe, C. K. Kingóndu and S. L. Suib, *Catal. Lett.*, 2010, **142**, 427–432.
42. N. Opembe, Y. Son, T. Sriskandakumar and S. L. Suib, *ChemSusChem*, 2008, **1**, 182–185.
43. R. Ghosh, X. Shen, J. Villegas, Y. Ding, K. Malinger and S. L. Suib, *J. Phys. Chem. B*, 2006, **110**, 7592–7599.

44. S. Sithambaram, Y. Ding, W. Li, Y. Shen, F. Gaenzler and S. L. Suib, *Green Chem.*, 2008, **10**, 1029–1032.
45. R. Ghosh, Y. Son, V. Makwana and S. L. Suib, *J. Catal.*, 2004, **224**, 288–296.
46. C. Jia, T. Kitamura and Y. Fujiwara, *Acc. Chem. Res.*, 2001, **34**, 633–639.
47. M. A. Campo and R. Larock, *Org. Lett.*, 2000, **2**, 3675–3677.
48. G. Qabaja and G. Jones, *Tetrahedron Lett.*, 2000, **41**, 5317–5320.
49. Y. Sprinzak, *J. Am. Chem. Soc.*, 1958, **80**, 5449–5455.
50. J. Long, L. Wang, X. Gao, C. Bai, H. Jiang and Y. Li, *Chem. Commun.*, 2012, **48**, 12109–12111.
51. B. Hereijgers and M. Weckhuysen, *J. Catal.*, 2010, **270**, 16–25.
52. J. Thomas, R. Raja, G. Sankar and R. Bell, *Acc. Chem. Res.*, 2001, **34**, 191–200.
53. G. Bellussi and C. Perego, *CATTECH*, 2000, **4**, 4–16.
54. H. Tsunoyama, Y. Liu, T. Akita, N. Ichikuni, H. Sakurai, S. Xie and T. Tsukuda, *Catal. Surv. Asia*, 2011, **15**, 230–239.
55. M. Cokoja, C. Bruckmeier, B. Rieger, W. Herrmann and E. Kühn, *Angew. Chem., Int. Ed.*, 2011, **50**, 8510–8537.
56. B. Hu, V. Stancovski, M. Morton and S. L. Suib, *Appl. Catal., A*, 2010, **382**, 277–283.
57. T. Leon, A. Correa and R. Martin, *J. Am. Chem. Soc.*, 2013, **135**, 1221–1224.
58. J. Spivey, E. Wilcox and G. Roberts, *Catal. Commun.*, 2008, **9**, 685–689.
59. T. Sakakura, J. Choi and H. Yasuda, *Chem. Rev.*, 2007, **107**, 2365–2387.
60. S. L. Suib, *Acc. Chem. Res.*, 2008, **41**, 479–487.
61. L. Jin, J. Reutenauer, N. Opembe, M. Lai, D. Martenak, S. Han and S. L. Suib, *ChemCatChem*, 2009, **1**, 441–444.
62. J. Armor, *Appl. Catal., B*, 1992, **1**, 221–256.
63. G. Centi, P. Ciambelli, S. Perathoner and P. Russo, *Catal. Today*, 2002, **75**, 3–15.
64. P. Forzatti, *Catal. Today*, 2000, **62**, 51–65.
65. K. Pirkanniemi and M. Sillanpa, *Chemosphere*, 2002, **48**, 1047–1060.
66. V. Polshettiwar and R. Varma, *Green Chem.*, 2010, **12**, 743–754.
67. Y. Wada, H. Yin and S. Yanagida, *Catal. Surv. Jpn.*, 2002, **5**, 127–138.
68. B. Hu, C. Chen, S. Frueh, L. Jin, R. Joesten and S. L. Suib, *J. Phys. Chem. C*, 2010, **114**, 9835–9844.
69. G. Xia, Y. Yin, W. Willis, J. Wang and S. L. Suib, *J. Catal.*, 1999, **185**, 91–105.
70. S. Auerbach, D. Bristol, J. Peckham, G. Travlos, C. Hébert and R. Chhabra, *Food Chem. Toxicol.*, 2010, **48**, 169–177.
71. M. Oz, D. Lorke and G. Petroianu, *Biochem. Pharmacol.*, 2009, **78**, 927–932.
72. M. Oz, D. Lorke, M. Hasan and D. Petroianu, *Med. Res. Rev.*, 2011, **31**, 93–117.
73. M. Rowley, K. Riutort, D. Shapiro, J. Casler, E. Festic and M. Freeman, *Neurocrit. Care*, 2009, **11**, 88–93.

74. A. Babuponnusami and K. Muthukumar, *Environ. Sci. Pollut. Res.*, 2012, 1–10.
75. H. Jiang, Y. Fang, Y. Fu and Q. Guo, *J. Hazard. Mater.*, 2003, **101**, 179–190.
76. R. Juang, W. Huang and Y. Hsu, *J. Hazard. Mater.*, 2009, **164**, 46–52.
77. W. Kujawski, W. Warszawski, W. Ratajczak, T. Porelbski, W. Capaa and I. Ostrowska, *Desalination*, 2004, **163**, 287–296.
78. F. Trabelsi, A. Lyazidi, B. Ratsimba, A. Wilhelm, H. Delmas, P. Fabre and J. Berlan, *Chem. Eng. Sci.*, 1996, **51**, 1857–1865.
79. T. Viraraghavan and A. Maria, *J. Hazard. Mater.*, 1998, **57**, 59–70.
80. R. Kerr, E. Kintisch and E. Stokstad, *Science*, 2010, **328**, 674–675.
81. J. Yuan, K. Laubernds, J. Villegas, S. Gomez and S. L. Suib, *Adv. Mater.*, 2004, **16**, 1729–1732.
82. H. Akimoto, *Science*, 2003, **302**, 1716–1719.
83. A. Nel, *Science*, 2005, **308**, 804–806.
84. H. Kim and J. Bernstein, *Curr. Allergy Asthma Rep.*, 2009, **9**, 128–133.

Subject Index

π-electrons 194
ζ potential 102

ABA copolymer 197
ABTA (2,2'-azinobis(3-ethylbenzo-thiazoline-6-sulfonic acid)) 198
Acetobacter xylinum 138
acetonitrile–water solvent 143
acetophenones, brominated 89, 91, 92
4-acetylphenylboronic acid 141, 142
4-acetylphenylchloride 137–138
acid chlorides 126
acrylic acid *n*-butyl ester 144
active site characteristics 7
active supports 161, 162
aerobic oxidation 159, 160, 165
air 207, 242, 246
alcohols 159, 160, 239
aldehydes 75–76, 93, 94, 96, 243
aliphatic aldehydes 93, 94, 96
alkenes *see* olefins (alkenes)
alkenyl halides 134
alkenylboranes 134
alkylamine ligands 19
alloys for templating synthesis 23
alumina membranes in two-compartment reactors 117, 118
alumina supports 70–71
aluminium oxides 36, 37, 161, 162
aluminosilicate frameworks 92, 93
  HZSM-5 192

ambient pressure X-ray photoelectron spectroscopy (APXPS) 13
amines
  aldehydes amination 94–96
  amine-borane 60–63
  amine-rich ionic liquids 129
  noble metal nanoparticle synthesis 59–60
  OMS-catalysed selective oxidation 239
aminoalcohol-stabilised ruthenium nanoparticles 50–51
p-aminophenol 23, 35
3-aminopropyltriethoxysilane (APTS) 60–63
ammonia 34–35
ammonium formate 93, 94
ammonium surfactant-capped rhodium nanoparticles 99–109
  asymmetric catalysis 105–108
  biphasic hydrogenation catalysis 100–105
anastase nanocrystals 19, 20, 34
aniline 211, 213
anisotropic shapes 172
applications of nanocatalysts
  clock reactions 216–217
  industrial catalysis 1, 2, 6, 31
  oil spill cleanup 237, 244–246
  overview 173
APTS (3-aminopropyltriethoxysilane) 60–63
APXPS (ambient pressure X-ray photoelectron spectroscopy) 13

arenes
- asymmetric nanocatalysis 105
- derivatives biphasic hydrogenation with HEA16Cl-capped rhodium colloids 103, 104
- hydrogenation reactions 99
- prochiral 107–108

aryl halides
- palladium-catalysed carbon–carbon coupling reactions 119, 121, 141
  - Heck coupling reaction 147
  - Stille coupling reaction 126–130, 132, 133
  - Suzuki coupling reaction 134, 136, 137, 138, 140, 141, 143

arylboronic acid 143
ascorbic acid 211
assembly-dispersion mechanism 36
asymmetric nanocatalysis *see* enantiocontrol/selectivity
atom transfer radical polymerisation (ATRP) 197
atom-leaching mechanism 116–126, 133, 141
ATRP (atom transfer radical polymerisation) 197
attenuated total reflectance (ATR) 185-7
2,2′-azinobis(3-ethylbenzothiazo-line-6-sulfonic acid) (ABTA) 198

bacterial cellulose (BC) 138
barium salts 159, 167
BE-O (binding energy for oxygen) 225, 226
bead templates 136, 174, 177, 178
Belousov–Zhabotinsky (BZ) reaction 205
benzene, biphasic hydrogenation 103
benzene/toluene/xylene (BTX) 242
benzyl alcohol oxidation 72–75, 76–78, 239–240
Berthollet's reversible chemical reactions 203
bimetallic systems 13, 123
BINAP (2,2′-bis(diphenylphosphino)-1,1′-binaphthyl) 40, 41–42
binding energy for oxygen (BE-O) 225, 226
biothiophene 135–136
biphasic hydrogenation 99–109
biphasic liquid–liquid hydrogenation 54
4-biphenylbenzoic acid 128
4-biphenylcarboxylic acid 126, 127, 128, 132
bisulfite/iodate clock reaction 204
block copolymers 135, 196, 197, 198
blue bottle experiment 207
Bohr exciton radius 14, 15
borane compounds 134
borohydride reduction
- eosin Y dye 182, 183, 184, 185
- 4-nitrophenol 180, 183–184, 185
- silver nitrate 209, 210

bottom-up approaches 48
see also organometallic approach
BP *Deepwater Horizon* oil spill 244–246
brain tissue 217, 236
Bray–Liebhafsky reaction 205–206
Briggs–Rauscher reaction 206–207
brominated acetophenones 89
1-bromo-2-methylnaphthylene 41
1-bromo-4-nitrobenzene 139
bromobenzene 134–135, 140, 142
4-bromobenzoic acid 133
bromotoluene 130, 139, 148
bulk semiconducting titanium dioxide 31, 32
n-butyl-formylcinnamate 144
n-butyl-*p*-iodocinnamate 146

Subject Index

n-butylacrylate 117, 118, 144, 145, 146
n-butylamine 22
t-butylstyrene 147

cadmium/selenium quantum dots 15
caesium salts 159, 168
cage effect 172–87
    active inner/inactive outer surfaces 180–182
    assembly on substrates 176–179
    hollow nanostructure characteristics 179–185
    hollow/solid activities comparison 182–183
    optical properties 184–185
    overview 172–174
    single shell hollow/double shell comparison 183–184
    spectroscopic studies 185–187
    synthetic approaches 174–176
calcium chloride 203
capillary inclusion method 21
capping agents
    see also individual capping agents; stabilising agents
    ammonium surfactants 99–109
    PVP-capped gold nanospheres 177, 178
    silver nanoparticles 209, 210
carbene-stabilised ruthenium nanoparticles 51–52
carbon dioxide 221–223, 228–230, 243–244
carbon material supports 75–78
carbon monoxide
    catalyst poison 14
    oxidation 72–75, 157–167
        on Au/LaPO$_3$ 162–163
        on gold 18–19, 162–163
    water-gas shift reaction 221–223, 228–230
carbon nanotubes 75–76, 198–199

carbonyl compounds hydrogen-transfer reduction 88–94
carbon–carbon coupling reactions
    applications 112
    bimetallic palladium–gold nanoparticle system 123
    colloidal catalysts 63–67
    greener methods 112
    Heck reaction 144–148
    mechanisms 116–126
    overview 3
    palladium catalysed 112–149
    Stille reaction 126–134
    Suzuki reaction 134–144
carbon–hydrogen bond activation 241–243
CASTEP numerical basis sets 220
catalysis definition 2
catalytic efficiency 179–185
    colloidal metallic hollow nanostructures
        determining factors 179
        hollow/solid similar shape comparisons 182–183
        inner active/outer inactive surafces 180–182
        metal–organic frameworks 180
        optical properties 184–185
        single shell hollow/double shell comparisons 183–184
        spherical shape efficiency 180
cellobiose 168
ceria-supported platinum 229
cerium oxide 228–230
cerium(IV) sulfate 205
cetyl trimethylammonium bromide (CTAB)-Pt nanocrystals 21
N-cetyl-ammonium derivatives 101
cetyltrimethylammonium chloride (CTACl) 198
chain length 59, 100
channel proteins 198

Subject Index

charge transfer excited states 32
charge trapping centres 20–21
Chaudret's decomposition of metalloorganic complexes 48
chemical reduction approach 115
chemical vapour deposition (CVD) 243
chiral nanocatalysis *see* enantiocontrol/selectivity
1-chloro-4-nitrobenzene 139, 148
chloroanisole 104
4-chlorophenol 137
4-chloropyridine 137
cinchona derivatives 105, 106
cinnamaldehyde 75–76
cinnamate 146, 147
cinnamyl methyl carbonate 146
circular dichroism (CD) 41
citric acid 205
citronellal 166, 167
click chemistry 140
clock reactions 203–217
    Belousov–Zhabotinsky reaction 205
    blue bottle experiment 207
    Bray–Liebhafsky reaction 205–206
    Briggs–Rauscher reaction 206–207
    definition 204
    history 204–207
    iodine clock reaction 204
    mechanistic approach 215–216
    methylene blue 210–215
CMC (critical micellar concentration) 101
colloidal catalysis
    carbon–carbon coupling reactions 63–67
    hydroformylation reactions 67–69
    metallic hollow nanostructures 172–187
        assembly on substrates 176–179
        behaviour in catalysis 179–185
        catalytically active inner surface 180–182
        hollow/solid nanocatalysts comparison 182–183
        optical properties of plasmonic nanocatalysts 184–185
        proposed mechanism for nanocatalysis 185–187
        single/double shell hollow nanocatalysts comparison 183–184
        synthetic approaches 174–176
    noble metal nanoparticles for 46–69
    packing on crystallographic faces 2–3
    precursors for size-controlled metal nanoparticloes 35–37
colorimetric oxygen indicators 217
computational approaches
    copper nanocatalysts for WGS reactions 221–223
    density functional theory 219–231
    metal (core)–platinum shell nanocatalysts for ORR in fuel cells 223–227
    metal nanocatalysts 220–227
    method 220
    overview 4–5
    shape effects 226–227
    size effects 224–226
    supported metal nanocatalysts 227–230
confinement properties 57–60, 182, 183, 184, 185
    see also cage effect
coordination numbers (CNs) 124
copolymers 135, 196, 197, 198

Subject Index 255

copper
    copper oxide nanoparticle surface restructuring 12
    copper/zinc oxide 229
    nanocatalysts for water-gas shift reactions 221–223
    nanocrystal formation 17
    nanoparticles in clock reactions 207, 209
    photo-assisted deposition 35
core–shell nanocatalysts
    hydrogen peroxide formation 42–44
    magnetic FePt@Ti-SiO$_2$ synthesis 38–39
    model 224
    nanoreactor catalysts 195
corner atoms 9, 174, 179
cost of transition metals 84
Cotton effect 41
Coulomb's interaction 209, 210
coupling reactions *see* carbon–carbon coupling reactions
critical micellar concentration (CMC) 101
cryptomelane-type manganese oxide octahedral porous molecular sieves 235–247
crystallinity
    copper-29 catalyst crystal structure 221, 223
    domain size/structure sensitivity 10
    exposed planes 173
    packing on crystallographic faces 2–3
    palladium nanoparticles 119–123
    surface free energy of facets 11
CTAB (cetyl trimethylammonium bromide)-Pt nanocrystals 21
CTACl (cetyltrimethylammonium chloride) 198
cubo-octahedron shape analysis 9–10, 11
cuprous oxide 212, 213, 214, 215

customisability 194, 195, 212
4-cyanophenylbenzoic acid 138
cyclodextrin (CD) 39–41, 143
1,3-cyclohexadiene (CYD) 60
cyclohexane 71–72, 165, 242, 243
cyclohexanone formation 104
cyclohexyl compounds 107
cyclooctene 39, 241
Cylindrotheca fusiformis R5 peptide 130, 131, 132

data reporting standards 8–9
decanoyl chloride 145
Deepwater Horizon oil spill 244–246
definition of nanocatalysis 6
dehydrogenation reactions 60–63, 243, 244
dendrimer-encapsulated nanoparticles (DENS) 126, 127, 128, 139
dendrimer-stabilised nanoparticles (DSN) 128, 140, 144–145
dendrimers 23, 193–194
dendritic bromobenzene 142
dendritic cages 141
dendritic structures 237, 238
dendritic-phosphine-based palladium nanoparticles 128
DENS *see* dendrimer-encapsulated nanoparticles (DENS)
density functional theory (DFT) 219–231
deposition-precipitation 160
deuteration experiments 89
2,6-di-*tert*-butyl-4-methylphenol (IONOL) 241
4,4'-di-*tert*-butylbiphenyl (DTBB) 85, 86
2,6-dichloropyridine 137
diffusion flow 174
1,4-difluorobenzene 137
dihydride-type transfer hydrogenation mechanism 90
p-diiodobenzene 146
diisopropylamine 148
2,4-dimethoxyphenylboronic acid 137

N,N-dimethyl-N-alkyl-N-(hydroxyalkyl)ammonium salts 100, 101, 103
N,N-dimethyl-N-cetyl-N-(3-hydroxypropyl)ammonium chloride (HPA16Cl) 100
dimethylamine-borane (DMAB) 60–63
dimethylphenylsilane 165
4-diphenylphosphinobenzoic acid 145
diphosphate-stabilised noble metal nanoparticles 51
diphosphines 54–57
dipping method of LB film transfer 178
dissolution potential 226
2,4-di(trifluoromethyl)phenylboronic acid 137
DMAB (dimethylamine-borane) 60–63
DMF (dimethylformamide) solvent 117, 118, 119
DMol3 numerical basis sets 220
double shell nanoparticles 180, 183–184
DSN (dendrimer-stabilised nanoparticles) 128, 140, 144–145
DTBB (4,4′-di-*tert*-butylbiphenyl) 85, 86
dual hydrogenation–oxidation reactions 78
dyes 180, 181, 203–217, 244

edge atoms 9, 174, 179
EDX-SEM (energy-dispersive X-ray spectroscopy-scanning electron microscopy) mapping 175, 176
electrochemical nanoparticle synthesis method 144
electron beam lithography (EBL) 177
electron microscopy 8, 175, 176, 237
electronic structure 14–15, 198–199
electrostatic stabilisation 210
Eley–Rideal mechanism 162, 215–216

emulsions 195–198
enantiocontrol/selectivity
  ammonium surfactant-capped rhodium nanoparticles for biphasic hydrogenation 105–108
  chiral diphosphite ligands 63–65, 68
  cis/trans isomerism 22, 23, 146, 147
  hydrogenation reactions 50–51, 88–89, 90, 106–107
  optically-active ammonium salts 106, 107, 108
  plasmonic nanocatalysts 184–185
  roll-over *trans* diastereoisomer formation 104
  stereoselective arylated alkene products 134
  steric hindrance 106, 107, 108, 141
encapsulation 21, 196
endocrine disruptors 104
energy-dispersive X-ray spectroscopy-scanning electron microscopy (EDX-SEM) mapping 175, 176
environmentally-friendly processes
  carbon–carbon coupling reactions 112
  environmental remediation 244–246
  green solvents 99
  oil spill cleanup 237, 244–246
enzymes 7, 194, 195, 197–198
eosin Y 180, 182, 183, 184, 185
epoxidation 39
equilibrium constant K 203–204
equilibrium shapes of nanocrystals 11
Et$_3$N (triethylamine) base 144
ethane 15–16, 21–22, 243, 244
ethyl-4-iodobenzoate 140
ethylene 14, 22, 243

Subject Index

ethylpyruvate
    hydrogenation 106–107
europium (Eu) 161, 162, 163
ex situ conditions 1–2
excitons 14, 15
extended X-ray absorption fine
    structure (EXAFS) 124

Fehling test 212
Fermi energy levels 173, 210
field emission scanning electron
    micrographs (FESEM) 237
finely-divided material structure
    sensitivity factors 9–16
Fisher carbene complexes 129
flavylium cations 37
9-fluorene 242
9-fluorenone 242
fluorine removal 19
fluorocarbon-hydrocarbon
    solvent 145
fluorohydrocarbon solvent 144
Fourier transform infrared (FTIR)
    spectra 185–187
free radical dye photodegradation
    mechanism 180–181
free-standing membranes 237
fuel cells 223–227
functional group transfer
    hydrogenation 83–96
        aldehydes reductive
            amination 94–96
        alkenes 86–88
        carbonyl compounds 88–94
        history 84–86
functionalised silica materials 125
furfural hydrogenation 78

galvanic replacement
    approach 174–175, 176
Gibbs law 101
gluconic acid 168, 169
glucose 168, 207
glucose oxidase 195
G_nDenP-Pd nanoparticles 140, 141, 142, 143

gold (Au) nanocatalysts
    bimetallic palladium–gold
        nanoparticle system 123
    carbon monoxide
        oxidation 18–19
    clock reactions 207, 209
    cost 84
    eosin Y dye reduction by
        borohydride 182, 183, 184, 185
    heteropolyacid salt-
        based 168–169
    hydroxyapatite-based 164–166
    hydroxylated fluoride-
        based 166–167
    metal salt-based 157–169
        metal carbonate-
            based 158–160
        metal phosphate-
            based 160–163
        metal sulfate-based 167
    micelle-based
        synthesis 196–197
    nanocages 175, 176, 178, 179, 184–185
    nanocrystals
        formation 17
        nitrobenzene reduction
            to aniline 211, 213
    overview 3–4
    PAD-PdAu/TS-1 33
    water-gas shift reaction 221, 222, 229
Grignard transmetalation
    reagents 114

halogenoanisoles 104
Hamiltonians 220
HAP (hydroxyapatite)-based gold
    catalysts 164–166
HEA (hydroxyethylammonium) 100, 101, 103
Heck coupling reaction
    palladium nanoparticles
        arylation of olefins 113
        dendrimer-based 144–145

Heck coupling reaction (*continued*)
    ionic polymer-ionic liquid-stabilised 147–148
        n-butyl acrylate/iodobenzene reaction 117, 118
        poly(propylene imine)-stabilised 145, 146, 147
helium 164
heterocoagulation technique 177
heterogeneous catalysis 1, 123–124, 125
    see also homogeneous/heterogeneous catalysis
heteropolyacid salt-based gold catalysts 168–169
hexacyanoferrate(III) 21, 185–187
N-hexadecyl-quicoridinium (QCD16)+ salt 105
hexadecylamine (HDA) 59–60
history of catalysis 7–8
    clock reactions 204–207
    palladium-catalysed carbon–carbon coupling reactions 112–114
    transfer hydrogenation of functional groups 84–86
hollow metallic nanostructures colloidal 172–187
    assembly on substrates 176–179
    catalytic proposed mechanism 185–187
    characteristic catalytic behaviour 179–185
    synthetic approaches 174–176
holmium (Ho) 161, 162, 163
homogeneous/heterogeneous catalysis
    carbon–carbon coupling using palladium nanoparticles 116–126
    colloidal nanocatalysis question 173, 185
    customised dendrimers housing metallic nanocatalysts 193
    palladium nanoparticle catalysis mechanism 125
homolytic pathways 241
horseradish peroxidase (HRP) 198
HPA (hydroxypropylammonium) 100, 101, 103
hybrid nanocrystal catalysis 23–24
hydroformylation reactions 67–69
hydrogen
    see also hydrogenation reactions
    adsorption on platinum or rhodium 14
    hydrogenolysis, ethane on rhodium 15–16
    production
        semi-conducting anastase-type titanium dioxide photocatalysis 34–5
        water–gas shift reaction 221–223, 228–230
    transfer reduction
        aldehydes 94–96
        alkenes 86–88
        carbonyl compounds 88–94
hydrogen peroxide
    Bray–Liebhafsky reaction 205–206
    Briggs–Rauscher reaction 206–207
    formation
        core–shell nanostructured catalyst 42–44
        cyclooctene epoxidation 39
        PAD-PdAu/TS-1 33
hydrogenation reactions 50–60
    alumina supports 70–71

ammonium surfactant-capped rhodium nanoparticles for biphasic hydrogenation 99–109
carbon materials as supports 75–76
dual hydrogenation–oxidation reactions 78
ethylpyruvate asymmetric hydrogenation 106–107
ligand-stabilised nanoparticles as colloidal catalysts 50–52
peptide-capped palladium nanoparticles 132–133
silica supports 71–72
surfactant-supported nanoreactor structures 196
transfer hydrogenation of functional groups 83–96
hydrophobic–hydrophilic structures 195–198
hydrothermal synthesis 19, 237–238
hydroxyapatite-based gold catalysts 164–166
2-hydroxychalcon derivatives 37
hydroxyethylammonium (HEA) 100, 101, 103
hydroxyl-terminated PAMAM dendrimers 126, 127, 139–140, 147
hydroxylated fluoride-based gold catalysts 166–167
4-hydroxyphenyl iodide 141
hydroxypropylammonium (HPA) 100, 101, 103

ICP-AES (inductively coupled plasma-atomic emission spectroscopy) 138
imidazolinium-based ionic polymers 129, 136
imines 95, 165, 240
immobilisation inside dendrimers 145, 146
in situ conditions 1–2, 193
inactive outer surfaces 180–182
inactive supports 161, 162
incipient wetness technique 31
inductively coupled plasma-atomic emission spectroscopy (ICP-AES) 138
industrial catalysis 1, 2, 6, 31
inner walls of hollow nanostructures 180–182
iodate/iodine clock reactions 204, 205–207
4-iodoacetophenone 128
iodoanisole 120, 121, 122, 123
iodobenzene
 n-butyl acrylate Heck coupling reaction 117, 118, 145
 methylacrylate Heck coupling reaction 146
 phenylboronic acid Suzuki coupling reaction 134, 135, 139–140
4-iodobenzoic acid
 Stille coupling reaction 126, 127
 4-biphenylbenzoic acid conversion 128
 phenyltin trichloride 131, 133
iodoferrocene 143
4-iodophenol 127
2-iodothiophene 135
p-iodotoluene 118
ion-exchange method 31
ionic bonds 145–6
ionic liquids (ILs) 57–60, 129, 136
ionic polymer-ionic liquid stabilised palladium nanoparticle (Pd-IP-IL) 147–148
IONOL (2,6-di-*tert*-butyl-4-methyl-phenol) 241
iridium (Ir)
 carbonyl compounds hydrogen-transfer reduction 88
 cost 84
 diphosphate-stabilised nanoparticles 51
 nanocrystal formation 17

iron (Fe)
    active supports for gold
        nanocatalysts 161, 162
    cost 84
    cyclodextrin-stabilised metal
        nanoparticles 39–41
    FePd magnetic nanoparticle
        modified with chiral BINAP
        ligand 41–42
    FePt@Ti-SiO$_2$ synthesis 38–39
    metal salt-based gold
        catalysts 162
    single crystal surfaces 11
isopropanol 89
isopulegol 166, 167
isotropic shapes 172

Keggin-type polyoxometallates 130, 138–139, 148
ketones 93, 94–96, 126
kinetics of nanocrystal formation 17
Kirkendall effect 174
Kumada reactions 113, 114

Landolt clock reaction 204
Langmuir isotherm 216
Langmuir–Blodgett (LB)
    technique 173, 178–179
Langmuir–Hinshelwood
    mechanism 162, 216
lanthanides (Ln) 161, 162, 163
lanthanum (La) 160, 161, 162, 163
leaching mechanisms 116–126, 133, 141
lead dioxide 20
leucomethylene blue (LMB) 210, 211, 213, 214
Lewis acid sites 241
ligand stabilisers 50–52, 115, 116, 144
    see also individual ligand
        stabilisers
limonene hydrogenation 60
lipophilic chain lengths 100, 101
liposomes 195–198
lithography techniques 177

localised surface plasmon resonance
    (LSPR) spectrum 184, 185
low-coordinated sites 221–223

magnetic nanoparticle-based
    multifunctional catalysts 38–44
manganese oxidation 237
manganese oxide octahedral
    molecular sieves 5, 235–247
Mars van Krevelen mechanism 240
melting points of nanomaterials 12
membranes 117, 118, 237
memory facilitation, brain
    oxygen 217
menthol 166, 167
mercury (Hg) 125
mesoporous aluminosilicate
    frameworks 92, 93
mesoporous silica supports 21–22
metal carbonate-based gold
    catalysts 158–160
metal (core)–platinum shell
    nanocatalysts for oxygen
    reduction reactions 223–227
metal nanoparticles (MNPs)
    see also individual metal
        nanoparticles; noble metal
        nanoparticles
    organometallic
        approach 47–79
    supported metal nanoparticle
        catalysts, new synthesis
        method 31–37
    theoretical aspects 219–231
metal phosphate-based gold
    catalysts 160–163
metal salt-based gold
    nanocatalysts 157–169
        heteropolyacid salt-
            based 168–169
        hydroxyapatite-based 164–166
        hydroxylated fluoride-
            based 166–167
        metal carbonate-
            based 158–160

Subject Index

metal phosphate-based 160–163
metal sulfate-based 167
metal sulfate-based gold
  catalysts 167
metallic hollow nanostructures,
  colloidal 172–187
metalloorganic complexes 48–49
metal–organic frameworks
  (MOFs) 180
metal–semiconductor hybrid
  nanocrystals 24
3-methoxyphenylboronic acid 138
methyl orange (MO) dye 180, 181
methyl phenyl sulfide oxidation 44
methyl-2-phenylacrylate 146, 147
methyl-4-bromobenzoate 128, 129
methylacrylate 146
3-methylanisole 107, 108
4-methylbiphenyl compounds 130
methylene blue
  ascorbate ion redox
    reaction 211, 212
  blue bottle experiment 207
  brain oxygen 217
  clock reaction 210–215
  OMS-2 degradation 244
N-methylephedrine 105
N-methylprolinol 105
micelles
  cationic/anionic/neutral 212
  emulsion interfaces 198
  formation 101
  nanoreactors 103, 195–198
Michaelis–Menton kinetics 198
microgels 194–195
microwave-assisted deposition
  method 34–35
mobility of surface adsorbates 14
molecular brush structure 195
molecular sieves 5, 235–247
molybdenum-doped catalysts 244
montmorillonite K10 87, 88, 92
morphology see shape
multifunctional magnetic
  nanoparticle-based catalysts 38–44
myrcene 71–72

N-heterocyclic carbenes (NHC) 51–52
nanocluster exclusion process 117,
  118, 119
nanocrystals 11, 16–17, 18
  see also crystallinity
nanogels 197
nanoparticle networks (NPNs) 130,
  131
nanoparticle-mediated clock
  reactions 203–217
nanoreactor catalysis 192–199
  absorbing nanocatalyst surface
    type 198–199
  block copolymers 196, 197, 198
  cage effect 172–187
  carbon nanotubes 198–199
  dendrimers 193–194
  emulsions 198
  micelles 198
  microgels 194–195
  polymer core–shell
    structures 195
  steric/structural
    effects 193–198
  surfactant-supported
    structures 195–198
nanosphere lithography (NSL) 177
naphthylboronic acid 41
natural catalysts 7
Negishi reactions 113, 114
NEt$_3$ base 135–136, 140
NHC (N-heterocyclic carbenes) 51–52
nickel hydride 85
nickel (Ni) nanocatalysts
  aldehyde hydrogen-transfer
    amination 94–96
  alkene hydrogen-transfer
    reduction 86–88
  carbonyl compound hydrogen-
    transfer reduction 88–94
  cores in silica shells 91, 92
  cost 84
  functional group transfer
    hydrogenation 83–96
  Ni@SiO$_2$ yolk–shell
    nanocatalysts 91–92

nickel (Ni) nanocatalysts (*continued*)
   NiCl$_2$-Li-DTBB 85
   NiCl$_2$.2H$_2$O-Li-arene 84, 85
   PAD-PdNi/Ti-HMS catalyst 343
para-nitro-substituted aryl bromides 145
nitrobenzene 211, 213
nitrogen 34–35
p-nitrophenol 23, 35
4-nitrophenol 180, 183–184, 185
p-nitrophenyl diphenyl phosphate 198
nobility of metals 207
noble metal nanoparticles
   see also individual noble metals
   organometallic synthetic approach 47–79
   overview 3
   for supported catalysis 69–78
   transfer hydrogenation 83–96
   tunability 172
NSL (nanosphere lithography) 177
numerical basis sets 220

octahedral molecular sieves (OMS-2) 235–47
octylamine 59–60
oil spill cleanup 237, 244–6
olefins (alkenes)
   cis-trans isomerism in platinum catalysts studies 22, 23
   hydrogen-transfer reduction 86–88
   hydrogenation using silica-supported noble metal nanoparticles 71–72
   palladium-catalysed arylation 113
   selective oxidation 239–241
oleic acid 40, 41, 42
oleylamine 40, 41, 42
OMS-2 (octahedral molecular sieves) 235–427
one-pot hydrogen peroxide formation 42–44

optical activity *see* enantiocontrol/selectivity
organic functional groups transfer hydrogenation 83–96
organic ligands for supported metal nanoparticle catalysts 31
organoboranes 134
organomagnesium 113
organometallic approach
   noble metal nanoparticle synthesis 47–79
      colloidal catalysis 49–69
      supported catalysis 69–78
organostannane reagents 126
organozinc 113
ORRs (oxidation reduction reactions) 183–184
oscillating chemical reactions 203–217
outer walls of hollow nanostructures 180–182
oxazoline-stabilised ruthenium nanoparticles 50–51
oxidation reactions
   alumina supports 70–71
   carbon materials as supports 76–78
   carbon monoxide 157–167
      oxidation on Au/LaPO$_3$ 162
   cyclohexane aerobic oxidation 165
   dual hydrogenation–oxidation reactions 78
   glucose 168
   selective with OMS-2 239–41
   silica supports 71–5
oxidative addition 121, 122, 123
oxide binary catalytic compounds 212
oxide layer formation in surface restructuring 12
oxime tandem synthesis with imines 165
4-oxo-2,2,6,6-tetramethylpiperidine-1-oxyl (TEMPO) 241

Subject Index

oxo-metal formation 241
oxygen
    binding energy for 225, 226
    colorimetric indicator 217
    consumption during memory formation 217
    metal (core)–platinum shell nanocatalysts 223–227
    reduction reactions 183–184

palladium (Pd) nanocatalysts
    benzyl alcohol oxidation 76–78
    carbon–carbon coupling reactions 3, 112–149
        Heck reaction 144–148
        overview 112–115
        reaction mechanism 116–126
        Stille reaction 126–134
        Suzuki reaction 134–144
    chiral diphosphite ligands 63–65
    core–shell structures 42–44, 195, 223–227
    cost 84
    crystal structure 119–123
    fabrication methods 115–116
    FePd magnetic nanoparticle modified with chiral BINAP ligand 41–42
    nanocages 175, 176
    nanocrystal formation 17
    PAD-PdNi/Ti-HMS catalyst 343
    Pd/SiO$_2$@TiMSS 43, 44
    Pd(core)–Pt(shell) NPs 223–227
    Pd(OAc)$_2$ 124–125, 128
    Pd–Pt alloy 183–184
    polymer core–shell structures 195
    polymer nanotubes with Pd–Rh nanoparticles 196
    PVP-stabilised 21, 22, 134
    pyrazole ligands 65–67
    reversible redox reactions 207
    silica-supported for myrcene hydrogenation 71–72
    supported 32, 33, 71–72

PAMAM (poly(amido amine)) dendrimers 126, 127, 139–140, 147
PANI (polyaniline) nanofibres 136–137, 138
paraformaldehyde 243
PC (propylene carbonate) 144
Pd-IP-IL (ionic polymer-ionic liquid stabilised palladium nanoparticle) 147–148
Pd4(TSNAVHPTLRHL) 12-mer peptide 132–133
PDMS (polydimethyl siloxane) polymer 245
PEG (polyethylene glycol) 129, 135–136, 147
peptide-capped palladium nanoparticles 132–133
perfluorinated polyether-derivatised PPI DENs 144
perfluoro-2,5,8,11-tetramethyl-3,6,9,12-tetraoxapentadecanoyl perfluoropolyether 146
perthiolated cyclodextrin (β-SH-CD) 143
petroleum industry 237, 241-3, 244-6
pH 36, 197
phage display 132
phenol removal from wastewater 244
phenylboronic acid
    carbon–carbon coupling reactions 134–135, 136, 138, 139, 140, 141
    iodoanisole reaction 120, 121, 123
    p-iodotoluene reaction 118
4-phenylnitrobenzene 139
4-phenylphenol 127
2-phenylthiophene 135
phenyltin trichloride 126, 127, 131, 132, 133
4-phenyltoluene 139
phenyltributylstannane 128, 129
phosphine-based dendrimers 140, 141

photo-assisted deposition (PAD) method 31–34, 343
photo-induced assembly-dispersion control 37
photocatalytic reactions 31–34, 180–181
photodegradation, dyes 180–181
pi-electrons 194
plane-wave basis sets 220
plasma jets 216
plasmon absorption 208
plasmonic nanocatalysts 184–185
platinum alloys 183–184
platinum dioxide layers 13
platinum fuel cells 223–227
platinum (Pt) nanocatalysts
    nanocages 175, 176
    nanoclusters 243
    nanocrystals
        cyclodextrin-stabilised 39–41
        FePt@Ti-SiO$_2$ synthesis 38–39
        formation 17
        shape control 21–23
        titanium dioxide 20
    nanocubes 178
    nanoparticles
        colloidal metallic hollow nanostructures mechanism 185–187
        Pd(core)–Pt(shell) 223–227
        PTA-stabilised 52–54
        titanium-containing mesoporous silica platform 34
    Pt/CeO$_2$ catalysed 228–230
    Pt(110) single crystals 12–13
    reversible redox reactions 207
PNIPA (poly(N-isopropylacrylamide)) 195
PNIPA-b-P4VP (poly(N-isopropylacrylamide)-b-poly(4-vinyl pyridine)) 196–197
polar carbon nanotube walls 199
polar heads 100

poly(amido amine) (PAMAM) dendrimers 126, 127, 139–140, 147
polyaniline (PANI) nanofibres 136–137, 138
polydimethyl siloxane (PDMS) polymer 245
polyethylene glycol (PEG) 129, 135–136, 147
polyhedra crystals 9–10
polymers
    see also dendrimers; *individual polymers*
    block copolymers 135, 196, 197, 198
    microgels 194–195
    polymer core–shell structures 195
    polymer templates 177
    polymer-stabilised palladium nanoparticles 123, 136
    polymersomes 197–198
poly(N-isopropylacrylamide) (PNIPA) 195
poly(N-isopropylacrylamide)-b-poly(4-vinyl pyridine) (PNIPA-b-P4VP) 196–197
poly(N-isopropylacrylamide-co-1-vinylimidazole) 198
poly(N-vinylcaprolactam-co-1-vinylimidazole) 198
polyoxometallates 130, 138–139
poly(propylene imine) (PPI) dendrimers 144–145, 146, 147
polystyrene polymer bead support 177, 178
polystyrene-b-poly(sodium acrylate) (PS-b-PANa) 135
polyvinylpyrrolidone (PVP) 21, 22, 134, 177, 178
pore sizes 181, 194
porous cryptomelane-type manganese oxide octahedral molecular sieves 5, 235–247
porous metal nanocatalysts see hollow metallic nanostructures

Subject Index

potassium bisulfate 167
potassium bromate 205
PPI (poly(propylene imine)) dendrimers 144–145, 146, 147
precatalysts for Stille coupling reaction 127–128
prices of transition metals 84
prismatic templates 177
prochiral arenes 107–108
2-propanol 86, 87, 88, 89, 90, 91, 95
propylene carbonate (PC) 144
protected catalytic environments *see* nanoreactor catalysis
protective agents *see* stabilising agents
proton pumps 198
PTA (1,3,5-triaza-7-phospha-adamantane) 52–54
push–pull reduction method 208
PVP (polyvinylpyrrolidone) 21, 22, 134, 177, 178
pyrazole ligands 65–67
pyridine *N*-oxides 165
pyridyl chloride 137
pyrrole hydrogenation 22

quantum confinement 14, 15
quantum dots 15
quantum mechanical methods 220
quaternary hydroxylated ammonium salts 105
quinoxalines 240

R5 peptide (*Cylindrotheca fusiformis*) 130, 131, 132
Raney nickel 86, 88, 95
rate/mass of catalyst relationship 7
reaction mechanisms
    carbon monoxide oxidation on Au/LaPO$_3$ 162
    clock reactions 215–216
    colloidal metallic hollow nanostructures 185–187
    micelle-based 196–197
    nanocrystal formation 17
reactive crystal facets 19–21

recyclability
    bacterial cellulose-stabilised palladium nanoparticles 138
    challenge 176–177
    colloidal metallic hollow nanostructures 178–179
    dendrimer-based materials 127, 140, 141, 145
    imidazolinium-based ionic polymer-stabilised palladium nanoparticles 130
    palladium-PANI nanoparticles 137–138
redox reactions 203–217
reducing agents 203–217
reduction pathways 182–183, 184, 185
reductive amination of aldehydes 94–96
reporting standards 8–9
reverse micelles 93, 197
reverse water-gas shift reaction (RWGS) 244
reversible redox reactions 203–217
rhodium (Rh) nanocatalysts
    ammonium surfactant-capped for biphasic hydrogenation 99–109
    carbonyl compounds hydrogen-transfer reduction 88
    cost 84
    diphosphate-stabilised nanoparticles 51
    ethane hydrogenolysis 15–16
    hydroformylation reactions 67–69
    nanocrystal formation 17
    particle atomic restructuring 14
    polymer nanotubes with Pd-Rh nanoparticles 196
    protective agents 100, 101, 103
    Rh/TPPTS catalyst 196
    rhodium/palladium systems 13, 196

rhodium (Rh) nanocatalysts (*continued*)
   rhodium/silicon dioxide catalyst for ethane hydrogenolysis 15–16
roll-over *trans* diastereoisomer formation 104
roughened surfaces 16, 179
ruthenium (Ru) nanocatalysts
   alumina as support for hydrogenation/oxidation reactions 70–71
   benzyl alcohol oxidation 76–78
   carbene-stabilised ruthenium nanoparticles 51–52
   carbon monoxide/benzyl alcohol oxidation 72–75
   carbonyl compounds hydrogen-transfer reduction 88
   confinement properties of ionic liquids 57–60
   cost 84
   diphosphate-stabilised nanoparticles 51
   nanocrystals
      alkylamine ligands 18
      formation 17
   oxazoline-stabilised nanoparticles 50–51
   PTA-stabilised 52–54
   Ru/APTS 60–63
   sulfonated diphosphine-stabilised 54–57
rutile nanocrystals 20
RWGS (reverse water-gas shift reaction) 244

SAED (selected area electron diffraction) experiments 121, 122
salts *see individual metal salts*; metal salt-based gold catalysts
scaffold-like nature of biomolecules 130, 131, 132
Schrödinger equation 220
SDBS (sodium dodecylbenzenesulfonate) 167
seed-mediated techniques 136, 174
seedless methods 174
selected area electron diffraction (SAED) experiments 121, 122
selenium nanowires 214, 215
semiconductors 2, 24
   see also individual semiconductors
shape
   analysis for catalyst development 7–8
   control/tuning
      advances 2–3
      colloidal metallic hollow nanostructures 172–187
      nanocrystal formation 17–18
      OMS-2 236–239
      platinum nanocrystals 21–23
   effects on nanoparticle catalyst properties 220, 226–227
   polyhedra crystals 9–10
silaffin proteins 130
silanes 165
silica (silicon dioxide)
   magnetic nanoparticles 38–39, 43, 44
   oxidation reactions 72–75
   palladium nanoparticle catalysis mechanism 125
   shells with nickel cores 91, 92
   supports for hydrogenation reactions 71–72
   titanium-containing 31–34
silver (Ag) nanocatalysts
   nanocrystal formation 17
   nanocubes 175, 176, 177, 178
   nanoparticles
      clock reactions 207, 208, 209, 210, 211
      reversible formation/dissolution 209, 210, 211
      size-controlled deposition 35–37

Subject Index

silver nitrate reduction by sodium borohydride 209, 210
silver oxide 181, 182
single metal shelled hollow nanoparticles 176, 183–184
single-site photocatalysts 31–34
size of particles 16, 220, 224–226
size of pores 181, 194
size-controlled metal nanoparticles
    advances 2
    deposition as colloidal precursors 35–37
    OMS-2 particles 238
    palladium nanoparticles by dendrimer encapsulation 127
sodium borohydride 209, 210, 214
sodium carbonate 203
sodium dodecyl-benzenesulfonate (SDBS) 167
sodium hydroxide 207
soil clean-up 244, 245, 246
sol–gel method 92
solids
    electronic structure change 14–15
    solid supports
        gold nanocatalysts 158
        palladium nanoparticle immobilisation 116
    solid/hollow nanocatalysts comparison 182–183
solvent-controlled swelling/ heterocoagulation technique 177
Sonogashira reactions 113, 114
spectroscopic studies
    ambient pressure X-ray photoelectron spectroscopy 13
    cage effect 185–187
    energy-dispersive X-ray spectroscopy-scanning electron microscopy mapping 175, 176
    Fourier transform infrared spectra 185–187
    inductively coupled plasma-atomic emission spectroscopy 138
    localised surface plasmon resonance spectrum 184, 185
    X-ray diffraction 8
stabilising agents
    3-aminopropyltriethoxysilane 60–63
    ammonium surfactant-capped rhodium nanoparticles for biphasic hydrogenation 100
    chiral diphosphite ligands 63–65
    electrostatic stabilisation 210
    noble metal nanoparticle synthesis 49, 59–60
    palladium
        nanoparticles 139–140
        synthesis 115, 116
    problem-solving 173
    steric stabilisation 209
    sulfonated diphosphine 54–57
    1,3,5,-triaza-7-phospha-adamantane 52–54
standards, data reporting 8–9
stereoselectivity *see* enantiocontrol/selectivity
steric effects
    hindrance 106, 107, 108, 141
    nanoreactor catalysis 193–198
    stabilisation 209
Stille coupling reaction
    palladium nanoparticles
        Cylindrotheca fusiformis R5 peptide-templated synthesis 130, 131, 132
        dendrimer-encapsulated 126, 127
        dendrimer-stabilised 128–129
        imidazolinium-based ionic polymer-stabilised 129–130

Stille coupling reaction (*continued*)
    Keggin-type
        polyoxometallates 130
    Pd4 peptide-
        capped 132–133
strong metal/support
  interaction 228
structural effects 3, 193–198
structure-sensitive/insensitive
  reactions 8–16
styrene
    bromotoluene reaction 148
    derivatives transfer-
      hydrogenation 86–87
    hydroformylation with
      diphosphite-stabilised
      rhodium nanoparticles 69
    hydrogenation 60
    ruthenium/sulfonated
      diphosphine aqueous
      colloidal solutions 56
    selective epoxidation 165
substrates for assembly of colloidal
  metallic hollow
  nanostructures 176–179
sulfate sources 167
sulfite 204
sulfonated diphosphines 52, 54–57
sulfoxides 165
sulfuric acid 211
supercritical carbon dioxide
  (scCO$_2$) 146
supported catalysis
    active supports 161, 162
    carbon material
      supports 76–78
    metal salt-based gold
      nanocatalysts 157–169
    organometallic approach to
      metal nanoparticle
      synthesis 48, 69–78
    theoretical aspects 227–230
surface aspects
    adsorbate mobility 14
    crystal facets
      atom arrangements 11
      free energy 11
      plane orientation in cubo-
        octahedron 11
    plasmon absorption 207, 208
    restructuring 12–14
    sites in platinum catalyst
      studies 22
    surface area in cage effect 181
    surface-to-volume ratio 179
surfactants 18, 99–109, 195–198
Suzuki coupling reaction
  palladium nanoparticles
    atom-leaching
      mechanism 123, 124
    bacterial cellulose-
      stabilised 138
    block copolymer-
      stabilised 135
    bromobenzene/
      phenylboronic
      acid 134, 135
    dendrimer-
      encapsulated 139
    dendrimer-
      stabilised 140–143
    imidazolium-based ionic
      polymer-stabilised 136
    iodoanisole/phenylboronic
      acid 120, 121, 123
    iodoferrocene/
      phenylboronic acid
      stable inclusion
      complexes 143
    Keggin-type
      polyoxometallates
      138–139
    PANI-stabilised 136, 137
    PEG-stabilised 135–136
    perthiolated cyclodextrin
      as surface passivant 143
    phenylboronic acid/*p*-
      iodotoluene
      reaction 118
    PVP-stabilised 134
    reaction
      mechanism 123–124
    seed-mediated nanoparticle
      synthesis methods 136

Subject Index

polymer core–shell
structures 195
single/double shell hollow
nanocatalysts
comparison 183–184
Suzuki–Miyaura coupling
reaction 41, 65
SXRD (UHV-high-pressure surface
X-ray diffraction) 13
synthesis methods
colloidal metallic hollow
nanostructures 174–176
micelle-based 196–197
multifunctional magnetic
nanoparticle-based
catalysts 38–44
noble metal nanoparticles,
organometallic
approach 47–79
overview 3
porous cryptomelane-type
manganese oxide
octahedral molecular
sieves 236–239
supported gold
nanocatalysts 158
supported metal nanoparticle
catalysts 31–37
templating methods 23
titanium dioxide nanocrystals
with reactive facets 19
well-defined
nanocrystals 16–18

templating methods
hollow metallic nanocatalyst
synthesis 174
nanocrystal synthesis 23
R5 peptide (*Cylindrotheca fusiformis*) 130, 131, 132
TEMPO (4-oxo-2,2,6,6-
tetramethylpiperidine-1-oxyl) 241
tetraalkylammonium salts 144
tetrabutylammonium salts 129, 134
tetrabutylphenylstannane 130
tetraheptylammonium bromide 130
tetrahydrofuran solution 49

tetrakis(triphenylphosphine) 134
tetraphenyltin 130
THEA (trishydroxyethylammonium)
100, 101, 103
theoretical aspects of metal
nanocatalysts 219–231
computational
method 219–231
copper nanocatalysts for
water-gas shift
reactions 221–223
metal (core)–platinum shell
nanocatalysts for oxygen
reduction reactions 223–227
supported
nanocatalysts 227–230
types of metal
nanocatalysts 220–227
thermolysis 144
thiazine groupd 210
thiols 239
2-thiopheneboronic acid 135, 139
thiosulfate 21, 185–187
time-of-flight (TOF) 15–16
tin reagents
C–C coupling reactions 113, 114
organostannane
reagents 126
phenyltin
trichloride 126, 127, 131, 132, 133
tetrabutylphenylstannane 130
tetraphenyltin 130
tributyl(phenyl)stannane 128, 129
titanium (Ti)
FePt@Ti-SiO$_2$ synthesis 38–39
mesoporous silicas 31–34
PAD-PdNi/Ti-HMS catalyst 343
Pd/SiO$_2$@TiMSS 43, 44
titanium dioxide 19–21, 31, 32, 229–30
zeolites 31–34
toluene 59–60, 145, 159
p-tolylboronic acid 139–140

transfer hydrogenation
    aldehydes 94–96
    alkenes 86–88
    carbonyl compounds 88–94
    overview 3
transition metals 84
    see also individual transition metals
transmission electron microscopy (TEM) 102, 118, 175, 176
trapped holes 20–21, 32
1,3,5-triaza-7-phosphaadamantane (PTA) 52–54
1,2,3-triazolylsulfonate dendrimers 140
tributyl(phenyl)stannane 128, 129
3,4,5-triethoxybenzoyl chloride 145
triethylamine (Et_3N) base 144
trimethyltetradecylammonium bromide (TTAB) 177
trishydroxyethylammonium (THEA) 100, 101, 103
TTAB (trimethyltetradecylammonium bromide) 177
tuning 194, 195, 212
turnover frequency (TOF)
    anomalous behaviour of ethane hydrogenolysis on rhodium 15–16
    carbon–carbon coupling using palladium nanoparticles 133
    dendrimer-stabilised palladium nanoparticle system 140
    palladium nanoparticle-catalysed Suzuki reaction 124
    Pd(core)–Pt(shell) nanoparticles 225
    reporting standards 8–9
turnover number (TON) 147
two-compartment reactors 117–126
two-electron reduction 182, 183, 184, 185

UHV-high-pressure surface X-ray diffraction (SXRD) 13
ultraviolet (UV)-activated colorimetric oxygen indicator 217

VASP plane-wave basis sets 220
voids in hollow nanostructures 180

wastewater treatment 244
water dissocation 229
water purification using clock reactions 216
water-based synthesis
    ammonium surfactant-capped rhodium nanoparticles for biphasic hydrogenation 99–109
    metal salt-based gold nanocatalysts 157–169
    microgels 194–195
    palladium nanoparticle catalysts 129
water–gas shift (WGS) reactions 164, 221–223, 228–230
water-soluble nanoparticles 39–41, 52
water–ethanol solvent 140, 148
well-defined nanocrystal synthesis/properties 16–18
wetness impregnation 16
Wulff construction 11

X-ray diffraction 8

yolk–shell nanocatalysts 91–92, 180–182

zeolites 31–4, 192–193
zeta potential (ζ) 102
zinc 161, 162, 229
zirconium phosphate 160